D1719020

Evolutionary Systems Design

Holden-Day Series in Explorations in Operations Research, Systems, and Artificial Intelligence
Melvin F. Shakun, Editor

Evolutionary Systems Design: Policy Making Under Complexity and Group Decision Support Systems
Melvin F. Shakun, New York University

Alternate Realities: Models of Nature and Man
John L. Casti, Institute for Econometrics and Operations Research, Vienna

Operations Research and Artificial Intelligence
Andrew Whinston and Suranjan De, Purdue University

Cabries: All by Melvin F. Shakun, 1980.

Evolutionary Systems Design

Policy Making Under Complexity and Group Decision Support Systems

Melvin F. Shakun
New York University

HOLDEN-DAY, INC., OAKLAND, CALIFORNIA

Production Management: Richard Mason, Bookman Productions
Copyeditor: Sylvia Stein
Compositor: Science Typographers, Inc.

EVOLUTIONARY SYSTEMS DESIGN
Policy Making Under Complexity
and Group Decision Support Systems

4432 Telegraph Avenue, Oakland, CA 94609

Library of Congress Catalog Card Number: 87-080645
ISBN: 0-8162-7819-9

Printed in the United States of America

To Norma Redstone Shakun,
which is to say to all (One).

Contents

Preface xi

1 Evolutionary Systems Design **1**

1.1 Introduction and Overview 1
1.2 The Evolutionary Systems Design Paradigm 3
1.3 Evolutionary Systems Design: Mathematical Development 9
1.4 Problem Representation in Decision Support Systems 20

2 Policy-Making as Evolutionary Systems Design **25**

2.1 Purposeful Systems, Policy-Making, and Conflict Resolution 25
2.2 Planning: Discussion 31
2.3 Planning: Mathematical Formulation 33
2.4 Act of Control 38
2.5 Consciousness 39
2.6 Conflict Resolution: Increase in Consciousness 41
2.7 Formalizing Conflict Resolution in Policy-Making 42
2.8 Goals/Values Referral: Additional Comments 50
2.9 Negotiations: A Mathematical Approach 51
2.10 Policy-Making and Meaning as Design of Purposeful Systems 51

3 Conflict Resolution as Evolutionary Systems Design: The Arab-Israeli Conflict **58**

3.1 Resolving the Conflict: June 1978, Before Camp David 58
3.2 A Postscript After Camp David 61
3.3 *Q*-Analysis of the Conflict 62

4 A Goals / Values Solution to Mandatory Retirement Age for Tenured Faculty: An Application of Situational Normativism **64**

4.1 Introduction 64
4.2 Nature of the Conflict: Goals/Values Analysis 65
4.3 Proposed Solution 67
4.4 Implementation 67

5 Structuring the International Marketplace for Maximum Socioeconomic Benefits from Space Industrialization **70**

5.1 Introduction 70
5.2 Economic Regulation 73
5.3 Feasible Solutions for Technology (Market Structure) 75
5.4 Goal Delivery Classification Analysis: Controls 76
5.5 Results 77
5.6 Conclusions 78

6 A Normative Model for Negotiations **81**

6.1 Introduction 81
6.2 Negotiation Behavior Concepts 84
6.3 Mathematical Model 86
6.4 Numerical Example 89
6.5 Extension to Multivariate Case 93
6.6 Concluding Remarks 93

7 Bivariate Negotiations as a Problem of Stochastic Terminal Control **96**

7.1 Introduction 96
7.2 A Two-Player Bivariate Negotiation Model 99
7.3 Numerical Example 107
7.4 Empirical Results 111
7.5 Concluding Remarks 112

8 Decision Support Systems for Semistructured Buying Decisions **115**

8.1 Introduction 115
8.2 Decision Making as Conflict Resolution 117
8.3 The Decision Support System 121
8.4 An Illustrative Example: The Decision to Purchase a Car 127
8.5 Concluding Remarks 131
8.6 Relaxing the Additivity and/or Piecewise Linearity Assumptions 132

9 Decision Support Systems for Negotiations **134**

9.1 The Case of One Decision Maker 135
9.2 Group Decision Making: Negotiations 136
9.3 Negotiations: Axioms and Concession Making 140
9.4 Concluding Remarks 149

10 MEDIATOR: Toward a Negotiation Support System **152**

10.1 Introduction 152
10.2 The Case of One Decision Maker 154
10.3 Group Decision Making: Negotiations 155
10.4 Using MEDIATOR: Application to Group Car-Buying Decisions 159
10.5 System Architecture for MEDIATOR 166
10.6 Concluding Remarks 179

11 A Decision Support System for Design and Negotiation of New Products **182**

11.1 Introduction 182
11.2 Controls, Goals, Criteria, Preferences: The Four Reference Spaces 183
11.3 Mapping Between the Four Spaces: Support for Negotiations 186
11.4 Implementation on Microcomputers 189
11.5 Concluding Remarks 194

12 Using MEDIATOR for New Product Design and Negotiation **196**

12.1 Introduction 196
12.2 Initial Design Phase 197
12.3 Design Evolution 202
12.4 Meeting with the President: Decision 212
12.5 Concluding Remarks 217

13 Negotiating to Free Hostages: A Challenge for Negotiation Support Systems **219**
13.1 Introduction 220
13.2 Case Presentation 220
13.3 Analysis 224
13.4 A Framework for Negotiation Support Systems: ESD 231
13.5 Applying the ESD Framework to the Case 233
13.6 Concluding Remarks 244

14 Effectiveness, Productivity, and Design of Purposeful Systems: The Profit-Making Case **247**
14.1 Introduction 247
14.2 Design of Purposeful Systems 248
14.3 Effectiveness 249
14.4 Physical Efficiency: Total Factor Productivity Analysis 252
14.5 Cost and Efficiency Controls 254
14.6 A Numerical Example: Producing Cadillacs and Chevrolets 254
14.7 A Second Example: Illustrating Profit-Pollution Control Trade-offs 260
14.8 Concluding Remarks 262

15 Irrationality and Effectiveness in Public Decision Making **264**
15.1 Introduction 264
15.2 Empirical Examples 265
15.3 Preferences, Bureaucratic Procedures, and Ambiguity 266
15.4 Coalition Formation and Model Dynamics 272
15.5 Concluding Remarks 275

16 An Evolving Conclusion and Wall As All **277**
16.1 Wall As All 278

Index **284**

Preface

In this book I develop evolutionary system design (ESD) as a methodology for policy-making under complexity involving multiparticipant, multicriteria, ill-structured, dynamic problems in self-organizing contexts. Examples include conflict resolution, negotiations, group decision making, and planning. As a methodology for problem definition and solution ESD provides an artificial intelligence (AI) framework for computer group decision support systems (GDSS) and negotiations—for policy sciences—that can aid the policy-making (design) process. In turn GDSS implement and provide concrete expression and operational meaning for ESD. Under ESD design of purposeful systems and policy-making are identified with each other.

Thus the ESD paradigm is a design methodology for policy-making under complexity and an AI framework for GDSS, but it is more than this. It is a model of nine mutually explanatory metaphors—design of purposeful systems, policy-making, meaning, cybernetics as cooperative control, dissipative self-organization, consciousness, problem solving, evolution, and wall as all. The ESD paradigm says these concepts are one and express One (all there is) in that they are mutually explanatory metaphors for the process of all there is—the process we term "reality" or "general systems." ESD is at once a practical, operational methodology for policy-making under complexity and a philosophic, experiential statement. Operationally, it is intended to aid policy-making through decision support in business, government, and nonprofit organizations, but it should be useful for systems design in general, as in engineering, architecture, the arts (painting, sculpture, music), writing, mountain climbing, and living.

ESD may also be viewed as evolutionary operations research (OR)/ management science (MS). Classical OR/ MS generally deals with well-structured problems having more than one feasible solution, choosing among them by some kind of optimization. The new evolutionary OR/ MS, while including the classical field, focuses on ill-structured (evolving) problems—which at least initially may have no feasible solution—and self-organizing phenomena.

The book is based on the author's research and teaching of graduate courses in OR/MS, policy sciences, and computer applications and information systems (recently, for the latter, a course on negotiation and group decision support systems) at the Graduate School of Business Administration, New York University, as well as at the University of Paris IX-Dauphine and the University of Aix-Marseille. The book can be used as a primary or supplementary text in graduate courses on OR/MS, systems design, policy sciences, computer decision support systems, and economics in schools of business, engineering, and public administration. The methodology and applications complement each other. The book is also suggested to practicing management scientists and systems designers and to those policymakers interested in using formalized methodology and group decision support (e.g., in office information systems and in negotiation) for policy-making under complexity. Some redundancy of material will be found between chapters, which helps make them relatively self-contained.

The book is in the nature of a partial retrospective and a prospective on my work, for I have always been and perhaps will always be working on the same problem—understanding the process of all there is cognitively, affectively, conatively, i.e., through increasing consciousness, which is tantamount to overcoming separateness from all there is. It is the same problem, I believe, on which everyone (everything) is explicitly or implicitly engaged. Formally, this work is expressed in ESD, which operationally is an AI framework for and, in turn, is implemented by GDSS in policy-making under complexity.

ACKNOWLEDGMENTS

By way of acknowledgment, I would like to thank my colleagues—in particular, John Casti, Alain Checroun, Guy O. Faure (Chapter 13), Françoise Fogelman-Soulie (Chapter 7), Jean-Louis Giordano (Chapter 11), William A. Good (Chapter 5), Eric Jacquet-Lagreze (Chapters 8, 11, and 12), Matthias Jarke (Chapter 10), M. Tawfik Jelassi (Chapter 10), Arie Lewin, Bertrand R. Munier (Chapters 7 and 15), Ambar G. Rao (Chapter 6), George S. Robinson (Chapter 5), Jean-Louis Rulliere (Chapter 15), and Ephraim F. Sudit (Chapters

4, 5, and 14)—who have contributed directly or indirectly to the ideas in this book. I would also like to thank my secretary, Ms. Eurita Brooks, for typing and administrative assistance.

Finally, I especially wish to thank my wife, Norma, for her love and support.

Melvin F. Shakun
Williamsville, Vermont

CHAPTER 1

Evolutionary Systems Design

1.1 Introduction and Overview

During the past three decades developments in management science/operations research, artificial intelligence, and cognitive/behavioral science have contributed to an understanding of procedural rationality—how decisions should be or are made in intelligent systems. These are information-processing (symbol/pattern manipulating) adaptive systems that use change (control) to cope with change in the environment or internally to attain goals that are operational expressions of values. Such systems have also been termed "artificial" (Simon 1980, 1981). They include manmade artifacts—thus a science of the artificial is a science of design (control)—and man himself, in terms of behavior. Simon notes three forms of adaptation corresponding to increasing time scales of observation: problem solving, learning, and evolution. Adaptive systems that can choose their own values and goals are purposeful (Ackoff and Emery 1972; Lewin and Shakun 1976). In this book we are concerned with design of purposeful systems that we identify with policy-making—they are mutually explanatory metaphors.

We develop evolutionary systems design (ESD) as a methodology for policy-making under complexity involving multiparticipant, multicriteria, ill-structured, dynamic problems in self-organizing contexts. Examples include conflict resolution, negotiations, group decision making, and planning. The ESD methodology with its expression in computerized group (coalition) decision support systems (GDSS) as an evolving group problem representation depicts formalized artificial intelligence (AI)—decision support for problem definition and solution in complex, self-organizing contexts. In other words ESD is an AI framework for GDSS. To achieve this decision support (realize the ESD framework), a GDSS may be viewed as integrating OR methods—mathematical programming, control theory, game theory, multicriteria decision-making methods, heuristics, etc.—and AI approaches—problem-solving strategies (search, heuristics, etc.), knowledge representation, and language processing.

We formalize design/policy-making as cybernetics/self-organization. This involves consideration of a cooperative control problem (mathematically designing and solving a dynamical system of values, goals, technology, and controls) and self-organizing phenomena. Under self-organization, open, nonequilibrium systems subjected to fluctuations can evolve to new, more complex structures having increased variety (adaptivity)—evolution. Design/policy-making also constitutes meaning, which we define as relation among values, goals, technology, and controls, in general, in a multiparticipant setting. Design, policy-making, meaning, cybernetics as cooperative control, self-organization, problem solving, and evolution are manifested by increasing consciousness regarded as self-organizing response capacity (awareness) and represented mathematically by sets of values, goals, technology, and controls whose expansions signify increase in consciousness. On a direct experiential basis, these metaphors are the metaphor wall as all. In this book we develop the ESD paradigm—a model of mutually explanatory metaphors and a design methodology for policy-making under complexity—and present applications.

As to the organization of the book, in Chapter 1, following this introduction and overview (Section 1.1), we discuss the ESD paradigm (Section 1.2), present the ESD mathematical development (Section 1.3), and discuss its use in problem representation in decision support systems (Section 1.4). Chapter 1 is thus an overview of the ESD paradigm.

In Chapter 2 we elaborate on the ESD framework presented in Chapter 1, discussing in more substantive detail such ideas as the goals/values referral process, conflict resolution and negotiations, planning, act of control, and consciousness. We formalize conflict resolution and illustrate goals/values analysis by two examples. We also reexplore policy-making, meaning, and the design of purposeful systems. Thus Chapters 1 and 2 present the basic ESD theory.

The next three chapters—3, 4, and 5—consider three applications of ESD involving goals/values analysis. The first (Chapter 3) deals with the Arab-Israeli crisis and also introduces the use of *q*-analysis for analyzing the goals/values matrix. The second (Chapter 4) develops a goals/values solution to the problem of mandatory retirement age for tenured faculty, as faced by the U.S. Congress. The last (Chapter 5) treats systems design for space industrialization involving consideration of values, goals, and controls in designing (choosing) market structure (the technology).

A fundamental process of ESD conflict resolution within a given goal space involves concession making through negotiations. Chapter 6 formalizes concession making on a single negotiation variable as a problem of stochastic terminal control that can be solved by dynamic programming to yield normative recommendations as to concession making. Chapter 7, also using dynamic programming, extends the work to the case of bivariate negotiations where local preference information is obtained interactively at each stage. Bivariate (in general, multivariate) utility functions are not assessed directly on the goal variables, as is required in the multivariate extension noted in Chapter 6.

The next six chapters utilize ESD as a methodology—an AI framework—for computer group decision support systems (GDSS). In turn, GDSS implement and provide concrete expression for ESD. In Chapter 8 a microcomputer-based DSS for multicriteria decision making involving one decision maker and applied to semistructured buying decisions, illustrated by car buying, is discussed.

Chapter 9 treats the case of two or more decision makers where DSS for negotiations—negotiation support systems (NSS)—are the central focus. Ways in which a DSS can support the negotiation process to resolve conflict are considered. Chapter 10 expands and implements these ideas through the NSS MEDIATOR. Chapters 11 and 12 operationalize the ESD, evolving group problem representation for the design and negotiation of new products at a multinational corporation. Drawing on a real case, Chapter 13 discusses negotiation support systems for freeing hostages. With such systems, we are moving toward DSS for multiplayer, multicriteria, ill-structured, dynamic problems that implement ESD.

Chapter 14 applies the ESD methodology to the problem of effectiveness and productivity and further discusses ESD in hierarchical systems. Chapter 15 treats irrationality and effectiveness in public decision making using the ESD methodology for case modeling. Chapter 16 presents an evolving conclusion and the wall-as-all metaphor.

1.2 The Evolutionary Systems Design Paradigm

The ESD paradigm is a model of mutually explanatory metaphors:

1. design of purposeful systems
2. policy-making, in particular, policy-making under complexity involving multiparticipant, multicriteria, ill-structured, dynamic problems (Examples include conflict resolution, negotiations, group decision making, and planning.)
3. meaning
4. cybernetics as cooperative control
5. dissipative self-organization
6. consciousness

7. problem solving
8. evolution
9. wall as all

The ESD paradigm relates these concepts—says they are one and express One (all there is) in that they are mutually explanatory metaphors for the process of all there is—the process we term reality or general systems.

The ESD paradigm is also a methodology for design (synthesis) of purposeful systems. It is scientific, philosophic, and experimental. ESD implies systematic application of all we know to the operational design of purposeful systems involving values, goals, technology, and controls in a self-organizing, evolutionary, cybernetic framework.

ESD involves the evolutionary design of systems and the design of evolutionary systems—in general, the evolutionary design of evolutionary systems. The design of a system is evolutionary in the short-term sense that problem solving is an evolving process that, in mathematical parlance, amounts to designing and solving a dynamical system involving values, goals, technology, and controls. The system itself is evolutionary if over the intermediate term it can retain adaptations for future use—learning—and over the long term it can change to more complex states (new structures)—evolution, biological, and cultural.

Policy-making may be viewed as the design of purposeful systems to deliver values to participants in the form of operational goals. Values are preferable (desired) modes of conduct (instrumental values) or end-states of existence (terminal values) (see Rokeach 1973). Maslow's (1954) needs (values) hierarchy involving safety, security, love, self-esteem, and self-actualization expresses terminal values. Operational goals are defined by specific, unambiguous operations and are characterized by performance measures. They are operational expressions of values.

Evolutionary system design is a design methodology for policy-making under complexity—the system is not completely known, or otherwise it would be merely complicated (Atlan 1981), involving multiparticipant, multicriteria, ill-structured, dynamic problems. An ill-structured problem (Simon 1973) is one in which the problem representation (structure) is not given but evolves. It is a game with incomplete information (Harsanyi 1977)—there is only partial knowledge of the rules of the game (structure)—whose information (structure) can change during the design process. As a methodology for problem definition and solution, ESD provides an artificial intelligence (AI) framework for computer group decision support systems (GDSS) and negotiations—for policy sciences—that can aid the policy-making (design) process. In turn, GDSS implement and provide concrete expression and meaning for ESD. With ESD we stretch the limits of bounded rationality (Simon 1983). We need to do this because the world today is more interconnected and complex—less factorable into separate problems—than was the world of our ancestors to whom we are biologically equivalent. However, cultural evolution can augment biological evolution to stretch the bounds of rationality.

In ESD decision makers (players) define and try to attain goals as operational expressions of underlying values. N players (who may change over time) are viewed as playing a dynamical (difference) game in which a coalition (group) C (which may also change over time) of the set of N players can form, provided it can deliver to itself (and hence to its members) a set of agreed-upon goals, thus defining and solving its policy-making/design problem. Formally this means that for each time period t the intersection of coalition goal target set $Y^C(t)$ and its technologically feasible performance set $y^C(t)$ is nonempty. For a given operational goal space, if the intersection $Y^C(t)$ and $y^C(t)$ is empty, then one or both of these sets may expand to give a nonempty intersection. Also the dimensions of the operational goal space itself may be redefined using a goal/values referral process (see Section 1.3 and Chapter 2). Within the new goal space—either originally or after goal target and/or technologically feasible performance expansion—the target-performance intersection may be nonempty, thus solving a redefined policy-making problem. More precisely, the nonempty intersections must be a single set or a single point to arrive at a solution (see Section 1.3). The problem-solving process is cybernetic: (1) the attempted mutual adaptation of multiple coexisting systems, i.e., the N players, which results in coalition C reducing the difference (distance) between its goal target and technologically feasible performance, is biological cybernetics, and (2) the use of feedback and feedforward for control of disturbances in the dynamical system is engineering cybernetics. These are discussed further in Chapter 2.

From this viewpoint (which we will expand) policy-making is a cybernetic (control) problem—an N-player planning game—in which nothing is considered necessarily to be fixed, i.e., it is a systems design problem. Because the game affords opportunities for cooperation among the players or a subset (coalition C) of them, the policy-making problem is one of cooperative control (cybernetics). The coalition designs a system technology and exercises controls (inputs) to deliver its values through operational goals or directly as in relationship-oriented systems, like the family. In this the dialectical process discussed by Churchman (1971) can be useful. Overcoming separateness from all there is (God, One, Tao, the absolute) is regarded as the ultimate value. However, the operational methodology itself does not depend on this assumption. Enlightened policy-making is at once pragmatic, operational control, and cooperative control to all there is.

As used here control is in the context of self-organization. An open system (open to energy/information exchange with the environment) in a state of nonequilibrium (so-called dissipative structure) can change to a new dynamic regime (structure) corresponding to an increased state of complexity if random fluctuations (external or internal) are introduced. Occurring far from equilibrium, this is the principle of order through fluctuation, and the process is called dissipative self-organization (Jantsch 1975, 1979, 1980; Jantsch and Waddington 1976; Nicolis and Prigogine 1977; Prigogine 1980; Prigogine and Stengers 1984). Evolution is viewed as an integral aspect of dissipative self-organization (Jantsch 1981).

Rosen (1978) associates complexity with error or deviation between actual system behavior and expected behavior based on some description of the system. Complex systems require multiple descriptions to account for system behavior. Points at which the system can pass from one description to another are bifurcation points. Casti (1984) defines system complexity in terms of the number of bifurcation points or the number of nonequivalent descriptions (models) that an observer (here designer, coalition C) generates for the system. Rosen (1978) relates this notion of complexity to that arising from a probabilistic description of a system that permits complexity to be defined as variety or uncertainty in the state of the system (Atlan 1981). Variety is measured by information content or negative entropy (Ashby 1956; Weaver and Shannon 1949). Variety is nonrepetitive order (disorder) associated with unknown constraints between system elements (Atlan 1981). Atlan has shown that self-organization requires redundancy or repetitive order that can be physically repetitive (the same element repeated) or deductively repetitive (knowing one element gives some information on others, i.e., there are known constraints between parts). Under self-organization redundancy decreases and complexity (variety) increases, allowing increased capability in regulatory performance and entropy production.

As the evolutionary design of evolutionary systems, ESD views participants in purposeful systems as codesigners (partial controllers) in dissipative self-organization. Purposefulness is choice of values, goals, technology, and controls within evolving variety. ESD is at once cybernetic self-organization (control in the context of self-organization) and self-organizing cybernetics (self-organization in the context of control). They are alternative descriptions. With the latter the viewpoint is that of the designer (coalition C) whose control efforts are subject to self-organizing phenomena. With the former the designer transcends himself and sees a self-organizing universe in which he exercises some control. The ESD paradigm includes at once both descriptions and one more. ESD is at once self-organizing, biological, engineering cybernetics; cybernetic self-organization; and—the additional description—simply, self-organization. With the latter the designer transcends himself still further and sees that he along with everything else is the self-organizing universe in which he is a catalyst enhancing creative processes. In the evolutionary design of evolutionary systems—ESD—self-organizing, biological, engineering cybernetics corresponds to Jantsch's (1980) rational attitude (being outside the stream), cybernetic self-organization to his mythical attitude (steering in the stream), and self-organization to his evolutionary attitude (being the stream). This is all to say that the three descriptions are one—cybernetics/self-organization. As one policy maker has put it: "My young friends, history is a river that may take us as it will. But we have the power to navigate, to choose direction and make our passage together. The wind is up, the current is swift and opportunity for a long and fruitful journey awaits us" (President Reagan at Fudan University in Shanghai, May 1, 1984).

Thus we consider policy-making to be design of purposeful systems—a process of evolutionary cooperative control (cybernetics) to all there is—dissipative self-organization—cybernetics/self-organization. The design and implementation of purposeful systems—dynamical systems of values, goals, technology, and controls—also constitutes meaning. (See Frankl 1962, 1969 on the will to meaning and values meaning of life and Simon 1981 on the notion of design as valued activity.) Meaning is relation (redundancy or deductively repetitive order) among elements (values, goals, technology, controls) defining the current problem representation as it relates to other representations of the same or different problems in long-term memory. (For related discussions see Mayer 1983 on meaning theory.) Meaning is also change in the current relation—evolution to a new relation—what Nozick (1981) refers to as transcending of limited orders. Mathematically, a relation is a subset of a Cartesian product of sets. Structure is also a relation (Atkins 1974), as is system. Structure (system, relation) follows from interactive processes whose self-organization is evolution to a new structure (Jantsch 1980, 1981). Meaning is relation (structure, system) and change in relation.

Thus meaning is design of purposeful systems is policy-making—mutually explanatory metaphors. They are evolutionary cooperative control (cybernetics) to all there is—dissipative self-organization—manifested by increasing consciousness. The latter is regarded as self-organizing response capacity (awareness) operating through cognition, affection, and conation and represented mathematically (cognitively) in the design problem by sets of values, goals, technology, controls whose expansions (mappings) signify increase (change) in consciousness (Chapter 2), which can also lead to a new coalition C. To the extent that decisions are made unconsciously, consciousness as sensor having cognitive, affective, and conative capabilities "reads" them more or less imperfectly in the design process. Simon (1983) suggests that affection (emotion) serves to focus attention in cognition.

To continue our discussion of the ESD paradigm, we consider hierarchy, control, evolution, and problem solving. The wall-as-all metaphor is discussed in Chapter 16. A hierarchy is a system composed of interrelated subsystems each of which in turn is composed of interrelated subsubsystems, etc., until some lowest level of elementary subsystem is reached (Simon 1981). The idea of control (and of communication or flow of information upon which control depends) is intimately involved with that of hierarchy and emergence in evolution. Checkland (1981) notes that these concepts are central to systems thinking—thinking about organized complexity. The model of organized complexity that arises from the systems approach is that of evolution of a hierarchy of levels of organization of increasing complexity—what Jantsch (1981) identifies as anagenesis. Hierarchy involves constraints (control) imposed on lower-level inputs resulting in higher-level outputs that have emergent properties. When we do not know the constraints (control) chosen, i.e., do not know the organizing principle, the system appears to be self-organizing. Consciousness as self-organizing response capacity (awareness)—see Chapter 2—may be regarded as an emergent property. Troncale (1982) stresses the need for research on the control process that forms hierarchical structure. We note that "control" here means constraints on inputs that themselves are commonly called "controls."

Simon (1981) emphasizes that the architecture of complexity is hierarchical—that the emergence of intermediate stable forms (subsystems) in a hierarchy greatly speeds up the evolution of a complex system from simple elements. His watchmaker's parable shows that it will take a watchmaker subject to interruptions a small fraction of the time to assemble a complex watch if he makes subassemblies. Simon suggests that problem solving is also hierarchical—a partial result representing progress toward goals is analogous to a stable subassembly. Problem solving toward goals involves selective trial and error or a generate-test search process. Alternatives are generated selectively based on heuristics or rules of thumb. This is followed by trial, outcome test for progress, and alternative selection. The process is evolutionary. In addition to generate-test search processes, problem solving involves pattern recognition. Simon (1983) identifies intuition as pattern recognition. He advocates a behavioral model of rationality (problem solving) involving bounded rationality and satisficing in which emotion focuses attention, and generate-test search and pattern recognition are complementary processes in problem solving. He also notes that generate-test search in the behavioral theory of rationality is the direct analogue of the variation-selection mechanism in Darwin's theory of evolution.

Bateson (1980) argues for a hierarchic structure of thought (problem solving) based on the idea of logical types (Whitehead and Russell 1910–1913) in which a class (output) is of a different logical type, higher hierarchically, than that of its members (input). Watzlawick et al. (1974) use the idea of logical types as a basis for change (reframing) involving emergence in problem solving. Bateson argues that thought and evolution follow the same stochastic process involving random generation of outcomes followed by outcome selection. Dissipative self-organization involves random generation and selection. In problem solving, as Simon points out, generation is not completely random but selective.

Citing supporting literature in several areas (including in the social systems area itself, see Hayek's [1973] concept of self-organizing spontaneous social orders), Malik and Probst (1982) discuss an evolutionary conception of management emphasizing self-organization. In this respect their evolutionary management is akin in spirit to ESD. In support of a self-organizing view of management, Malik and Probst mention the following. (1) A polycentric, self-organizing system displays greater adaptability in complex situations than a hierarchical command structure given limited system information-processing capability. (2) Human behavior, although directed by goals, is more fundamentally based on tacit, specific evolved rules of behavior—self-organizing norms subject also to conscious modification. (3) Unintended (self-organizing) side effects are prevalent. (4) Self-organizing relationship-oriented systems (not primarily goal driven), like the family, exist. An important implication of self-organization is that management is capable of only partial design (control) of organization. The rest is left to self-organization. Attempted full control is neither possible nor desirable.

We also note McKelvey's (1982) discussion of organizations from a biological evolutionary perspective, but one that does not focus on dissipative self-organization.

1.3 Evolutionary Systems Design: Mathematical Development

The mathematics of ESD formalizes cybernetics/self-organization involving the three descriptions noted in Section 1.2: (1) self-organizing, biological, engineering cybernetics, (2) cybernetic self-organization, and (3) self-organization. The difference between these three descriptions rests on the degree of control exercised by the designer (coalition C) in comparison to self-organizing forces. Description (1) involves the most control and (3) the least. Because they all involve at least some minimal control (in the limit no control—letting go, which paradoxically is the ultimate in control) on the part of the designer, we can draw on the mathematics of control and interpret it to accommodate cybernetics/self-organization.

Toward this end problem solving is viewed as a cooperative control problem within an N player planning game in which nothing is considered necessarily to be fixed. A subset of the set η of N players can cooperate (with agreements binding and enforceable) and form a coalition $C(C \subseteq \eta)$, which can be η itself, the grand coalition. $\overline{C}$ is the set of all other players not in coalition C who can form one or more coalitions, with one coalition being the most difficult for coalition C. The game between C and $\overline{C}$ is played noncooperatively—otherwise if C and $\overline{C}$ cooperate, they can form the grand coalition $C = \eta$. We note that N, C, and $\overline{C}$ can change over time. Mathematically speaking, coalition C's problem amounts to the design and solution at present time τ of a dynamical system such as (1), (2) below (involving goals, technology, and controls) with associated dynamical values. The mathematics of control are as follows:

$$x_s^C(t+1) = f_s^C\left[x_s^C(t), u_s^C(t), u_s^{\overline{C}}(t), t\right], \; x_s^C(\tau) = x_{Cs\tau} \tag{1}$$

where $x_{Cs\tau}$ is the known present state of coalition C's system at hierarchical level s at present time τ;

$$y_s'^C(t) = g_s^C\left[x_s^C(t), u_s^C(t), u_s^{\overline{C}}(t), t\right] \in Y_s^C(t) \tag{2}$$

C = a coalition or subset of the set η of N players ($C \subseteq \eta$) including η itself, the grand coalition.

$\overline{C}$ = set of all other players not in C who can form one or more coalitions.

τ = present time with $\tau = 0, 1, 2, \ldots$, representing a moving present time τ.

$t = \tau, \tau + 1, \tau + 2, \ldots, \tau + T(\tau)$ represents time periods over the planning horizon of length $T(\tau) + 1$.

$s = 1, 2, \ldots, S$ are descending hierarchical levels. Higher level, more general goals are delivered by (inclusive of) lower level, less general goals.

$\mathrm{x}_s^C(t) = (x_{1s}^C(t), \ldots, x_{n_s s}^C(t))$ is a n_s-dimensional vector that represents the state of coalition C's system at hierarchical level s at time t.

$y_s'^C(t) = (y_{1s}'^C(t), \ldots, y_{p_s s}'^C(t))$ is a p_s dimensional vector of systems outputs (goals) at level s for coalition C at time t. $Y_s^C(t)$ = coalition C's goal target, i.e., set of desired outputs at level s for coalition C at time t. $Y_s^C(t) = \cap Y_s^j(t)$ where $Y_s^j(t)$ is player j's ($j \in C$) goal target, a set of desired outputs at level s for player j at time t. If $Y_s^C(t)$ is empty, the $Y_s^j(t)$ can expand to give a nonempty intersection, $Y_s^C(t)$, a necessary condition for coalition C formation. $Y_s^C(t)$ and $Y_s^j(t)$ are subsets of coalition C's goal space (normally, R^{p_s}, the p_s-dimensional real vector space). The latter is defined to include all the goal dimensions over the planning horizon of interest to any player $j \in C$. Also the dimensions of this goal space can be redefined—see discussion of goals/values relation below and in Chapter 2.

$\mathrm{u}_s^C(t)$ = a c_s-dimensional vector of system inputs (controls) for level s for coalition C at time t.

$u_s^{\bar{C}}(t)$ = a $\bar{c}_s$-dimensional vector of system inputs (controls) for level s for players $\bar{C}$ at time t.

$u_s(t) = (u_{1s}(t), u_{2s}(t), \ldots, u_{m_s s}(t)) = (u_s^C(t), u_s^{\bar{C}}(t))$ = an m_s-dimensional vector of system inputs (controls) for level s partitioned between controls $u_s^C(t)$ and $u_s^{\bar{C}}(t)$ assigned respectively to C and $\bar{C}$. Note that $c_s + \bar{c}_s = m_s$.

Equation (1) expresses state transition function f_s^C. Equation (2) gives the output function g_s^C and, as a necessary condition, constrains outputs $y_s'^C$ to a set of desired outputs $Y_s^C(t)$. Normally $Y_s^C(t) \subset R^{p_s}$, as noted. With R^{c_s} and $R^{\bar{c}_s}$ partitions of R^{m_s}, control space, we have constraints $x_s^C(t) \in X_s^C(t) \subset R^{n_s}$, $u_s(t) \in U_s(t) \subset R^{m_s}$, $u_s^C(t) \in U_s^C(t) \subset R^{c_s}$, $u_s^{\bar{C}}(t) \in U_s^{\bar{C}}(t) \subset R^{\bar{c}_s}$.

In general the last constraint is not known by coalition C which necessarily must predict it. $U_s^C(t)$ is a set of c_s-dimensional technologically feasible controls available to coalition C. Similarly we have $U_s^{\bar{C}}(t)$ for players $\bar{C}$. $U_s^C(t)$ and $U_s^{\bar{C}}(t)$ are projections of $U_s(t) \subset R^{m_s}$ onto R^{c_s} and $R^{\bar{c}_s}$ where $U_s(t)$ is a set of m_s-dimensional technologically feasible controls available to C and $\bar{C}$. Coalition C's control space R^{c_s} is defined to include all the control dimensions over the planning horizon available to any player $j \in C$ (similar definitions for $R^{\bar{c}_s}$ and R^{m_s}). Coalition C's set of technologically feasible controls

$U_s^C(t) \supseteq \cup U_s^j(t)$ where $U_s^j(t) \subset R^{c_s}$ is player j's set of technologically feasible controls.

We let $\bar{C}(\nu)$ for $\nu = 1, 2, \ldots$ represent coalitions of $\bar{C}$ that coalition C may choose to predict. If not, coalition C may design for the worst case—that players $\bar{C}$ form a single coalition.

If $\bar{C}$ consists of more than one coalition, $u_s^{\bar{C}}(t)$ is partitioned into $\bar{c}(\nu)$-dimensional controls $(u_s^{\bar{C}(\nu)}(t))$, i.e., $u_s^{\bar{C}}(t) = (u_s^{\bar{C}(\nu)}(t))$, where $\Sigma_\nu[\bar{c}(\nu)]_s = \bar{c}_s$.

Each coalition $\bar{C}(\nu)$ chooses controls $u_s^{\bar{C}(\nu)}(t) \in U_s^{\bar{C}(\nu)}(t)$, coalition $\bar{C}(\nu)$'s set of technologically feasible controls. We have $U_s^{\bar{C}(\nu)}(t) \supseteq \cup U_s^{\bar{j}(\nu)}(t)$, where $U_s^{\bar{j}(\nu)}(t) \subset R^{c(\nu)_s}$ is the set of technologically feasible controls for player $\bar{j}(\nu) \in \bar{C}(\nu)$. Thus the expression $u_s^{\bar{C}}(t) \in U_s^{\bar{C}}(t)$ represents the set of constraints $\{u_s^{\bar{C}(\nu)}(t) \in U_s^{\bar{C}(\nu)}(t)\}$ where $U_s^{\bar{C}(\nu)}(t)$ is the projection of $U_s^{\bar{C}}(t)$ onto $R^{\bar{C}(\nu)_s}$, i.e., $U_s^{\bar{C}(\nu)}(t) \subset R^{c(\nu)_s}$.

We note that cooperation is possible only within coalitions, so between coalitions—the coalitions $\bar{C}(\nu)$ and coalition C—the game is played noncooperatively.

Inputs $u_s^C(t)$ and $u_s^{\bar{C}}(t)$ carry the subscript s to denote they are inputs (controls) for outputs $y_s'^C(t)$, but hierarchically they are drawn at level $(s + 1)$, as shown in Figure 1.1. In (1), (2), and Figure 1.1 the constraint (control) on inputs (controls) is expressed by $f_s^C(t)$ and $g_s^C(t)$, together comprising the technology (structure) that in general is not fixed but evolves. The technology constrains the inputs, producing the emergent properties of the outputs that signify the transition from level $(s + 1)$ to level s. Mathematically we have f_s^C: $R^{n_s} \to R^{n_s}$ and g_s^C: $R^{m_s} \to R^{p_s}$. The former expresses the transition (mapping) from the system state $x(t)$ at time t to the state $x(t + 1)$ at time $(t + 1)$; the latter expresses the mapping from control space, R^{m_s} to goal space, R^{p_s}.

When the system is Markovian, as is (1), (2), in that only the present state $x_{Cs\tau}$ (and not past states) enters into the determination of change of state and output, we may simplify the time indexing by defining present time τ always to be 0 with the understanding that there is a moving present, i.e., a moving 0.

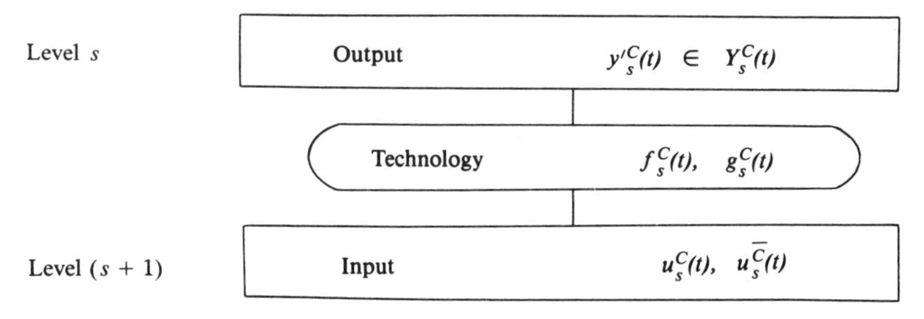

Figure 1.1. **Hierarchical relationship between output and input for coalition C's system**

In this case $t = 0, 1, 2, \ldots, T$ where $T = T(0)$, with the moving present τ symbolized by $t = 0$. Further, if we understand that (1), (2) applies to a hierarchical level s, we can also simplify the notation by dropping the subscript s. We also omit the superscript C when it is not needed for clarity. Then we rewrite (1), (2) as follows:

$$x(t+1) = f\left[x(t), u^C(t), u^{\bar{C}}(t), t\right], \; x(0) = x_0 \tag{3}$$

$$y'(t) = g\left[x(t), u^C(t), u^{\bar{C}}(t), t\right] \in Y(t) \tag{4}$$

where $t = 0, 1, 2, \ldots, T$.

We note that dynamical models other than (1), (2) can be used—e.g., ones incorporating state memory, anticipation (disturbance feedforward), and output feedback (see below and Chapter 2).

Let $\mathscr{R}(t)$ be the set of states in R^n reachable by coalition C from x_0 at time t by application of $U^C(\ell)$ and $U^{\bar{C}}(\ell)$ in (3) where $\ell = 0, 1, \ldots, t-1$. $\mathscr{A}(t)$ is the intersection of $\mathscr{R}(t)$ with $X(t)$, the set of admissible states, i.e., $\mathscr{A}(t) = \mathscr{R}(t) \cap X(t)$. $\mathscr{A}(t)$ is the set of admissible reachable states. For details on the characterization and computation of $\mathscr{R}(t)$, see Casti (1977) and Hermann and Krener (1977). However, coalition C does not control the input choice from $U^{\bar{C}}(\ell)$—players $\bar{C}$ do. Because of this, coalition C can control only to a set of states and not to a single state. For each control sequence $u^C(\ell) \in U^C(\ell)$ for $\ell = 0, 1, \ldots, t-1$ available to coalition C, there is a set of admissible reachable states, $\mathscr{A}(t)|u^C(\ell) = (\mathscr{R}(t)|u^C(\ell)) \cap X(t)$ associated with $U^{\bar{C}}(\ell)$, i.e., with $\bar{C}$'s possible use of any of the $u^{\bar{C}}(\ell) \in U^{\bar{C}}(\ell)$, where $\mathscr{R}(t+1)|u^C(\ell) = f(\mathscr{A}(t)|u^C(\ell), |u^C(t), U^{\bar{C}}(t), t)$. When $t = 0$, ℓ is undefined, $\mathscr{A}(0) = x_0$, and $\mathscr{R}(1)|u^C(0) = f(x_0, u^C(0), U^{\bar{C}}(0), 0)$.

Using (4) for each control $u^C(t) \in U^C(t)$ available to coalition C there is a set of reachable outputs $y(t)|u^C(t) = g(\mathscr{A}(t)|u^C(\ell), u^C(t), U^{\bar{C}}(t), t)$ associated with $U^{\bar{C}}(t)$, i.e., with coalition $\bar{C}$'s possible use of any of the $u^{\bar{C}}(t) \in U^{\bar{C}}(t)$. Then $y(t)$ is the set of reachable sets $y(t)|u^C(t)$ for $u^C(t) \in U^C(t)$; i.e., $y(t) = (y(t)|u^C(t))$ represents technologically feasible performance (output). In choosing a control $u^C(t)$, coalition C can reach the set of outputs $y(t)|u^C(t)$ regardless of which of the controls in $U^{\bar{C}}(t)$ players $\bar{C}$ choose. Coalition C can attempt to control to a single output by predicting $u^{\bar{C}}(\ell)$ for here, $\ell = 0, 1, \ldots$ t, by analyzing the game between C and $\bar{C}$, but it can be wrong. However, only the present control $u(0) = u^C(0), u^{\bar{C}}(0)$ is implemented. There is a moving present and a redesign and resolution of the system by coalition C at the next present, one time period later.

If for $t = 0, \ldots, T$ the intersection of the goal target $Y^C(t)$—what coalition C wants—and the technologically feasible performance $y^C(t) = y(t)$—what it can get—is a single set, i.e., one of the sets $y(t)|u^C(t)$, then, as a necessary and sufficient condition, coalition C has defined and solved its problem. Otherwise these sets may change, i.e., expand or contract until a single-set intersection is found. In general, there is a mapping from a current set ($Y^C(t)$ or $y^C(t)$) to a new set by which the current set is redefined (see

Chapters 2 and 9). This can include a goals/value referral process whereby the dimensions of the operational goal space R^P can be redefined (see below and Chapter 2).

Alternatively, if coalition C predicts $u^{\bar{C}}(\ell)$ for $\ell = 0, 1, \ldots, t$ by analyzing the game between C and $\bar{C}$, the solution (intersection of $Y^C(t)$ and $y^C(t)$) can be defined as a single point rather than a single set, but if the prediction is wrong, the solution so defined will not be realized. Also a solution can be defined as a subset of a set $y(t)|u^C(t)$ if coalition C predicts that $u^{\bar{C}}(\ell)$ for $\ell = 0, 1, \ldots, t$ will fall in a subset of $U^{\bar{C}}(\ell)$. If $C = \eta$, the grand coalition, the solution can be defined and realized as a single point.

If coalition C correctly predicts $u^{\bar{C}}(t)$ for $t = 0, 1, \ldots, T$ and maximizes a coalition utility function defined over $t = 0, 1, \ldots, T$, and vice versa for other players $\bar{C}$, then the noncooperative equilibrium solution of classical game theory is obtained (see Kinberg, Shakun, and Sudit 1978). If for each $u^C(t) \in U^C(t)$ a coalition C utility function is defined on outcomes $y'(t) \in y(t)|u^C(t)$ for $t = 0, \ldots, T$, then by choosing $u^C(t)$, coalition C can maximize its minimum utility, i.e., a maximin utility solution for coalition C playing against $\bar{C}$.

In our mathematical development here we assume the controls available to coalition C is a set of technologically feasible controls, $U^C(t)$. In some cases (see Chapter 9) the controls available to coalition C may be a set of a priori admissible controls, $U_0^C(t) \subset U^C(t)$. Then $U_0^C(t)$ can be used in the mathematics presented instead of $U^C(t)$. Thus, in general, only some of the technologically feasible controls $U^C(t)$ may be a priori admissible for a given player $j \in C$, i.e., a set of controls $U_0^j(t)$ yielding outputs (goals) belonging to a set of a priori specificed outputs (goals). Then $U_0^C(t) = \cap U_0^j(t)$ is the coalition C joint set of a priori admissible controls.

Recalling that goals are operational expressions of values, related to the goals $y_s'^C(t)$ in (2) or $y'(t)$ in (4) are a set of K values $v(t) = v_s^C(t) = (v_{ks}^C(t))$ at levels s for coalition C at time t ($k = 1, 2, \ldots, K$). Here $s = \ldots, -3, -2, -1, 0$ are descending hierarchical value levels with higher level, more general values being inclusive of lower level, less general values. If we do not distinguish among these levels, then all values are at level $s = 0$. As noted above, descending goal levels are denoted by $s = 1, 2 \ldots, S$, so that the highest level goals are at $s = 1$. In the following discussion we shall relate values all considered at level $s = 0$ to highest level goals at $s = 1$. For development of the values/goals hierarchy, see Chapter 14.

Consider a set of p goal dimensions $\mathscr{y}^C(t) = \mathscr{y}(t) = (\mathscr{y}_i(t))$ for $i = 1, 2, \ldots, p$ comprising the dimensions of coalition C's goal space R^p. (In later chapters the ith goal dimension is sometimes denoted by g_i.) We define relations $\lambda(t) = (\lambda^j(t))$ from a set of values $v^C(t) = v(t)$ to a set of goal dimensions $\mathscr{y}(t)$, which shows which values are delivered by which goal dimensions for each player $j \in C$. We note that sets $v(t)$ and $\mathscr{y}(t)$ include all the values and goal dimensions of interest to any player $j \in C$. The relations $\lambda(t)$ can be represented as a J-matrix where J is the number of players in coalition C. For example, for $J = 2$ we would have a bimatrix—at each row-column intersection (k, i) we would have two λ entries corresponding, respectively, to $j = 1, 2$. The goals/values matrix gives the relations

$$\lambda^j(t) \subset v(t) \times \mathscr{y}(t) \quad j \in C \tag{5}$$

and is shown by

	$\mathscr{y}(t)$
$v(t)$	$(\lambda^j_{ki}(t))$

where $\lambda^j_{ki} = 1$ indicates player j is "for" value v_k being delivered by goal dimension $\mathscr{y}_i$ (i.e., he favors both the value v_k and the goal dimension $\mathscr{y}_i$ as an operational expression of this value; $\lambda^j_{ki} = 0$ indicates player j is against value v_k being delivered by goal dimension $\mathscr{y}_i$; $\lambda^j_{ki} = X$ indicates player j is neutral or does not perceive value v_k as being delivered by goal dimension $\mathscr{y}_i$. If rows and columns are interchanged, we use the inverse relation $\lambda^{-1j}(t) \subset \mathscr{y}(t) \times v(t)$. In Chapter 2 we discuss the use of the λ relation formalization of what we describe there also as the goals/values referral process by which the relations $\lambda(t)$ can change, including change in the dimensions of the operational goal space $\mathscr{y}(t)$ and values $v(t)$. This can lead to goal dimension agreement—agreement among the players $j \in C$ on a common set of goal dimensions—which is a necessary condition for goal target agreement, i.e., a nonempty coalition goal target $Y^C(t) = \cap Y^j(t)$. The possibility of the latter depends on players $j \in C$ agreeing to negotiate on the same set of goal dimensions, i.e., within the same goal space. We may expand Figure 1.1 into Figure 1.2, where goals are at level $s = 1$. We have a controls/goals/values relation, and can use goals/values, controls/goals, and controls/values referral processes operating between or within hierarchical levels (see Chapter 2). In addition to goals, in effect at level $s = 1$ we have norms—tacit, specific, evolved rules of behavior. Norms reveal themselves by mismatch with current operational goals or values. For relationship-oriented systems (not primarily goal driven), like the family, controls and technology deliver values directly and norms tacitly (goals may be imputed).

Suppose some players $j \in C$ are uncertain about their future values and goals, i.e., are uncertain about their $\lambda^j(t)$ matrices for $t > 0$. If these matrices, although uncertain, can be identified, then contingency planning (Chapter 2) can be used where here the contingency involves uncertain goals/values matrices. For example, one can determine—assuming a present control $u^C(0)$ were to be implemented in (3), (4)—whether, starting from the set of admissible reachable states $\mathscr{A}(1) = \mathscr{R}(1) \cap X(1)$, alternate sets of future ($t > 0$) goals deriving from alternate $\lambda^j(t)$ matrices can be realized. For a substantive discussion of the problem of uncertain future preferences that draws on behavioral studies and considers normative implications, see March (1978).

Recall that the search for a solution (a single set or single point intersection between $Y^C(t)$ and $y^C(t)$) involves mapping from a current set ($Y^C(t)$ or $y^C(t)$) to a new set by which the current set is redefined. This mapping may be

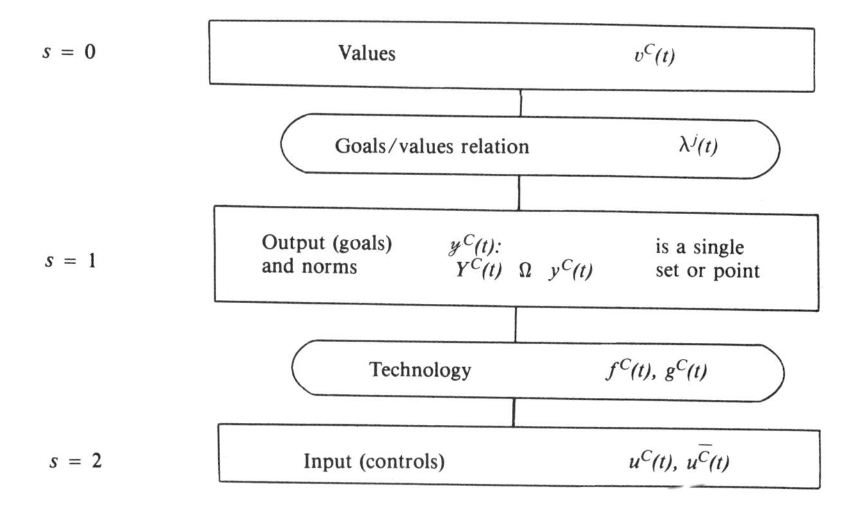

Figure 1.2. **Hierarchical relation between controls, goals and values (For an application see Figure 14.1)**

facilitated by the use of individual preference structures $\mu_j(t)$ at time period t and μ_j over all time periods for player $j = 1, 2, \ldots, J$ $(j \in C)$ where $\mu_j(t) = \mu^j(y'^C(t), t)$ and $\mu_j = \theta^j(\mu_j(t))$ for $t = 0, 1, \ldots T$. For example, μ^j and θ^j could each be a utility function or an ordering. Preference structures can be defined not only on single point outputs $y'^C(t)$ but also on single set outputs $y(t)|u^C(t)$; i.e., we can have $\mu_j^*(t) = \mu^{j*}(y(t)|u^C(t), t)$ where the μ^{j*} on sets would be a different function than the μ^j on points. We also have $\mu_j^* = \theta^{j*}(\mu_j^*(t))$. Of course, preferences can be defined on points $Y'^C(t)$ in the goal target $Y^C(t)$ as well, i.e., we have $\mu_j(t) = \mu^j(Y'^C(t), t)$. The individual preference structures μ_j can be represented in preference space R^J (e.g., utility space where utilities are used for μ_j or an ordering space where, say, rank orders are represented by positive integers). In Chapters 9 and 10, we discuss how a decision support system can show the decision problem—graphically or as relational data in matrix form—in three spaces as a mapping from control space to goal space to preference (there utility) space. Within each of these spaces there is a mapping from a current set (which can be a target set or a feasible set, e.g., $Y^C(t)$ or $y^C(t)$ in goal space) to a new set by which the current set is redefined.

The current-to-new set mapping in any one space has corresponding mappings in the other two because of the mapping from control space to goal space to preference space. Figure 1.3 shows the hierarchical relation between controls, goals, and preferences. In some cases a coalition preference structure μ_C is defined on individual preference structures (μ_j) as shown, i.e., $\mu_C = f(\mu_j)$.

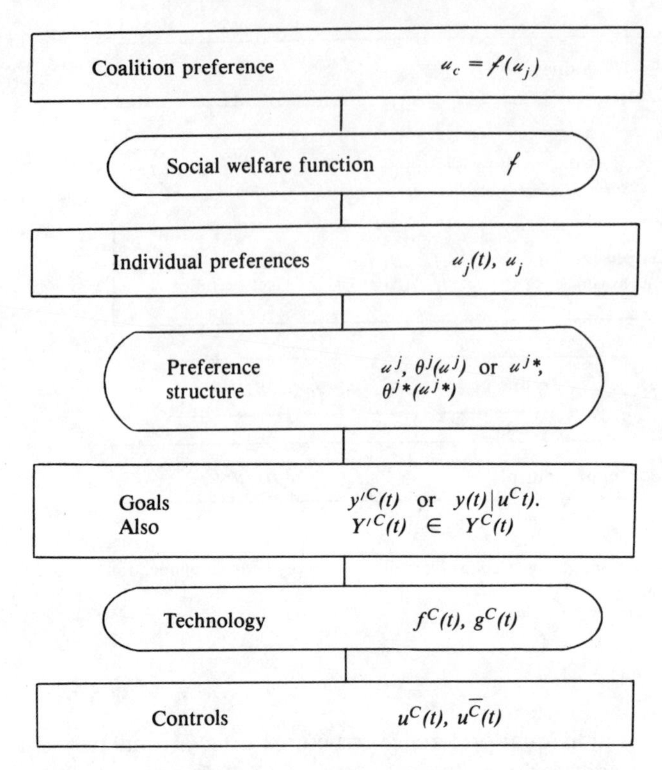

Figure 1.3. **Hierarchical relation between controls, goals, and preferences**

In sum, preference structures are represented by:

$$\mu_j(t) = \mu^j(y'^C(t), t) \quad \text{and} \quad \mu_j = \theta^j(\mu_j(t));$$

similarly defined on $Y'^C(t) \in Y^c(t)$.

$$\mu_j^*(t) = \mu^{j*}(y(t)|u^C(t), t) \quad \text{and} \quad \mu_j^* = \theta^{j*}(\mu_j^*(t)) \tag{6}$$

$$\mu_C = f(\mu_j) \quad \text{for} \quad j \in C \quad \text{and} \quad t = 0, 1, \ldots T.$$

The act of control for coalition C is the design and solution at present time $t = 0$ of a dynamical system such as (3), (4), (5), (6); it includes the implementation of present control $u^C(0)$. After operation in present period $t = 0$, a redesign and resolution of the system may be undertaken at the next present, one time period later. Only the present control $u^C(0)$ is implemented, and there is a moving present. A sequence of acts of control is called a process of control.

Sometimes, e.g., in Chapter 7 on bivariate negotiations as a stochastic terminal control problem, we have a given input producing a risky output characterized by a probability distribution. In this situation the preference

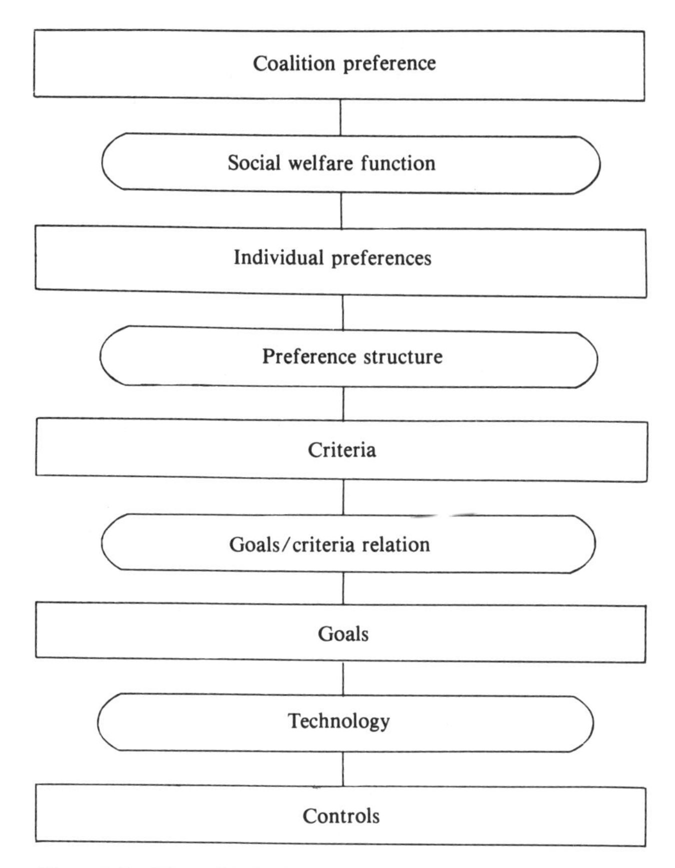

Figure 1.4. **Hierarchical relation between controls, goals, criteria, and preferences**

structure $u_j(t)$ can be defined on risky outputs. Various multicriteria methods employing such criteria as means, standard deviations, probabilities of ruin, etc., can be used, e.g., the UTA utility method described in Chapter 8. Such criteria can be used to define an output criteria space. For risky outputs the goal target $Y^C(t)$ is defined in the same output criteria space as the technologically feasible performance $y^C(t)$, and a single set or single point intersection in criteria space can be obtained. This intersection maps into a corresponding intersection in preference space. To modify Figure 1.3 for risky outputs, a criteria level is drawn between goals and individual preferences, giving Figure 1.4. The goals/criteria relation in Figure 1.4 expresses which criteria (means, standard deviations, etc.) are used to characterize risky outputs. Individual preferences are defined on these criteria. Otherwise the mathematics in Figure 1.4 is the same as in Figure 1.3.

Although a redesign of the system may be undertaken one time period later (moving present), formally system (1), (2) or its equivalent (3), (4) does not incorporate output feedback or disturbance feedforward (anticipation) loops. Hence it does not allow for negative feedback or feedforward—mainstays of cybernetic thinking—to stabilize structures and outputs, nor positive feedback or feedforward to permit cybernetic control to new structures. Self-organization, in fact, requires feedback or feedforward loops (Allen 1981; Jantsch 1980; Nicolis and Prigogine 1977). Such loops can be integral to the system dynamics (1), (2) or (3), (4), and/or we can add output feedback and disturbance feedforward.

Negative feedback or feedforward is associated with autopoiesis (organizational closure), a self-referential process within self-organization for system self-renewal. Positive feedback or feedforward is associated with self-organization to new structures (evolution). (In chemistry positive feedback corresponds to auto- or crosscatalytic reactions.) We may extend (1), (2) to include state memory, anticipation (disturbance feedforward), and output feedback. (For detailed discussion see development of [19], [20] in Chapter 2.)

$$x(t+1) = \mathscr{F}\left[\underline{\overline{x}}(t), u^{C}(t), u^{\overline{C}}(t), t\right], x(\tau) = x_{\tau} \tag{7}$$

$$y'(t) = \mathscr{G}\left[\underline{\overline{x}}(t), u^{C}(t), u^{\overline{C}}(t), t\right] \in Y(t) \tag{8}$$

where
$$u^{C}(t) = u^{C}\left[\underline{\overline{x}}(t), E(t-1), t\right] \tag{9}$$

$\underline{\overline{x}}(t)$ = vector of known and predicted states using predicted future disturbances (feedforward). For detailed discussion see Chapter 2.

$E(t-1) = \|\tilde{y}(t-1) - y_0^*(t-1)\|$ = feedback error or minimum distance between actual output $\tilde{y}(t-1)$ and goal target $Y(t-1)$. Thus $y_0^*(t-1)$ is the element of $Y(t-1)$, which minimizes the distance $\|\tilde{y}(t-1) - Y(t-1)\|$.

and $u^{\overline{C}}(t)$ includes disturbance, $u^0(t)$, as predicted by coalition C.

Equations (5), (6), (7), (8), (9) can model cybernetics/self-organization. System state x and output y' can have bifurcation points at which there is a choice of branch (structure). With description (1)—self-organizing, biological, engineering cybernetics—the designer, coalition C (subject to $\overline{C}$'s actions), can control the system to a bifurcation point. Then, at the bifurcation, a new branch (structure), e.g., a new technology $\mathscr{F}$, $\mathscr{G}$ in (7), (8), may be selected by the designer, associated with positive feedback or feedforward in (9). The goals/values and preference structures (5) and (6) can also be redesigned. The designer may also use negative feedback or feedforward in (9) to maintain the present structure to deliver goals and values, in which case he would, better, avoid the bifurcation point in the first place. Self-organizing phenomena under description (1) are relatively weak. Under description (2)—cybernetic self-organization—self-organizing forces are stronger. However, again the designer is basically able to control the system to a bifurcation point, but there fluctuations determine the branch the system will follow. In other words, under description (2) the branch (structure) chosen is not predictable but depends on chance fluctuation, i.e., on self-organization. Finally, with description (3), self-organization, the designer does not drive the system to a bifurcation point—self-organizing chance fluctuations do this and also choose the branch the system will follow. Here the designer—along with everything else—is self-organization.

The dog run conflict described in Chapter 2 is an example of description (3), self-organization, where the goals/values structure (5) is driven to a new structure by self-organizing forces. Suddenly a new value—minority rights—is introduced, and the goals/values structure evolves to a new one leading to conflict resolution.

An example of description (2), cybernetic self-organization, is also given in Chapter 2 in the bank–women's group conflict. There the designer uses the goals/values referral process to bring the system to a bifurcation point. Then a new goal is generated by self-organization, which changes the goals/values structure (5) leading to conflict resolution.

Nicolis and Prigogine (1977), Prigogine (1980), Jantsch (1980), and Allen (1981) present a number of models of self-organization in physics/chemistry, biology, ecology, and social systems, including, in the latter, marketing and urban evolution.

We may relate the previous models to catastrophe theory as follows: Suppose in (8) or (2) there is only one output $y'(t)$—a so-called "potential" function (not necessarily known)—that the system seeks to minimize so that $Y(t) = \min y'(t)$. Then, if the system follows gradient-type dynamics, i.e., if expressed as a differential equation, (7) or (1) can be written as:

$$\frac{dx_n}{dt} = -\frac{\partial y'(t)}{\partial x_n} \tag{10}$$

where $n = 1, 2, \ldots n$, then the steady-state solutions of the x_n in (10), which depend on the control parameters of u, show catastrophes (Casti 1979; Thom 1975; Zeeman 1977)—bifurcations—at which there is discontinuous change in the steady-state values of x_n associated with a continuous change in the control parameters.

Resilience is the capability of a system to persist under unknown (broad class) external or internal disturbance (Casti 1979; Holling 1976) or the capability to adapt to surprise because of past experiences of instabilities (Casti 1982; Holling 1981). In our terms resilience is the capability of a system (5), (6), (7), (8), (9) to continue to give solutions as present time τ moves forward. For qualitative comments see Chapter 2 and Holling (1976, 1981); for mathematical approaches to resilience see Casti (1979, 1982).

In sum, as design methodology, ESD—the evolutionary design of evolutionary systems—is cybernetics/self-organization, which can be modeled by (5), (6), (7), (8), (9).

1.4 Problem Representation in Decision Support Systems

Collectively, relations (5), (6), (7), (8), (9) represent coalition C's problem representation, which can serve as a basis for computer decision support systems (DSS) and, in particular, for negotiation support systems (NSS) involving multiple players. A basic idea in DSS (NSS) is to show the evolving group (coalition C) problem representation (5), (6), (7), (8), (9) graphically or as relational data in matrix (spreadsheet) form as relations between values and goal dimensions (5), preferences and goals or criteria (6), and goals and controls (7), (8), (9). Thus in applications we show the goals/values relation and mappings from control space to goal space to criteria space to preference space.

In DSS applications the problem representation for coalition C is a joint or combined view of the design problem attributed to all players $j \in C$. We mention two cases of interest in using DSS:

1. monouser case. One or more independent users in coalition C. Each user in coalition C develops a private joint problem representation. By this we mean he builds his own individual problem representation (which is supported by the DSS), and he imagines (either unsupported or supported) the problem representations for all other players in coalition C.
2. multiuser case. This case adds to the monouser case having two or more users the facility for building a public, explicit, common, joint problem representation.

At each stage in the design process the joint problem representation for coalition C shows the acknowledged degree of consensus (or conflict) among the players, i.e., at each stage players may show different individual problem representations. The evolution of problem representation (problem solving) can be described as a process of consensus seeking—through sharing of views, which constitutes exchange of information—within which compromise is possible. Thus at any stage of consensus seeking (problem representation) players can accept (suggest) compromise based on axiomatic solution concepts or engage in concession making to arrive at a compromise (see Chapter 9 for decision support on these). Computer display of the evolving problem representation can be used to support continued consensus seeking. Computer implementation of these ideas in the NSS MEDIATOR is discussed in Chapter 10. For methodological development and a DSS application to design and negotiation of new products see Chapters 11 and 12. For negotiation support in freeing hostages see Chapter 13. With such systems we are moving toward decision support systems for multiplayer, multicriteria, ill-structured, dynamic problems, thus implementing the ESD methodology in a decision support context.

Some examples will illustrate ESD problem representation in DSS. For example, at any stage of the design process the goals/values matrix (5), which shows which values are delivered by which goal dimensions, can be displayed by the DSS. To illustrate, in the dog run conflict described in Chapter 2, Figure 2.14 shows the initial goals/values matrix. As noted above and discussed in Chapter 2, this initial goals/values matrix evolves to a new structure, Figure 2.15, as a result of self-organizing forces. The new goals/values matrix is displayed by the DSS. The process is an example of self-organization (description 3 above) in DSS.

The bank–women's group conflict in Chapter 2 is an example of cybernetic self-organization (description 2 above). The initial goals/values matrix (Figure 2.16) is displayed by the DSS. Although there is goal dimension agreement, negotiations become deadlocked. Using the goals/values referral process, players pose the following question: Given the value "being for women's rights," is there any other goal dimension in addition to the number of women vice-presidents that delivers this value? The system has thus been brought to a bifurcation point. A second goal dimension—number of women's scholarships—is then generated by self-organization, which changes the goals/values structure to Figure 2.17 displayed by the DSS. If at the bifurcation point the second goal dimension is chosen from a set of goal dimensions built into the DSS knowledge base, then we have an example of self-organizing cybernetics (description 1). Another example of cybernetic self-organization (description 2) is the generation of a new goal dimension, profit sharing, in negotiation support to free hostages (Chapter 13).

We mention three additional ideas discussed and applied to design and negotiation of new products in Chapters 11 and 12. As another illustration of ESD problem representation in DSS, consider the situation in which each of two players has different individual mappings from control space to goal space, in particular with respect to goal i. The DSS uses the union of the two

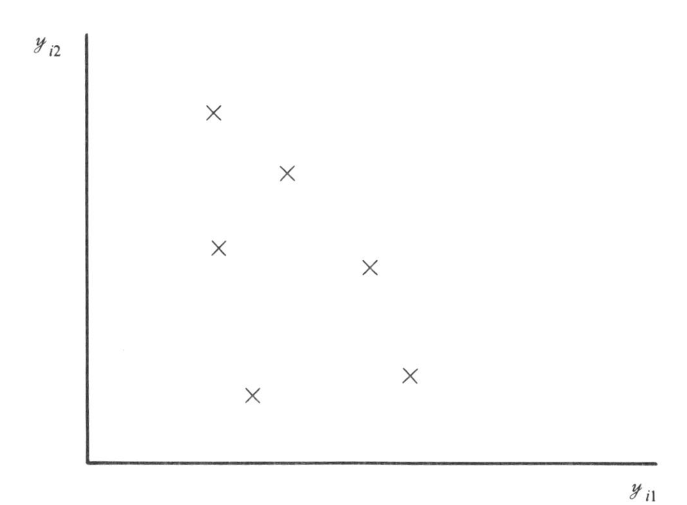

Figure 1.5. **Goal *i* representation in coalition goal space**

goal dimensions y_{ij}(y_{i1} and y_{i2} for players $j = 1, 2$, respectively) in the coalition goal space. Figure 1.5 shows the (y_{i1}, y_{i2}) plane projection that can be displayed by the DSS.

Each point in Figure 1.5 represents the mapping made by players 1 and 2 from a given point in control space. Evidently there is considerable conflict in players' mappings, which they can discuss. If points in Figure 1.5 fell along a 45^0 line from the origin, then players would fully agree on their mappings. In assessing players' preferences—utilities by PREFCALC (Chapters 8 through 12)—both dimensions y_{i1} and y_{i2} are included, but a player may choose to give zero weight to the other player's goal i dimension.

If the mappings from control (input) space are risky, i.e., a given input produces risky goals (outputs), the points in Figure 1.5 could be means of a bivariate probability distribution that maps into criteria space and then into preference space (see Figure 1.4).

If controls themselves are risky (e.g., controls on a risky technological frontier) in that at least one player estimates there is a probability less than one of realizing some given control (input), then these probabilities for the various players are dimensions in goal space and criteria space that express the probability of realizing goals and criteria due to risky controls.

In Chapter 2 we elaborate on the ESD framework presented in Chapter 1.

REFERENCES

Ackoff, R. L., and Emery, F. E. 1972. *On Purposeful Systems*. Aldine-Atherton, Chicago.

Allen, P. M. 1981. "The Evolutionary Paradigm of Dissipative Structures." In Jantsch, E. (ed.). *The Evolutionary Vision*. Westview Press, Boulder, CO.

Ashby, W. R. 1956. *An Introduction to Cybernetics*. Wiley, New York.

Atkin, R. H. 1974. *Mathematical Structure in Human Affairs*. Heinemann, London.

Atlan, H. 1981. "Hierarchical Self-Organization in Living Systems." In Zeleny, M. (ed.). *Autopoiesis*. North Holland, New York.

Bateson, G. 1980. *Mind and Nature: A Necessary Unit*. Bantam Books, New York.

Casti, J. 1977. *Dynamical Systems and Their Application: Linear Theory*. Academic Press, New York.

______. 1979. *Connectivity, Complexity and Catastrophe in Large Scale Systems*. Wiley, New York.

______. 1982. "Topological Methods for Social and Behavioral Systems." *International Journal of General Systems*, 8, pp. 187–210.

______. 1984. *On System Complexity: Identification, Measurement and Management*. International Institute of Applied Systems Analysis, Vienna.

Checkland, P. 1981. *Systems Thinking, Systems Practice*. Wiley, New York.

Churchman, C. W. 1971. *The Design of Inquiring Systems*. Basic Books, New York.

Frankl, V. E. 1962. *Man's Search for Meaning: An Introduction to Logotherapy*, rev. ed. Beacon Press, Boston.

______. 1969. *The Will to Meaning: Foundations and Applications to Logotherapy*. World Publishing, New York.

Harsanyi, J. C. 1977. *Rational Behavior and Bargaining Equilibrium in Games and Social Situations*. Cambridge University Press, Cambridge.

Hayek, F. A., von. 1973. *Law, Legislation and Liberty*, vol. 1, *Rules and Order*. University of Chicago Press, Chicago.

Hermann, R., and Krener, A. 1977. "Nonlinear Controllability and Observability." *IEEE Transactions on Automatic Control*, AC-22, pp. 728–740.

Holling, C. S. 1976. "Resilience and Stability of Ecosystems." In Jantsch, E., and Waddington, C. H. (eds.). *Evolution and Consciousness*. Addison-Wesley, Reading, MA.

______. 1981. "Resilience in the Unforgiving Society." Report R-24, Institute of Resource Ecology, University of British Columbia, Vancouver.

Jantsch, E. 1975. *Design for Evolution*. George Braziller, New York.

______. 1979. "The Unifying Paradigm Behind Dissipative Structures, Autopoicsis, Hypercycles and Ultracycles." Center for Research in Management Science, University of California, Berkeley.

______. 1980. *The Self-Organizing Universe*. Pergamon Press, New York.

______. 1981. "Unifying Principles of Evolution." In Jantsch, E. (ed.). *The Evolutionary Vision*. Westview Press, Boulder, CO.

Jantsch, E., and Waddington, C. H. (eds.). 1976. *Evolution and Consciousness*. Addison-Wesley, Reading, MA.

Kinberg, Y., Shakun, M. F., and Sudit, E. F. 1978. "Energy Buffer Stock Decisions in Game Situations." In Aronofsky, J. S., Rao, A. G., and Shakun, M. F. (eds.). *Energy Policy*. North Holland, Amsterdam.

Lewin, A. Y., and Shakun, M. F. 1976. *Policy Sciences: Methodologies and Cases*. Pergamon Press, New York.

Malik, F., and Probst, G. J. B. 1982. "Evolutionary Management." *Cybernetics and Systems*. 13, pp. 153–174.

March, J. G. 1978. "Bounded Rationality, Ambiguity and the Engineering of Choice." *Bell Journal of Economics*, 9, No. 2, Autumn, pp. 587–608.

Maslow, A. G. 1954. *Motivation and Personality*. Harper and Row, New York.

Mayer, R. E. 1983. *Thinking, Problem Solving, Cognition*. W. H. Freeman, New York.

McKelvey, B. 1982. *Organizational Systematics*. University of California Press, Berkeley.

Nicolis, G., and Prigogine, I. 1977. *Self-Organization in Nonequilibrium Systems: From Dissipative Structures to Order Through Fluctuation*. Wiley Interscience, New York.

Nozick, R. 1981. *Philosophical Explanations*. Harvard University Press, Cambridge, MA.

Prigogine, I. 1980. *From Being to Becoming*. W. H. Freeman, San Francisco.

Prigogine, I., and Stengers, I. 1984. *Order Out of Chaos*. Bantam Books, New York.

Rokeach, M. 1973. *The Nature of Human Values*. Free Press, New York.

Rosen, R. 1978. *Fundamentals of Measurement and Representation of Natural Systems*. Elsevier North-Holland, New York.

Simon, H. A., 1973. "The Structure of Ill-Structured Problems." *Artificial Intelligence*, 4, pp. 181–202.

______. 1980. "Cognitive Science: The Newest Science of the Artificial." *Cognitive Science*, 4, pp. 33–36.

______. 1981. *The Science of the Artificial*, 2nd ed. MIT Press, Cambridge, MA.

______. 1983. *Reason in Human Affairs*. Stanford University Press, Stanford, CA.

Thom, R. 1975. *Structural Stability and Morphogenesis*. W. A. Benjamin, Reading, MA.

Troncale, L. R. 1982. "Some Key Unanswered Questions About Hierarchies." In Troncale, L. R. (ed.). *A General Survey of Systems Methodology*, vol. 1, *Proceedings of the 26th Annual Meeting of the Society for General Systems Research*. Washington, DC, January 5–8.

Waltzlawick, P., et al. 1974. *Change*: *Principles of Problem Formation and Problem Resolution*. W. W. Norton, New York.

Weaver, W., and Shannon, C. E. 1949. *The Mathematical Theory of Communication*. University of Illinois Press, Urbana.

Whitehead, A. N., and Russell, B. 1910–1913. *Principia Mathematica*, 2nd ed. Cambridge University Press, Cambridge.

Zeeman, E. C. 1977. *Catastrophe Theory*: *Selected Papers 1972–1977*. Addison-Wesley, Reading, MA.

CHAPTER 2

Policy-Making as Evolutionary Systems Design*

2.1 Purposeful Systems, Policy-Making, and Conflict Resolution

Following some definitions, in Section 2.1 we discuss policy-making and conflict resolution as design of purposeful systems. The role in systems design of values and goals and changes in these through the goals/values referral process is brought out. In Sections 2.2 and 2.3, we consider planning and control in dynamical systems. Then in Section 2.4 we define the act of control to launch our discussion of policy-making as a process of cooperative control to all there is—dissipative self-organization—manifested by increasing consciousness (Sections 2.4 through 2.10). In this, Section 2.5 treats consciousness and Section 2.6 conflict resolution as increase in consciousness. Section 2.7 formalizes conflict resolution and, with Section 2.8, cites applications of the goals/values referral process to expand consciousness represented mathematically by sets. Section 2.9 notes the role of negotiations, in particular of mathematical models of negotiation within the systems design framework. Section 2.10 presents concluding remarks.

We begin with some definitions. A *system* is a "set of interrelated elements each of which is related directly or indirectly to every other element, and no subset of which is unrelated to any other subset", (Ackoff and Emery 1972). The environment of a system consists of elements that are not part of the system but can affect it.

*This chapter is based on Shakun (1981a, 1981b).

An *adaptive* system is one that *reacts* or *responds* to *change* to attain goals. Thus, an adaptive system uses change (i.e., the reaction or response is the use of change) to cope with change. The *change* may be *internal* (within the system) or *external* (in its environment). The *reaction* or *response* may be *passive* (the system changes itself—i.e., how it behaves) or *active* (the system changes its environment—i.e., how the environment behaves).

A *reactive* system can exhibit only one structural behavior in any one environment (e.g., a thermostatically controlled heating system). A *responsive* system can do more—it can learn from its performance, and can choose its means and increase its efficiency. Both are adaptive systems. However, an *adaptive system in policy-making* can do still more—it can also choose its goals and is thus *purposeful* (Ackoff and Emery 1972).

Shakun (1976, 1977) views policy-making and conflict resolution as the design of purposeful systems to deliver values $v = (v_k)$ for $k = 1, 2, \ldots, K$ to participants. Values are viewed as nonoperational goals that are delivered in the form of operational goals $g = (g_i)$ for $i = 1, 2, \ldots, p$. Operational goals are goals defined by specific, unambiguous operations and are characterized by performance measures. Values and operational goals express meaning (Frankl 1962, 1969). Purposefulness is meaning at the operational goal level. The operational goals are expressed in a mathematical model of a purposeful system (with minor notational changes) (Lewin and Shakun 1976):

$$g_i(u_1, u_2, \ldots, u_m) \geqq b_i \quad \text{for} \quad i = 1, 2, \ldots, p \tag{1}$$

or in vector notation

$$g(u) \geq b$$

where $u = (u_1, u_2, \ldots, u_m)$ is a vector of control (decision) variables controlled by various participants (policymakers) in the system. The operational goals g (including all goals of all participants) are included in the set of constraints (1) to be satisfied at least at aspiration level $b = (b_i)$. Other constraints in (1) involve technological, resource, etc., limitations. Nothing in system (1) is considered to be fixed. The b_i are subject to change. The functional forms in general may be controlled by various groups of policymakers. The set of policymakers in the system may change. The vector of control variables $u = (u_1, u_2, \ldots, u_m)$ is not fixed—e.g., some new variables $u_{m+1}, u_{m+2}, \ldots$ may be added. The number of goals (constraints) is not fixed—e.g., $g_{p+1}, g_{p+2}, \ldots$ may be added. Also the values v are not fixed. If one of the b_i is chosen to be as large as possible, i.e., one of the $g_i(u)$ becomes an objective function, then (1) is a basic mathematical programming problem.

At any stage in the design of system (1) we have a particular formulation for system (1). We also have a relation λ from the set of values v to the set of goals g that shows which values are delivered by which goals. The relation λ can be represented by incidence matrix:

λ	g
v	(λ_{ki})

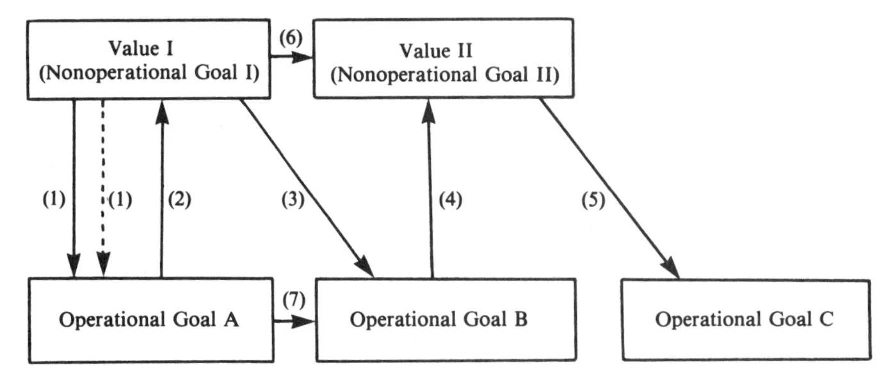

Figure 2.1. **The goals / values referral process**

where $\lambda_{ki} = 1$ indicates the relation λ is true, i.e., value v_k is delivered by the goal g_i and $\lambda_{ki} = 0$ indicates the relation λ is false, i.e., value v_k is not delivered by goal g_i. The associated inverse relation λ^{-1} from g to v shows which goals deliver which values.

λ^{-1}	v
g	(λ_{ik})

where $\lambda_{ik} = 1$ or 0 indicates the relation λ^{-1} is true (g_i delivers v_k) or false (g_i does not deliver v_k), respectively. The structure of a relation may be studied by methods of polyhedral dynamics or Q-analysis (Atkins 1974; Casti 1979).

Operational goals and related values may be redefined by a *goals/values referral process* (Figure 2.1) whereby values (nonoperational goals) are referred to operational goals and vice versa. Values may also be referred to other values and goals to other goals. The referral process provides for purposeful search and adaptation whereby values and operational goals are identified and changed. Summarizing Shakun (1975), this *referral process* goes (1) from nonoperational goal (value) I, "to be outdoors in nature in the winter," to operational goal A, "to do downhill skiing at least 20 days per year," which subjectively satisfies it. Or the process may begin with operational goal A, with nonoperational goal I implicit—represented by dotted (1) in Figure 2.1; (2) from operational goal A, which is no longer being satisfied (there is a gasoline shortage, and the mountains are far away) back to value I; (3) then from value I to new operational goal B, "to do cross-country skiing (available nearby) at least 20 days per year," which now satisfies value I. Failure to satisfice on goal B (the skier injures himself) can lead to referral (4) to value II, "to overcome separateness," and then (5) to operational goal C, "to take up pottery making." The path (6) from nonoperational goal I to II and then to C is also possible, as are others, e.g., (7). Thus, the referral process provides a mechanism for purposeful search and adaptation whereby operational goals and

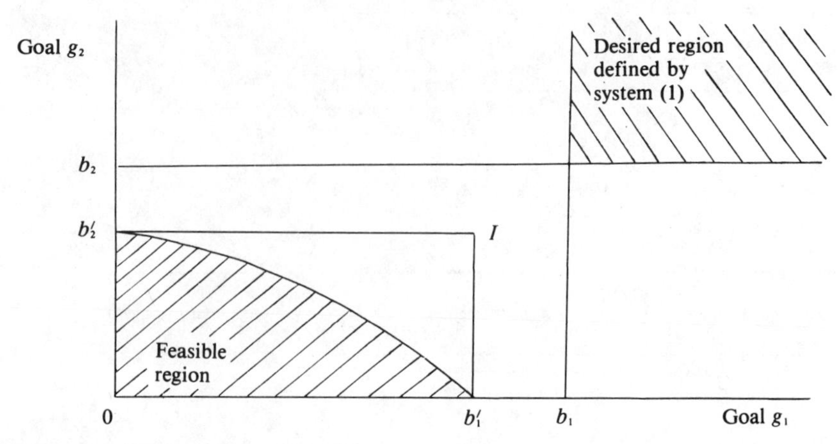

Figure 2.2. **The geometry of conflict**

values can be changed. For a discussion of goal changing in political systems see Deutsch (1966).

Although values are general concepts, there can be various levels of generality. More general values are inclusive of less general values. For example, the value "to overcome separateness" is more general than "to be outdoors in nature in the winter." The values and goals are hierarchically related (Kenney and Raiffa 1976, Ch. 2) in a goals/values tree or more general relation within which the referral process operates.

In policy-making often there is initially no feasible solution u satisfying the system of constraints (1) as defined by the participants, i.e., there is conflict. The technologically feasible g_i values lie below the desired region defined by (1). The geometry of conflict is shown in Figure 2.2. (Shakun 1978).

Conflict is resolved if the desired region and/or the feasible region is expanded so that they have at least one point in common. For example, if aspiration levels b_i are revised downward, the desired region will expand.* Expansion of the feasible region is possible if the parameters and/or the functional forms of the g_i can be redesigned. Both of these methods proceed within the operational goal space as defined.** However, an important aspect of conflict resolution is to redefine the operational goal variables themselves.

The geometry of conflict can radically change if the operational goal variables are changed through the goals/values referral process. Thus if goals

*For example, b_1 and b_2 may be reduced to b_1' and b_2' on the basis of considering the feasible region one dimension at a time. This gives rise to the concept of an ideal point I (Figure 2.2). A compromise point in the feasible region could be chosen as close as possible to the ideal, which itself could be displaced in the decision process (Zeleny 1977). Such a compromise point implies, of course, still further lowered levels of aspiration. Compromise points may also be arrived at through negotiation.

**For the use of mathematical programming in purposeful search and adaptation within a defined operational goal space, see Lewin and Shakun (1976).

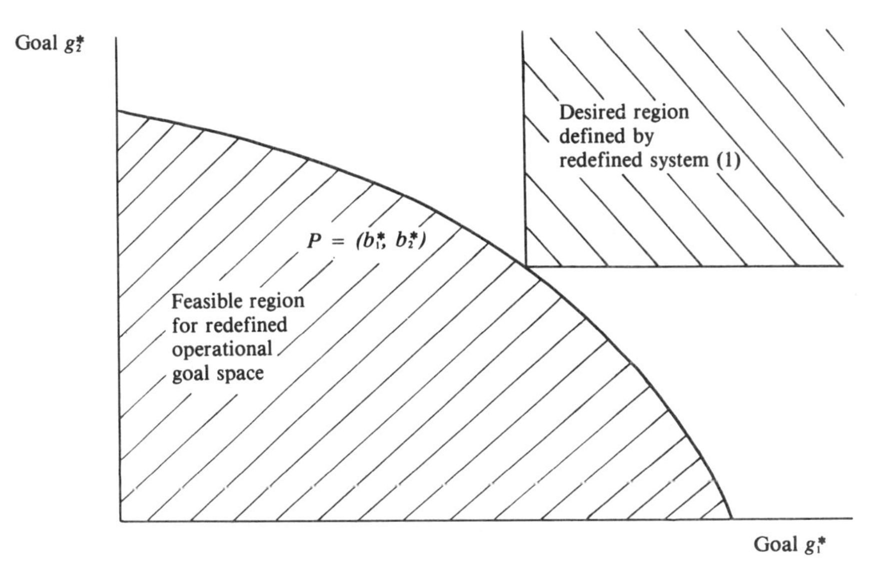

Figure 2.3. **Conflict resolution**

g_1 and g_2 in Figure 2.2 are changed to goals g_1^* and g_2^*, the geometry of conflict could become Figure 2.3, wherein conflict is resolved at point P.

Interpreting system (1) and Figures 2.2 and 2.3 further, we note that conflict arises from the difference between what participants want and what they can get, i.e., between target (values, goals—the desired region in Figures 2.2 and 2.3) and performance (the feasible region in Figures 2.2 and 2.3).* Conflict resolution is concerned with overcoming the difference between target and performance. It is thus cybernetic involving design of purposeful systems.

In control theory the basic cybernetic paradigm with feedback control involves three steps (Lewin and Shakun 1976):

1. *Specification* of goals and performance measures influenced by feedback from step 3
2. *Operation* in pursuit of the goals through a course of action selected in step 3
3. *Evaluation* of performance resulting in respecification of goals and reselection of a course of action for further operation

In control theory this cybernetic cycle is expressed at the operational level as the pursuit of operational goals. These are goals and means to attain them that are defined by specific, unambiguous operations and performance measures. Thus this cybernetic paradigm is normally viewed as an operational

*This view of conflict accommodates situations involving single or multiple decision makers and single or multiple goals. Thus problems in multicriteria decision making are problems in conflict resolution (see also Zeleny 1977).

model. It is, however, also applicable to the nonoperational or values level as well. The values and operational levels are hierarchically related at various levels of generality in the goals/values referral process discussed previously.

Thus participants choose targets at various levels of generality, the most specific being operational goals—the desired region in Figures 2.2 and 2.3. At the operational level, participants also choose performance—the feasible region in Figures 2.2 and 2.3—by designing the system (mathematically, the forms of the g_i) for delivering the operational goals. Conflict arises from the difference between target and performance. Policy-making means designing (redefining) system (1) and related values so as to reduce the difference between target and performance. The target and performance regions must have at least one point in common for conflict to be resolved—see Figure 2.3.

As noted, a conflict is a problem that, as defined, has no feasible solution, i.e., conflict is characterized by a difference (distance) between target and performance. If this distance is zero, then for the system, as defined, there is no conflict. However, the system, as defined, may not overcome separateness from the ultimate target. What is the ultimate target—the most general value? Perhaps the ultimate value is to overcome separateness from all there is. All there is is also called God, One, Tao, the absolute. The process of all there is (of the absolute) is the relative—what we term reality or general systems. While the ultimate target is all there is, in practice we use as intermediate targets less general values and associated operational goals chosen by participants through the goals/values referral process. The more participants as codesigners are in tune with the process of all there is, the better they can choose values and operational goals consonant with it; thereby conflict is less likely. Values and goals in turn are targets to finding this process. Shakun (1979) discusses some metaphors for the process of all there is (general systems)—religion, evolution, cybernetics, and policymaking.

The design of system (1) is cybernetic, as illustrated in Figure 2.4. The controller of the system (policymaker) specifies b. The regulator knows b and g and tries to choose u to satisfy (1), in which case the problem is solved. If (1) is not satisfied, the system (1) is redesigned as previously discussed.

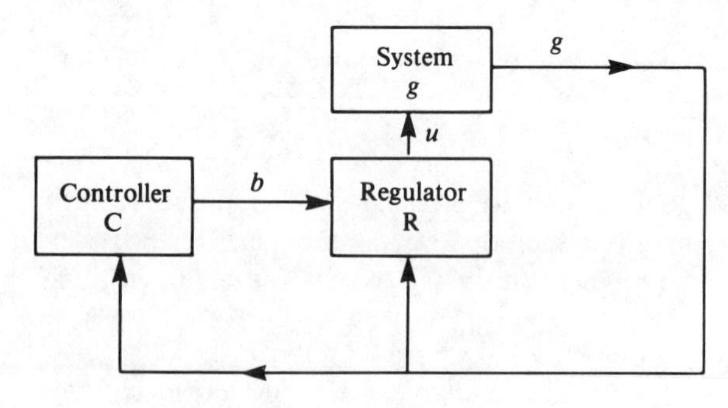

Figure 2.4. **Cybernetic design of system (1)**

After considering planning and control in dynamical systems (Sections 2.2 and 2.3), I discuss policy-making as a process of cooperative control to all there is manifested by increasing consciousness (Sections 2.4 through 2.9).

2.2 Planning: Discussion

Planning is policy-making and conflict resolution over future time. It is the process of designing and bringing about a desired future for system (1) (Shakun 1975). Thus for planning we need to subscript the system (1) on future time. We need to satisfy not only a present system (1) but also future systems (1). We must establish the interrelationships between present and future systems (1) and deal with a multiperiod constraint system problem.

The realization of a desired system (1) at a particular future time may be compatible only with certain present systems (1). For example, suppose a desired future system (1) may be attained if certain decisions, u values, are selected now. If these u values are not feasible, i.e., do not satisfy the present system (1), then the present system can be adapted until these u values are feasible. As another possibility, u's feasible within the present system that will allow satisficing in a different future system (1) can be chosen.

The design of system (1) over time (planning) is also cybernetic. Further, as the system operates and actual outputs are observed, the system may be redesigned, if needed. The essentials of a cybernetic (adaptive purposeful) system in operation are diagrammed in Figure 2.5 and discussed qualitatively by Lewin and Shakun (1976). For a quantitative presentation, see equations (15), (16) below.

Lewin and Shakun (1976) point out that two policymakers, 1 and 2, each viewed as an adaptive purposeful system, may be coupled to form the composite policy-making system shown in Figure 2.6. If Figure 2.5 represents policymaker 1, then one part of the Disturbance D in that figure, policymaker

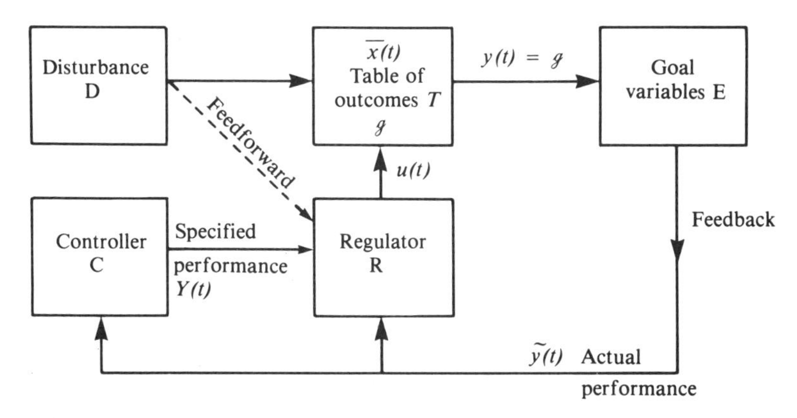

Figure 2.5. **An adaptive purposeful system**

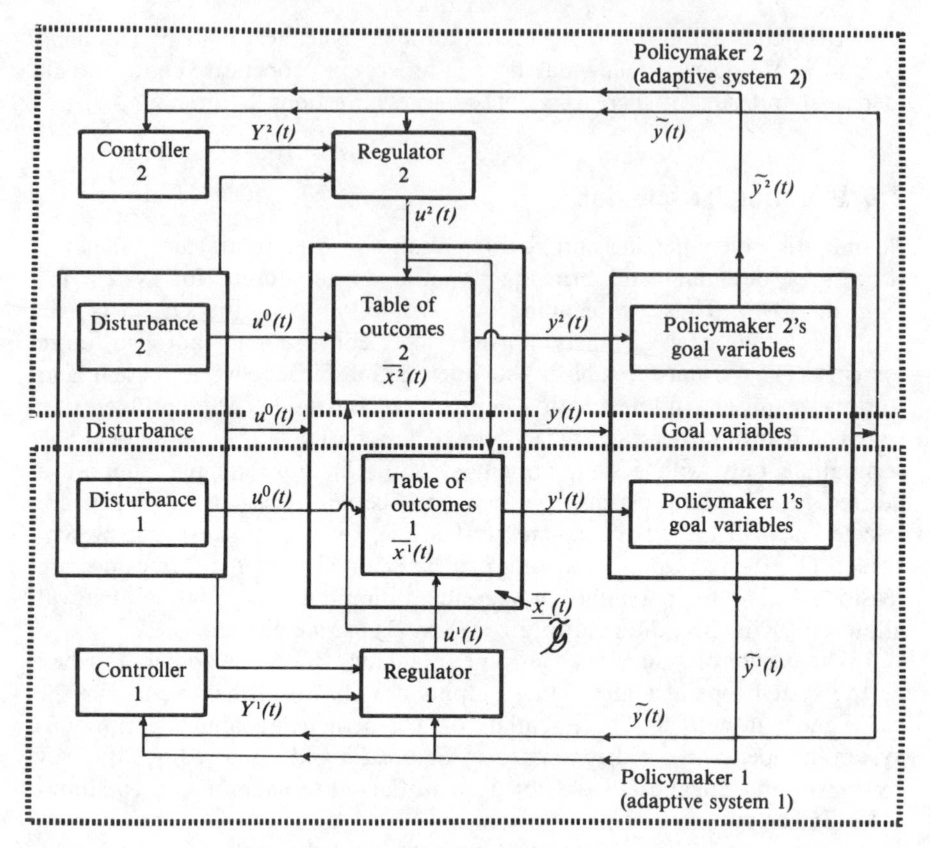

Figure 2.6. **A policy-making system with two policy makers**

2, is represented explicitly as a system in Figure 2.6. Thus the environment of each policymaker includes the other. For a quantitative presentation see equations (19), (20) below.

Effective planning requires feedforward (anticipatory) control in which future disturbance is predicted based on past disturbances. Thus effective planning requires uncovering future inevitabilities (Ackoff 1979a) and introducing future uncertainties. These uncertainties include ones identifiable by such techniques as statistical forecasting, scenario writing, Delphi, etc., for which contingency planning (Ackoff 1979a) involving the use of simulations, etc., may be employed. For example, the computer-based planning system of Hamilton and Moses (1973, 1974) and Moses (1975) permits the evaluation of alternative scenarios prior to the incidence of change (as well as *ex post*). In addition to uncertainties in interest rates, taxes, foreign exchange rates, etc., scenarios for this kind of computerized planning system can include such discontinuities as acquisitions-divestments, corporate takeover threats, political expropriations, commodity embargoes, etc. Purposeful search and adaptation in anticipatory planning could then follow the ideas presented in Section 2.1.

A recent mathematical approach, catastrophe theory (Casti 1979; Thom 1975; Zeeman 1976, 1977) permits modeling discontinuous change in a dependent variable associated with continuous change in independent variables. Thus catastrophe theory offers possibilities for predicting discontinuity in future disturbance.

Perhaps the most difficult discontinuities for which to plan involve qualitative uncertainties, i.e., ones of which we are qualitatively ignorant, so that their possibilities cannot specifically be considered. Formally, such uncertainties may be viewed as a very broad class of disturbance. Effective response to such discontinuities generally depends on early detection and on the flexibility with which a system's resources can be reorganized, and these depend in part on information capability. Consequently, responsiveness planning (Ackoff 1979a) for qualitatively unknown discontinuities could include a lookout capability and a computer-based planning system. Fundamentally, planning for qualitatively unknown (broad class) disturbance requires designing resilient systems. Resilience is the capability of a system to persist under unknown disturbance, a concept first identified in ecology by Holling (1976). Resilience is enhanced by flexibility in reorganizing a system's resources.

2.3 Planning: Mathematical Formulation

In planning we are interested in dynamical systems, ones that evolve over time. Such systems may be described by various mathematical descriptions: internal, external, finite-state, potential and entropy functions, sets and relations (Casti 1979). An internal description in continuous time may be written using a differential equation and has the following general form:

$$\dot{x}(t) = f(x(t), u(t), t), \qquad x(0) = x_0 \tag{2}$$

$$y(t) = g(x(t), u(t), t), \qquad 0 \leqq t \leqq T \tag{3}$$

where $x(t) = (x_1(\mathrm{t}), \ldots x_n(t))$ is an n-dimensional vector that represents the state of the system at time t, $y(t)$ is a p-dimensional vector of system outputs or goals (intended or unintended); $u(t)$ is an m-dimensional vector of system inputs, and x_0 is the initial system state. In discrete time these dynamics may be written using a difference equation:

$$x(t+1) = f(x(t), u(t), t), \quad x(0) = x_0 \tag{4}$$

$$y(t) = g(x(t), u(t), t), \qquad t = 0, 1, 2, \ldots, T \tag{5}$$

A diagram relating input, output, and state of the system is given in Figure 2.7.

System (1) and associated values v may be generalized for planning as a control problem over time. We consider the design of purposeful systems to deliver values $v(t) = (v_1(t), \ldots, v_k(t))$ to N participants (players) at time t

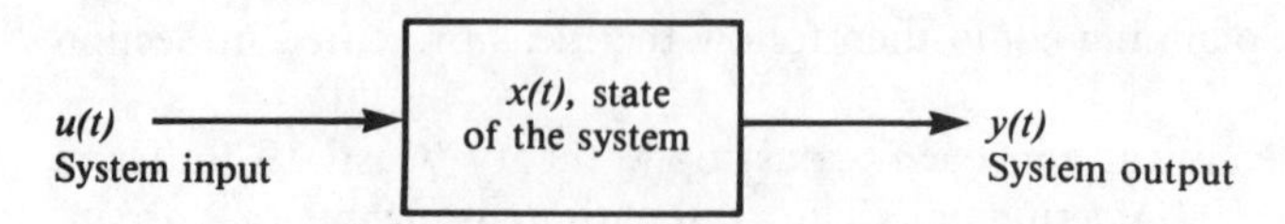

Figure 2.7. **Illustrating system input, output, and state**

$(t = 0, 1, 2, \ldots, T)$. Values are delivered in the form of operational goals $g(t) = (g_1(t), \ldots, g_p(t))$. There is a relation λ from the set of values $v(t)$ to the set of goals $g(t)$ that shows which values are delivered by which goals. The relation λ can be represented by an incidence matrix:

λ	$g(t)$
$v(t)$	$(\lambda_{ki}(t))$

where $\lambda_{ki}(t) = 1 (k = 1, 2, \ldots, K;\ i = 1, 2, \ldots, p)$ indicates the relation λ is true, i.e., value $v_k(t)$ is delivered by goal $g_i(t)$, and $\lambda_{ki}(t) = 0$ indicates the relation λ is false, i.e., value $v_k(t)$ is not delivered by goal $g_i(t)$.

Using (4) and (5), for planning we wish to find input or control $u(t)$ satisfying

$$x(t+1) = f(x(t), u(t), t), \qquad x(0) = x_0 \tag{6}$$

$$y(t) = g(x(t), u(t), t) \in Y(t), \qquad t = 0, 1, 2, \ldots, T \tag{7}$$

Equation (6) expresses the state transition function and is the same as equation (4). Equation (7) is the same as (5) except that outputs $y(t)$ are constrained to a set of admissible outputs $Y(t)$. Equation (7) restates equation (1) at a particular time t and also generalizes it by replacing the idea of outputs (goals) exceeding aspiration levels b with outputs belonging to a set of admissible outputs, $Y(t)$. Normally, $Y(t) \subset R^p$, i.e., $Y(t)$ is a subset of R^p, the p-dimensional real vector space. We can also have constraints $x(t) \in X(t), u(t) \in U(t)$ where $X(t) \subset R^n, U(t) \subset R^m$ (Bensoussan, Hurst, and Maskind 1974; Casti 1979).

Equations (6) and (7) formulate planning as a control problem formally generalizing system (1) to a dynamical system. We note that if there is only one time period, then (6) is unnecessary, and (7) may be written $y = g(u) \in Y$ or, as a special case, as a system of constraints (1). In planning (designing the future), as discussed previously for designing system (1), nothing is considered fixed, i.e., nothing is fixed in dynamical system (6), (7), and associated values $v(t)$.

Mathematically $u(t)$ will satisfy (6) and (7) if $u(t)$ can be found so that the "distance" between $y(t)$ and $Y(t)$ is zero for all t. Therefore mathematically suppose we

$$\min_{u(t)} Z = \sum_{t=0}^{T} \|y(t) - Y(t)\| \tag{8}$$

where $\|y(t) - Y(t)\| = d(y(t), Y(t))$ = distance between $y(t)$ and $Y(t)$. In

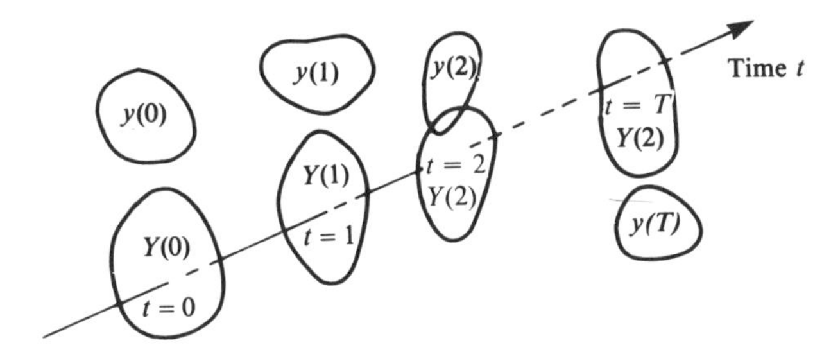

Figure 2.8. **The geometry of conflict**

Euclidean space

$$\|y(t) - Y(t)\| = \sqrt{\sum_{i=1}^{p} [y_i(t) - y_i^*(t)]^2}$$

where $y^*(t) \in Y(t)$ is the closest point to $y(t)$ and where subscript i denotes the ith component in the y and y^* vectors. If the minimum Z is zero, then we have satisfied (6) and (7). Otherwise there is no feasible solution, and by definition there is conflict.

In system theoretic terms, this is the problem of reachability. If $\mathscr{R}(t)$ is the set of states in R^n reachable from x_0 at time t by application of admissible inputs from R^m, then find the intersection $\mathscr{A}(t)$ of $\mathscr{R}(t)$ with $X(t)$. $\mathscr{A}(t)$ is the set of admissible reachable states. Then $g(\mathscr{A}(t), U(t), t)$ is the set of reachable outputs at time t. If the intersection of $Y(t)$ and $g(\mathscr{A}(t), U(t), t)$ is not empty, then $Z = 0$ and conversely. For details on the characterization and computation of $\mathscr{R}(t)$ see Casti (1977) and Hermann and Krener (1977).

Alternatively, if $Z \neq 0$, by the distance measure Z used we have found $u(t)$ to be as close as possible to satisfying equations (6) and (7) as specified—a close-as-possible solution. However, in planning, because (6) and (7) are not fixed, they can be changed until a minimum of zero is obtained in the changed system. This is the idea of cybernetic design and conflict resolution discussed above where performance $y(t)$ and/or target $Y(t)$ are changed. In other words, in the changed system, for all t the intersection of performance $y(t)$ and target $Y(t)$ can be nonempty (distance $Z = 0$) and conflict be resolved. The geometry of conflict is shown in Figure 2.8. Conflict resolution is depicted in Figure 2.9.

Formally, system (6), (7) can represent N players in a difference game*—now $u(t) = (u_1(t), \ldots, u_m(t)) = (u^1(t), \ldots, u^N(t))$ is a partition of m-dimensional input $u(t)$ among controls $u^1(t), \ldots, u^N(t)$ assigned respectively to the N players denoted by the superscripts. Here we consider joint problem solving by all N participants. In Section 2.7 we treat coalitions smaller than N.

*Dynamical systems in continuous (discrete) time with two or more players are called differential (difference) games (Blanquiere 1973; Friedman 1971; Kuhn and Szego 1971; Rufus 1965).

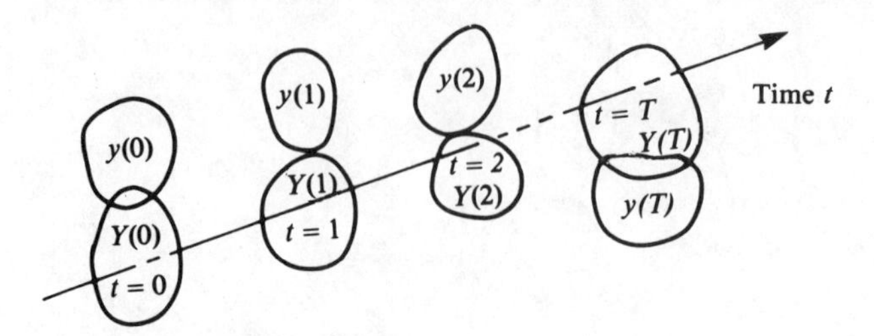

Figure 2.9. **Conflict resolution**

The system change required for conflict resolution can involve continuous expansion of $y(t)$ and $Y(t)$ until they have one point in common—a unique solution point in output goal space. The $u(t)$ that give this point are controls satisfying (6), (7). If the change in $y(t)$ and/or $Y(t)$ is discontinuous, they could have more than one point in common. By definition all of these points are admissible, and all the $u(t)$ that give these points are control solutions satisfying the new (6), (7) and giving $Z = 0$ in (8). Therefore any one of these points can be used as the solution if participants agree. Participants could agree to choose among these points at random. They could agree to choose among these points by selecting $u(t)$ to optimize a criterion

$$J = J(x(t), u(t), t) \tag{9}$$

for example, minimize

$$J = \Sigma\Theta(x(t), u(t), t) \tag{10}$$

where Θ is the cost at time t of the system being in state $x(t)$ when control $u(t)$ is being applied. A special case of (9) is where a terminal cost is minimized, i.e., minimize

$$J = \Theta(x(T)) \tag{11}$$

If participants agree to maximize the sum discounted over time of one of the output goals $y_i = g_i$ (e.g., profit) while satisfying (6) and (7), then $u(t)$ is chosen to maximize

$$J = \sum_t \alpha_t g_i(x(t), u(t), t) \tag{12}$$

where α_t is a discount factor.

If participants cannot agree on choosing among these points, the adaptation process continues using any of the processes discussed until $y(t)$ and $Y(t)$ have only one point in common (geometrically speaking, the expansion process in this case becomes negative, i.e., contraction). The $u(t)$ that give this point are controls satisfying new (6), (7).

Equations (6) and (7) are Markovian in nature in that only the present state $x(0)$ enters into the determination of change of state and output of the system. If past states are included, we have a dynamical system with state memory. We need a more general system of time notation.

Let τ = present time with $\tau = 0, 1, 2, \ldots$, representing a moving present time τ. $t = \tau, \tau + 1, \tau + 2, \ldots, \tau + T$ represents time periods over a planning horizon of length $T + 1$. Also let $T' = \tau + T$.

With $x(t) = (x_1(t), \ldots, x_n(t))$ a n-dimensional state vector at time t, define a $(t + 1)n$ dimensional vector $\bar{x}(t) = (x(0), x(1), \ldots, x(t))$. Then for a system with state memory we have:

$$x(t + 1) = F(\bar{x}(t), u(t), t) \tag{13}$$

$$y(t) = G(\bar{x}(t), u(t), t) \in Y(t) \tag{14}$$

which is mathematically the same form as (6) and (7).

Next we consider a system with state memory and output feedback. First, find a point in $Y(t - 1)$ so that the distance between actual output $\tilde{y}(t - 1)$ and that point is minimum, i.e., find

$$min\|\tilde{y}(t - 1) - Y(t - 1)\|$$

and let $y_0^*(t - 1)$ be the element of $Y(t - 1)$ that minimizes this distance. Then a system with state memory and output feedback may be represented by:

$$x(t + 1) = f(\bar{x}(t), u(t), t) \tag{15}$$

$$y(t) = g(\bar{x}(t), u(t), t) \in Y(t) \tag{16}$$

where $u(t) = u(\bar{x}(t), \|\tilde{y}(t - 1) - y_0^*(t - 1)\|, t)$

System (15), (16) is shown diagrammatically in Figure 2.5 where the feedforward dotted line from D to R in that figure is not operative so that the Disturbance input D is not known. The Table of Outcomes T is modeled by in (16), and that box in the figure contains $\bar{x}(t)$. According to the model (15, 16), the input $u(t)$ from the Regulator R in the figure into g produces output $y(t)$; however, the actual output is $\tilde{y}(t)$ as shown. The difference between $y(t)$ and $\tilde{y}(t)$ is due to unknown disturbance D. The difference between $\mathcal{F}$ and $\mathcal{G}$ in (13), (14) and f and g in (15), (16) is due to inclusion of the output error $\tilde{y}(t - 1) - y_0^*(t - 1)$.

If we can formalize a disturbance input, then feedforward control, which can be more effective than feedback, is possible. Thus in addition to state memory we now formalize a disturbance vector input from the environment and use it and past states to make predictions of upcoming states. If both past states and predicted future states of the system enter into the determination of change of state and output, we have a system with memory and anticipation (Rosen 1974) (disturbance feedforward). We define $w(\tau)$ as a q-dimensional disturbance vector at present time τ. We define a $(T' + 1)$ q-dimensional disturbance vector $\underline{\overline{w}}(\tau) = (\overline{w}(\tau), \underline{w}(\tau))$ where $\overline{w}(\tau) = (w(0), \ldots, w(\tau))$ is a $(\tau + 1)$ q-dimensional past-present disturbance vector (known), and $\underline{w}(\tau) = (\hat{w}(\tau + 1), \ldots, \hat{w}(T'))$ is a $(T' - t)$ q-dimensional predicted future disturbance vector dependent on past disturbances—i.e., $\underline{w}(\tau) = \psi(\overline{w}(\tau), \tau)$. We define a $(T' + 1)n$ vector $\underline{\bar{x}}(t) = (\bar{x}(t), \underline{x}(t))$ where $\bar{x}(t) = (x(0), \ldots, x(t))$ is a $(t + 1)n$ dimensional state vector (known), and $\underline{x}(t) = (\hat{x}(t + 1), \ldots, \hat{x}(T'))$ is a $(T' - t)n$ dimensional predicted state vector dependent on states $\bar{x}(t)$ and past, present and predicted future disturbances, $\underline{\overline{w}}(\tau)$, i.e., $\underline{x}(t) =$

$\phi(\overline{x}(t), \overline{w}(\tau), t)$. We note that because predicted future disturbances $\underline{w}(\tau)$ are dependent on past-present disturbances $\overline{w}(\tau)$ through ψ, we could also write $\underline{x}(t) = \Phi(\overline{x}(t), \overline{w}(\tau), t)$. Then for a system with memory and anticipation (disturbance feedforward):

$$x(t+1) = \mathscr{F}(\underline{\overline{x}}(t), u(t), t) \tag{17}$$

$$y(t) = \mathscr{G}(\underline{\overline{x}}(t), u(t), t) \in Y(t) \qquad t = 0, \ldots, T \tag{18}$$

where $u(t) = (u_1(t), \ldots, u_m(t)) = (u^0(t), u^1(t))$ is a partitioning of m-dimensional input $u(t)$ between disturbance $u^0(t) = \hat{w}(t)$ and control $u^1(t)$. We note that (17), (18) is of the same mathematical form as (6), (7).

If we formally represent N players in a dynamic planning game, then a system with N players, memory and anticipation (disturbance feedforward), may also be represented by (17, 18) where now $u(t) = (u_1(t), \ldots, u_m(t)) = (u^0(t), u^1(t), \ldots, u^N(t))$ is a partitioning of m-dimensional input $u(t)$ among disturbance $u^0(t) = \hat{w}(t)$ and controls $u^1(t), \ldots, u^N(t)$ assigned respectively to the N players denoted by the superscripts.

A system with N players, memory, anticipation (disturbance feedforward), and output feedback may be represented by:

$$x(t+1) = \mathscr{F}(\underline{\overline{x}}(t), u(t), t) \tag{19}$$

$$y(t) = \mathscr{G}(\underline{\overline{x}}(t), u(t), t) \in Y(t) \tag{20}$$

where $u(t) = u(\underline{\overline{x}}(t), \|\tilde{y}(t-1) - y_0^*(t-1)\|, t)$

As before, $u(t) = (u^0(t), u^1(t), \ldots, u^N(t))$ and $u^0(t) = \hat{w}(t)$. System (19), (20) is shown diagrammatically in Figure 2.6 for the case of $N = 2$ players. In the figure $y(t) = (y^1(t), y^2(t))$, $\tilde{y}(t) = (\tilde{y}^1(t), \tilde{y}^2(t))$, $Y(t) = (Y^1(t), Y^2(t))$, $\overline{x}(t) = (\underline{\overline{x}}^1(t), \underline{\overline{x}}^2(t))$ where these respective vector partitionings partition model output, actual output, admissible output, and state of the system for the composite system (19), (20) into corresponding quantities for component systems 1 and 2. The disturbance u^0 is common to the composite system and the component systems.

A number of applications involving dynamical systems such as (19), (20) appear in the literature (Bensoussan, Hurst, and Maslund 1974; Bensoussan, Kleindorfer, and Tapiero 1978; Blaquiere 1973; Casti 1977, 1979; Deal 1979; Feichtinger and Jorgesen 1983; Friedman 1971; Ghosal and Shakun 1980; Kuhn and Szego 1971; Negoita 1979; Rufus 1965; Tapiero 1977).

2.4 Act of Control

We have taken the view that planning is a control problem. Control is sometimes regarded as more general than planning in that in control we always include implementation. *The act of control* is the design and solution at present time τ of a dynamical system such as (19), (20) with associated dynamical values; it includes the implementation of control $u(\tau)$. The act of control always occurs at present time τ. After operation in period τ, a redesign and resolution of the system may be undertaken at the next present, one time period later. Only the present control $u(\tau)$ is implemented, and there is a moving present. A sequence of acts of control is called a *process of control.*

We have noted that dynamical system (19), (20) with associated dynamical values represents N players (participants) in a dynamic planning game. The act (and process) of control thus affords opportunities for cooperation among the players, or a subset (coalition) of them, i.e., cooperative control. For example, society may be viewed in terms of coalitions. There is evidence that values embodying cooperation among participants are fundamentally in tune with evolution (Wilson 1975). Using the Prisoner's Dilemma game, Axelrod and Hamilton (1981) show how cooperation based on reciprocity can get started (initial viability), can thrive in a variegated strategy environment (robustness), and can defend itself, i.e., resist invasion by mutuant strategies once fully established (evolutionary stability). Following the goals/values referral process, identification of cooperative values may proceed by referral to operational goals and other values (Shakun 1976). Thus cooperative control signifies opportunities for cooperation among a coalition of system participants who design a system and exercise controls to deliver values (generally through operational goals, although also directly, as in relationship-oriented systems (Checkland 1972).

As used in this chapter control is in the context of self-organization. Open (open to energy/information exchange with the environment) systems in a state of nonequilibrium (so-called dissipative structures) can change to a new dynamic regime corresponding to an increased state of complexity (nonrepetitive order) if fluctuations (external or internal) are introduced. This is the principle of "order through fluctuation," and the process is called dissipative self-organization (Jantsch 1975, 1979, 1980; Jantsch and Waddington 1976; Nicolis and Prigogine 1977; Prigogine 1980). Evolution is viewed as an integral aspect of dissipative self-organization (Jantsch 1981). Such irreversible processes imply time has an arrow (Layzer 1975)—it goes forward.

Autopoiesis (Zeleny 1977) is a process of control for system self-renewal —a self-referential maintenance process within the process of dissipative self-organization. Such self-production is deviation reducing proceeding by negative feedback or feedforward. The process of control advances to a new order by deviation amplification—by positive feedback or feedforward.

We consider policymaking to be a process of cooperative control to all there is—dissipative self-organization—manifested by increasing consciousness. We now consider consciousness.

2.5 Consciousness

Jantsch (1978) defines consciousness as the "autonomy gained by a self-organizing system in its coevolution with the environment; it brings into play different properties at each new level of evolutionary processes (or autopoietic existence)" (also see Jantsch 1980).

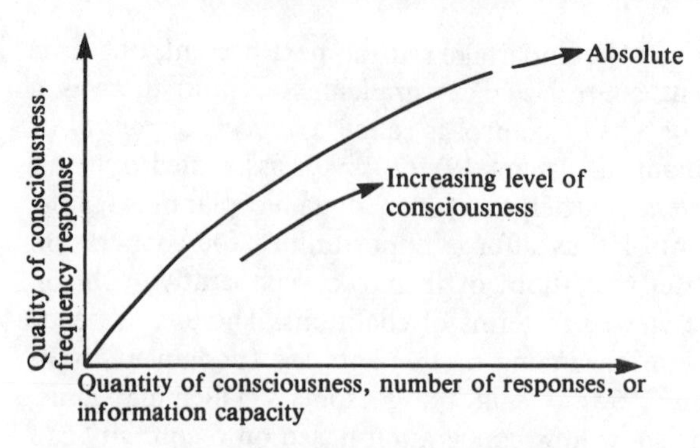

Figure 2.10. **Consciousness as response capacity**

Bentov (1977) considers consciousness to be "the capacity of a system to respond to stimuli." He characterizes response capacity in terms of vibration frequency response (quality of consciousness) and number of responses, i.e., information capacity (quantity of consciousness). Quality and quantity of consciousness thus defined are related by a monotonically increasing curve depicting increasing levels of consciousness—Figure 2.10 based on Bentov (1977). Quality of consciousness broken up into various bands (not shown on Figure 2.10) corresponds to Tart's (1975) discrete states of consciousness and Lilly's (1972) levels of consciousness. Bentov views the absolute as pure consciousness plus intelligence. Pure consciousness means infinite response capacity in frequency and number. Intelligence means self-organizing capability. The absolute has infinite response capacity and has self-organizing capability. Under self-organization the absolute (what I call all there is) manifests itself through the big bang in the relative (what I call the process of all there is) as vibratory energy—as holograms or interference patterns of information-carrying, coherent electromagnetic radiation. Our physical bodies—all matter—may be viewed as four-dimensional holograms that change with time (Bentov 1977) or whose change relative to consciousness (continuum of awareness) we experience as time (Woodhouse 1980). The relative increases complexity (response capacity or consciousness) through dissipative self-organization—the process of cooperative control—evolving toward the absolute.

Drawing on Jantsch and Bentov, we shall regard consciousness as autonomy or self-organizing response capacity (awareness). In Section 2.7 we shall represent consciousness mathematically by sets. For humans, drawing on Jaynes (1976), we may view this consciousness (autonomy) as a metaphor-generated model (analogue) of the world operating by constructing an analogue "I" of ourselves, "I" or self. "I" or self implies separateness from all there is. The analogue "I" allows self-transcendence—the capability of a system to represent itself (Jantsch and Waddington 1976).

Consciousness is thus self-reflective perception or apperception and operates through cognition, affection, and conation. The active stabilization of metaphors constitutes what Tart (1975) calls a discrete state of consciousness, i.e., awareness and a system of psychological structures experienced as a gestalt. Ornstein (1977) views consciousness as a stabilized personal construction amounting to an analogue of the world. He emphasizes two modes of consciousness: the analytic, rational, sequential mode associated with the left hemisphere of the brain and the intuitive, holistic, simultaneous, gestalt mode associated with the right hemisphere. Consciousness as self-organizing response capacity is not confined to what is sensed as rational with respect to the absolute (all there is).

Shakun (1979), drawing on Sagan (1977), discusses the Fall from Eden as a metaphor for a discontinuous change in man's consciousness—the development of self implying separateness from God (all there is)—associated with evolution of the neocortex. Eden is of course the site of the first sin. In the paradise of Eden before the Fall, man knew God—was at one with Him. There was no sin. As discussed by Shakun (1977), the Hebrew word most often used for sin in the Yom Kippur (Day of Atonement in Judaism) liturgy is *het*, which means "a miss"—missing the target as an archer might miss it. In Judaism the target is to find God. The means is through right deeds. When Adam and Eve ate the fruit, they missed the target—they sinned. Man, through the growth of the neocortex, sensed "self"—a separateness from God; he knew he had sinned (missed the target) in eating the fruit.

With self, implying separateness from all there is (God), arises the ultimate value (target) to overcome this separateness. This is the origin of the will—the self with the imperative to overcome separateness from all there is—which acts as cybernetic steersman—a codesigner—in the process of cooperative control. The act of will is the act of control. Although the ultimate target is all there is, in practice the will chooses as intermediate targets less general values and operational goals through the goals/values referral process. Thus the will chooses targets (involving self-transcendence) and corrects the error with respect to a given target—in general, it designs and solves a dynamical system such as (19), (20) with associated dynamical values. Its exercise thus defines meaning (Frankl 1962, 1969) and purposefulness (Ackoff and Emery 1972). Through the act of control the will participates as codesigner in dissipative self-organization—the evolution of the relative to the absolute manifested by increasing consciousness.

2.6 Conflict Resolution: Increase in Consciousness

Practically, we can define conflict resolution as the design of purposeful systems for reducing the difference between chosen target and chosen performance—thus target and/or performance must change. Change in target/performance would seem to rest on understanding human consciousness and evolving systems (Jantsch 1975, 1979; Jantsch and Waddington 1976). The design process should afford opportunities for mutual influence and change in consciousness among the system's participants consonant with evolution.

Target change can occur within the goals/values hierarchy following the goals/values referral process. These target changes may be interpreted as discontinuous changes in values and operational goals in system (1) or in dynamical versions as (19), (20)—redefining the operational goal space as discussed above—associated with discontinuous increase in consciousness. Performance change—systems redesign, i.e., rechoosing the functional forms of the g_i and u in system (1) or $\mathscr{F}$, $\mathscr{G}$ and u in (19), (20)—may also be associated with discontinuous increase in consciousness; also redefinition of coalition C.* Such discontinuities can follow from positive feedback within the cybernetic process and can be related to Nystrom's (1974) model of cognition as discussed by Shakun (1976). The cybernetic process through the use of negative feedback also allows for continuous changes, as when aspiration levels b_i or the goal target $Y(t)$ are revised, e.g., through negotiations. Parameter changes—e.g., due to technology changes—in given functional forms $g_i(u_1, u_2, \ldots, u_m)$ are also considered continuous changes. It is the discontinuous change in consciousness—represented mathematically (cognitively) by the goals/values relation λ, (19) and (20)—leading to new coalitions, values, goals, technologies, and controls that permits defining and solving difficult problems. The new systems design (pattern) is now in long-term memory.

The referral process appears to be a basic problem-solving process facilitating increase in consciousness and may be related to the left and right hemispheres of the brain in the following sense: Search at more general levels of values would appear to involve largely right hemisphere thinking describable as intuitive, holistic, simultaneous, and gestalt. At the more specific value and operational goal levels, opportunity for effective left hemisphere thinking —analytic, rational, and sequential—increasingly come into play.

2.7 Formalizing Conflict Resolution in Policy-Making

As indicated, a conflict is an N-player problem that, as defined, has no feasible solution. Conflict resolution involves redefining the problem so that for the new problem definition there is a solution. Operational goals and related values may be redefined by a goals/values referral process (Figure 2.1) whereby values (nonoperational goals) are referred to operational goals and vice versa. The operational goals may be expressed in a mathematical model of a purposeful system. Here we take (6), (7) with N players as that model. (Other models as (19), (20) can be used.) Let $U^C(t)$ be a set of c dimensional admissible controls available to a coalition C of the set η of N players

*Adams (1974) discusses how understanding conceptual blocks and techniques for overcoming them can lead to discontinuous change in consciousness in creative systems design.

Figure 2.11. **Goal target $Y^C(t)$ for coalition C**

($C \subset \eta$). We note that coalition C may be the grand coalition of all N players. Let $Y^C(t)$ be a set of p-dimensional admissible outputs (goals) defined by the coalition C as the intersection of admissible outputs $Y^j(t)$ for all players $j \in C$. The nonemptiness of $Y^C(t)$ is a necessary condition—called "goal target agreement"—for problem definition and solution by coalition C. $Y^C(t)$ is the goal target for coalition C. If $Y^C(t)$ is initially empty, it may become nonempty by expansion of the $Y^j(t)$ through negotiation among coalition members within a given operational goal space. However, an important approach to conflict resolution is to redefine the dimensions of the operational goal space itself (e.g., by using the goals/values referral process (Figure 2.1) as formalized below. Within the new goal space, either originally or after negotiation to expand individual coalition members' sets of admissible goals $Y^j(t)$, $Y^C(t)$ may be nonempty. See Figure 2.11.

Let $\mathscr{R}^C(t)$ be the set of states in R^n reachable from x_0 at time t by application of $U^C(\ell)$ and $U^{\bar{C}}(\ell)$ for $\ell = 0, 1, \ldots,\ t-1$ where $U^{\bar{C}}(\ell)$ is a set of $(m-c)$ dimensional admissible controls available to players not in coalition C who can form one or more coalitions $\bar{C}$. Find the intersection $\mathscr{A}(t)$ of $\mathscr{R}^C(t)$ with $X(t)$, the set of admissible states. $\mathscr{A}(t)$ is the set of admissible reachable states. Then, using (7), $y^C(t) = g(\mathscr{A}(t), U^C(t), U^{\bar{C}}(t), t)$ representing technologically feasible performance is the set of outputs reachable by coalition C regardless of the controls $U^{\bar{C}}(t)$ exercised by $\bar{C}$. If the intersection of target $Y^C(t)$ and performance $y^C(t) = g$ is not empty, then coalition C has defined and solved its problem; otherwise target-performance conflict remains. Thus conflict resolution (problem definition and solution) for coalition C (see Figure 2.12) requires both goal target agreement by its members and target-performance conflict resolution. See Section 1.3 for a more complete formulation.

Suppose target-performance conflict remains. For a given operational goal space, if the intersection of sets $Y^C(t)$ and $y^C(t)$ is empty, one or both of these sets may expand to give a nonempty intersection to resolve conflict. However, the dimensions of the operational goal space itself may be redefined. Within the new goal space—either originally or after target and/or performance expansion—the target-performance intersection may be nonempty. Both of these approaches—expansion of target and/or performance in a given goal space and redefinition of the dimensions of the goal space itself—require a player to expand his consciousness (self-organizing response capacity) as

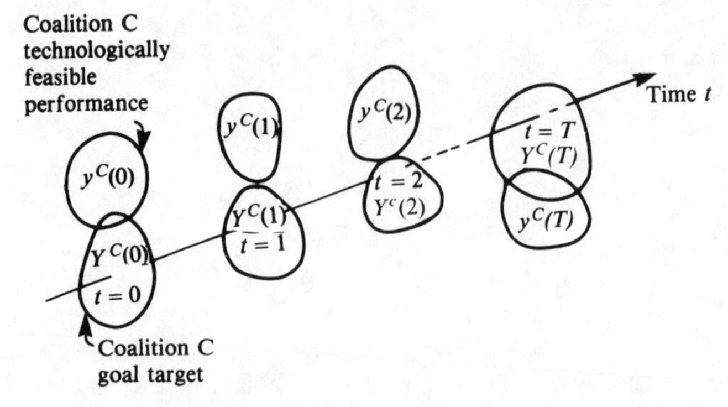

Figure 2.12. **Conflict resolution for coalition *C***

represented formally in the problem by his values, goals, controls, and technology. Mathematically each of these is a set—conflict resolution requires expansion of these sets, which represents increase in consciousness.

The goals/values referral process stimulates expansion of the sets of values and goals, which facilitates agreement on goal dimensions or agreed revision of goal dimensions should the coalition fail to achieve goal target agreement or nonempty target-performance intersection. For given goal dimensions the process also facilitates goal target agreement and nonempty target-performance intersection because players better sense expanded targets within the goal space that deliver identified values. In any case conflict must finally be resolved at the operational level, i.e., it is the nonempty intersection of operational quantities—goal target $Y^C(t)$ and technologically feasible performance $y^C(t) = g$—that finally resolves conflict.

We shall now develop a more formal treatment of goals/values referral using two applications.

DOG RUN CONFLICT

A conflict over a dog run in New York City provides an interesting setting for the goals/values referral process. A fenced-in dog run took up a large part of the area of a large vacant lot owned by New York University. Pending construction on the site, which took over ten years to begin, the run was allowed by the university in response to requests by faculty and local community members who owned dogs. After about a year, a meeting of the faculty tenants association was called by members not owning dogs who strongly objected to the dog run and demanded it be closed down. Their reasons for closing the run were (1) the run was unclean and unhealthy for children who played, if not in, then at least adjacent to the run, where dogs also sometimes ran free and soiled the grass—the whole area was soiled by the dogs and unhealthful; (2) the area available for play by children was smaller than that for the dogs; (3) children come before dogs. Thus, according to dog run opponents, the value that was not being delivered because of the run was a large, healthful play area for the children. Run supporters claimed that dogs should have a place too (a second value). They offered to clean up the run, keep it clean, and let dogs run free inside the run only. The run opponents, many of whom had children, were unmoved. They called for a vote to close down the run. It was clear from the discussion that the dog owners would lose the vote.

Suddenly there was an increase of consciousness (self-organizing response capacity) by one member of the group. He said he was opposed to the dog run for the reasons being cited. However, he said he was concerned about minority rights—the dog owners were in the minority. It was clear from the expressions on the faces of many of those previously opposed to the run that the new value of minority rights was associated with an increase in consciousness. This new value was immediately seized upon by the present writer, who was a dog owner. I had been attempting through use of the goal/values referral process to increase my own consciousness in order to create goals/values change. Suddenly another person had come up with the kind of value change I was seeking. I now emphasized that there were three values involved: (1) a large, healthful play area for children, (2) a play area for dogs, and (3) minority rights. I proposed that these values could be delivered by reducing dog run size (lowering aspiration level for dog owners), organizing by dog owners of a cleanup and ongoing maintenance program, and restricting dogs off the leash to the run area only. I proposed a three-month trial period. I emphasized that dog owners were a minority who ought to be given a chance to improve the situation. The group voted by a large majority to adopt this proposal. The proposal was implemented. The dog run was reduced in size, the owners successfully self-organized for cleanup and maintenance, and they kept dogs on the leash outside the run area. The run continued to operate until new construction on the site began about ten years later.

We now present a more formal treatment of the goals/values referral process for the dog run problem. The goals/values hierarchy appears in Figure 2.13:

The relation of the set of values to the set of goals (goal dimensions) for each player is shown by a goals/values matrix. For two players we use a bimatrix with the first entry associated with dog run opponents (player $j = 1$) and the second entry with dog owners (player $j = 2$). The entries in the bimatrix (Figure 2.14) are λ_{kij} where:

$\lambda_{kij} = 1$ indicates player j is "for" value v_k being delivered by goal g_i (he favors both the value v_k and the goal g_i as an operational expression of this value).

$\lambda_{kij} = 0$ indicates player j is against value v_k being delivered by goal g_i.

$\lambda_{kij} = X$ indicates player j is neutral or he does not perceive value v_k as being delivered by goal g_i.

Goal dimension agreement, i.e., agreement between the players on a common set of goal dimensions, requires that for each goal dimension column

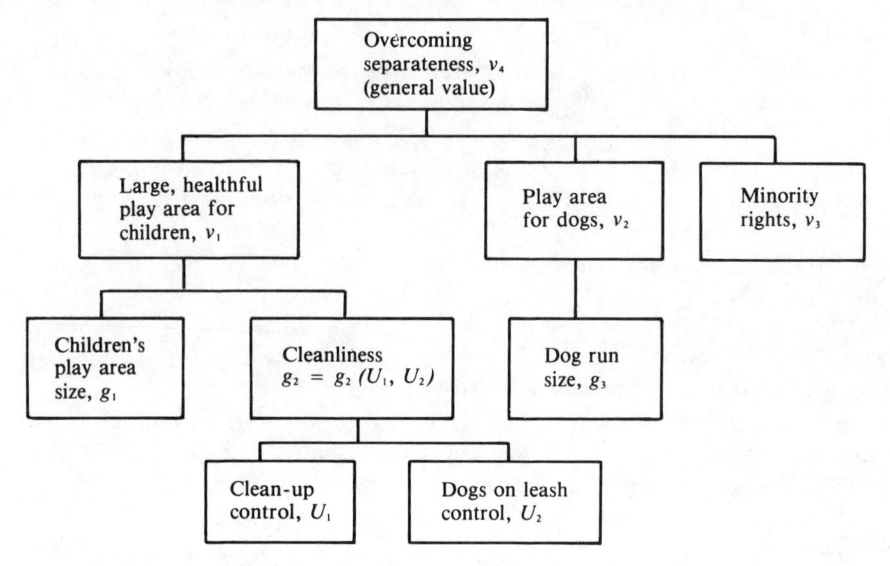

Figure 2.13. **Goals / values hierarchy for dog run conflict**

in which one or both players have an entry of 1 there be no 0 entry. We note that Figure 2.14 does not exhibit goal agreement because in column g_3 player 2 has an entry of 1 and player 1 has an entry of 0.

Failure to achieve goal dimension agreement suggests players expand the sets of goals and values. For example, for any given entry in the matrix players can try to generate additional g_i, which deliver the v_k passing through that entry, and additional v_k, which are delivered by the g_i passing through that entry. In other words, for any given entry players can try to move horizontally, generating new columns $g_i|v_k$ by referring values to goals, and vertically,

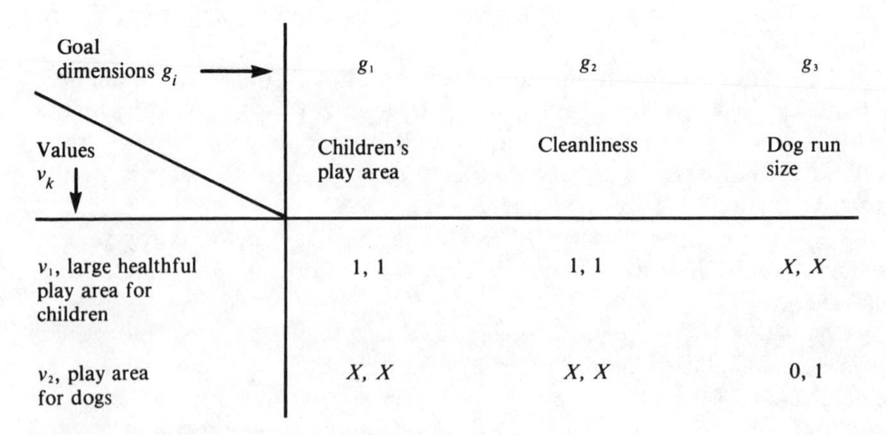

Goal dimensions g_i → / Values v_k ↓	g_1 Children's play area	g_2 Cleanliness	g_3 Dog run size
v_1, large healthful play area for children	1, 1	1, 1	X, X
v_2, play area for dogs	X, X	X, X	0, 1

Figure 2.14. **Initial goals / values matrix for dog run conflict**

Goal dimensions g_i → / Values v_k ↓	g_1 Children's play area	g_2 Cleanliness	g_3 Dog run size
v_1, large healthful play area for children	1, 1	1, 1	X, X
v_2, play area for dogs	X, X	X, X	X, 1
v_3, minority rights	X, X	X, X	1, 1
v_4, overcoming separateness	1, 1	1, 1	1, 1

Figure 2.15. **New goals / values matrix for dog run conflict**

generating new rows $v_k|g_i$ by referring goals to values. Values may also be referred directly to other values and goals to other goals to expand these sets—see discussion of the referral process in Section 2.1. Thus new goals and values are generated, and these in turn can generate still other goals and values, i.e., the sets of goals and values expand. For the enlarged matrix players enter new λ_{kij} values (players are allowed to change previous λ_{kij} entries if they desire).

For the dog run problem goals/values generation operated as follows: Focusing on the (v_2, g_3) entry in Figure 2.14 and thinking of what other values are delivered by g_3, dog run opponents (player 1) generated a new value v_3, minority rights. The bimatrix entry corresponding to (v_3, g_3) was $(1, 1)$. Also in the new goals/values matrix (Figure 2.15) player 1 changed his entry for (v_2, g_3) to X so that the double entry for (v_2, g_3) became $(X, 1)$. Figure 2.15 exhibits goal dimensional agreement because in each goal dimension column in which one or both players have an entry of 1, there is no 0 entry.

Why did player 1 change his (v_2, g_3) entry from 0 to X after the value v_3 was generated? He still did not favor the value of a play area for dogs. However, because he favored v_3 (delivered by g_3) very strongly, he entered a neutral X at (v_2, g_3) instead of 0, indicating that he was willing to accept g_3 because it delivered v_3. Evidentally, the more general value v_4—overcoming separateness (see Figures 2.13 and 2.15)—now included minority rights. In overcoming separateness player 1 weighed his support of v_3 more heavily than his opposition to v_2 because he accepts g_3 by changing his (v_2, g_3) entry from 0 to X.

As noted, goal dimension agreement was now achieved. Goal dimension agreement is a necessary condition for goal target agreement, i.e., the possibility of nonemptiness of $Y^C(t)$—called "goal target agreement"—depends on players agreeing to negotiate on the same set of goal dimensions, i.e., within the same goal space. In the dog run case goal target agreement followed from goal dimension agreement as dog owners lowered their aspiration level on the goal dimension of dog run size so that the aspiration level on the goal dimension of children's play area size could be raised. There was no goal conflict on the aspiration level for cleanliness. Technological performance with respect to cleanliness was achieved by dog owners engaging in cleanup U_1 and dogs on the leash U_2, the latter two being the control variables.

BANK-WOMEN'S GROUP CONFLICT

In this case there are two participants, a bank and a women's rights group, and the underlying value is "being for women's rights." Negotiations become deadlocked on the operational goal "the number of women the bank should promote to vice-president." Through the goals/values referral process an increase in consciousness is facilitated. A second operational goal, which also delivers the value "being for women's rights," is introduced—the amount of money the bank will give toward women's scholarships at a local graduate school of business administration. The operational goal space is thereby redefined—it now has two dimensions—affording an opportunity for renewing negotiations that can lead to a feasible solution. Thus in this case there was agreement on the value "being for women's rights." It was a matter of finding an additional operational goal to be able to deliver this value acceptably to both parties.

We now develop a more formal treatment of goals/values referral for the bank-women's group application. The initial goals/values matrix is given in Figure 2.16 (player 1 is the women's group and player 2 is the bank). There is goal dimension agreement in the initial matrix. However, negotiations on this dimension become deadlocked, so there is no goal target agreement. This initiates the goals/values referral process, which generates (v_1, g_2) in Figure 2.17. The latter, in turn, possibly might generate (v_2, g_2). The new goals/val-

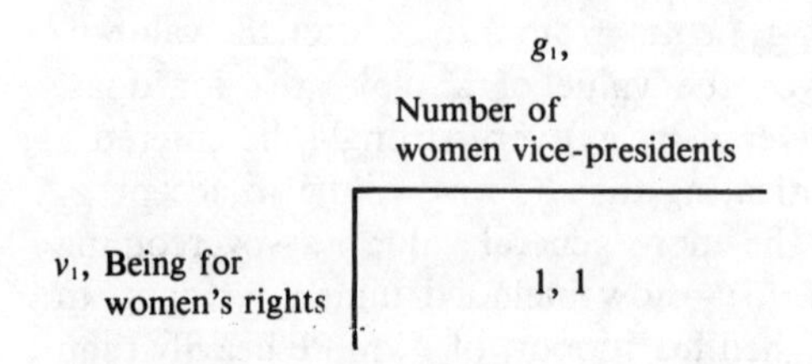

Figure 2.16. **Initial goals / values matrix for bank-women's group conflict**

	g_1, Number of women vice presidents	g_2, Amount of women's scholarships
v_1, Being for women's rights	1, 1	1, 1
(v_2, Support universities—possible generation of a second value)	(X, X)	(X, 1)

Figure 2.17. **New goals / values matrix for bank-women's group conflict**

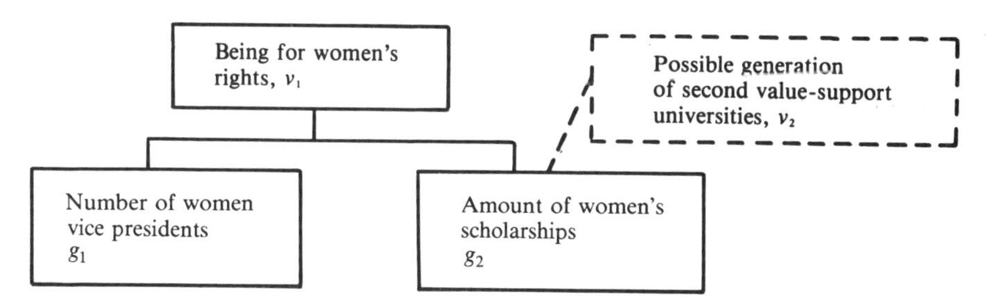

Figure 2.18. **Goals / values hierarchy for bank-women's group conflict**

ues matrix also shows goal dimension agreement. Now there are two goal dimensions that deliver the value of women's rights. The goals/values hierarchy is given in Figure 2.18. Within the new goal space, after negotiations, (see Section 2.9), players can come to an agreement.

We have illustrated the use of the goals/values referral process to facilitate increase in consciousness (self-organizing response capacity) expressed formally as expansion of the sets of goals and values. Values are intermediate between operational goals and the ultimate (most general) value regarded as overcoming separateness from all there is. The players, the sets of goals g/values v, and their relation (as given in the goals/values matrix), together with controls u and technology f and g in (6), (7), constitute a system—a local space/time expression of all there is that involves its direct experience.

We have formalized conflict resolution mathematically in terms of the ability of a coalition C of the set of N players to attain agreed goals for itself. In other words coalition C can form if it can resolve conflict for itself—formally, if after values, goals, controls, and technology sets expand, the intersection of goal target $Y^C(t)$ and technologically feasible performance $g(\mathscr{A}(t), U^C(t), U^{\bar{C}}(t), t)$ is nonempty. In principle there might be more than one coalition C that can feasibly resolve conflict for itself. We do not fully consider this question here. However, a very important coalition in applications is the grand coalition of all N players—stability and questions of ultimate value suggest that conflict resolution involving agreement of all N players is desirable, if achievable. Therefore we suggest both behaviorally and normatively that players try to choose coalition C to be as large as possible. Thus a systematic search would start with $C = \eta$, the set of all N players, and then consider successively coalitions C of size N-1, N-2, ... etc. Using this systematic search, we suggest that the first coalition C that can formally resolve conflict for itself will form.

2.8 Goals / Values Referral: Additional Comments

Invoking the referral process facilitates increase in consciousness (autonomy or self-organizing response capacity), which in turn provides for change in values and/or goals (hopefully in tune with the process of all there is) needed for conflict resolution.

The bank-women's group situation illustrates a change in operational goal space with no necessary change in underlying values. The dog run application involves a change in values with no change in the operational goal space. In other cases both values and goals change, e.g., the skiing example of the goals/value referral process described in Section 2.1. If neither changes, conflict remains unless within the unchanged goal space aspiration level b or the technology represented by $g(u)$ in system (1) changes; or in terms of (19) and (20) the set of admissible outputs $Y(t)$ or the technology represented by $\mathscr{F}$, $\mathscr{G}$ and u changes.

In Shakun and Sudit (1978b)—and Chapter 4 here—involving mandatory retirement age for tenured faculty, the authors, using no change in values and operational goal space, find a solution that delivered the goals at slightly lowered aspiration levels. They proposed this to members of the House Senate Conference Committee. The solution actually voted by Congress and signed into law by the president does not deliver certain goals/values for some of the participants. Shakun and Sudit discuss this in relation to their proposed solution.

From the point of view of methodology, we note that often one begins with identification of the participants, their values, and associated operational goals and then proceeds to search for a feasible solution. However, in cases such as the Arab-Israeli conflict, where there are many participants and many goals/values, it may be easier to begin the iterative search process with a proposed solution and then systematically identify the operational goals and values satisfied by it as a starting point for the next iteration in the search process. This was the approach taken by Shakun and Sudit (1978a)—see Chapter 3—in proposing a 14-component solution (with identified goals and values) to the Arab-Israeli conflict as a starting point for an iterative solution search to be undertaken by the participants themselves.

2.9 Negotiations: A Mathematical Approach

As noted above, a fundamental process of conflict resolution within a given operational goal space involves concession making through negotiations.

Rao and Shakun (1974)—see Chapter 6—mathematically treat the process of concession making as a sequential decision problem where concessions made alternately by participants depend on the current state of negotiations and the behavior concepts being employed (as a player maximizing his expected utility, maximizing his minimum utility–maxmin behavior, or using minimax risk comparison). The problem is formulated as one of stochastic terminal control that can be solved by dynamic programming to yield normative recommendations as to concession making. Rao and Shakun place their negotiation model within the above systems design framework.

Shakun (1976) extends the thinking of the Rao-Shakun model to incorporate cooperative behavior. He shows that various final negotiated agreements that may be reached dynamically using a cooperative behavior concept employing a cooperative *behavior* coefficient are also given by a proposed bargaining solution involving a cooperative *solution* coefficient.

Fogelman-Soulie, Munier, and Shakun (1980)—see Chapter 7—also using dynamic programming, extend the Rao-Shakun work to the case of bivariate negotiations where preference information is obtained interactively at each negotiation stage. They treat in detail the bivariate negotiations phase of the bank-women's group conflict.

2.10 Policy-Making and Meaning as Design of Purposeful Systems

The notion of cybernetics presented in this chapter—implied in Figure 2.6 and (19), (20)—includes (1) engineering cybernetics involving system control of known disturbances through feedback and feedforward; (2) biological cybernetics—e.g., see Ashby (1952), Beer (1972), and Vickers (1965)—involving mutual adaptation (conflict resolution) of multiple coexisting systems that plan for and adapt to qualitatively unknown (formally, very broad class) disturbances—see last paragraph of Section 2.2; (3) self-organizing evolutionary cybernetics (Jantsch 1975, 1979, 1980, 1981; Jantsch and Waddington 1976; Nicolis and Prigogine 1977), here, policy-making as design of purposeful systems—a process of cooperative control to all there is—dissipative self-organization—manifested by increasing consciousness. The latter is regarded as self-organizing response capacity (awareness) and represented mathematically by sets.

Thus viewed, policy-making—which includes conflict resolution and planning—is an adaptive control (design) problem in multicriteria decision making involving multiple participants (players) in which nothing is considered fixed. Mathematically, this control (design) problem is represented by a dynamical system such as (19), (20) with N players, memory, anticipation (disturbance feedforward), and output feedback, with associated dynamical values. Sometimes this model is only conceptual. This control problem is thus cybernetic—the design of purposeful systems to overcome the difference between chosen target (what participants want) and chosen performance (what they can get). Choice of target involves a goals/values referral process wherein participants can identify and change goals and values, as well as a negotiation process. Choice of performance to deliver the target involves choice by participants of system structure and technology. The ultimate value (target) is regarded as overcoming separateness from all there is. This separateness is experienced as uncertainty and can be overcome by faith (control)—letting go, the act of control into uncertainty,* wu wei (perfect action, nonaction) of Tao (Lao Tzu 1961; Merton 1969), meditation of Krishnamurti (1979).

Faith (control) is taking a horizontal step; faith is stepping up one toehold higher in rock climbing, noting ahead of time where the next handholds are; faith is stepping up in rock climbing without seeing any handholds above but anticipating that you'll find one after you've stepped up.

Control (faith) always means designed letting go. The more uncertainty (turbulence), the more we necessarily must let go, the greater must be our faith and the less is our influence (control) as codesigners in determining outputs (outcomes). The act of control is always an act of faith. In turbulence we necessarily must accept less control over outcomes and see what emerges. Lessened control requires more faith, cooperation, acceptance of vulnerability, love, and compassion (caring) in living (Michael 1983). Notably, near-death survivors speak of these same values (Ring 1980, 1982).

As a designer I always have a choice, i.e., control of my values, goals, technology, and controls. I always need faith. After one time period I have another choice. With death I accept ejection out of the current system—dissipative self-organization to a new dynamic regime. Life is an act/process of control into less or more uncertainty, i.e., it is an act/process of faith. It is expressed in policy-making where participants as codesigners choose as intermediate targets less general values and associated operational goals hopefully in tune with the process of all there is, as well as technology and controls. Thus enlightened policy-making is at once pragmatic, operational control, and cooperative control to all there is.

The design and implementation now of a dynamical system of goals, controls, technology, and associated dynamical values constitutes an act of cooperative control (faith)—a unique meaning of the moment (Frankl 1969)—an expression now of all there is that involves its direct experience. To my consciousness at this time the meaning of life and the universe (Frankl's (1962, 1969) will to meaning) is design and implementation of purposeful

*Expressed beautifully in Hesse's (1970) novella, *Klein and Wagner*.

systems—a process of cooperative control to all there is—dissipative self-organization—manifested by increasing consciousness.* It is policy-making.

Otherwise put, meaning at the moment (now) is an act of cooperative control (faith) to all there is and over time a process or sequence of such acts. Meaning is design of purposeful systems is policy-making—they are one and express One (all there is). At the moment (now) they are a local space/time expression of all there is that involves its direct experience. Over time, policy-making, meaning, and design of purposeful systems express evolution to all there is. They are manifested by increasing consciousness and characterized as evolutionary cooperative control to all there is—dissipative self-organization.

Finally, with regard to the management sciences, we note that the methodology presented encompasses the official, reformist, and revolutionary paradigms of operations research discussed by Dando and Bennett (1977). It is consonant with Checkland's (1972, 1977, 1980) soft systems methodology, Ackoff's (1974, 1979b, 1981) design approach to mess management and planning, Landry's (1979, 1980) view of problem definition and solution, Churchman's (1971) design of inquiring systems, and Mason's (1969) and Mitroff and Mason's (1980a, 1980b) dialectical approach to problem definition. As a methodology for problem definition and solution in complex contexts involving multiparticipant, multicriteria, ill-structured, dynamic problems, it can provide a basis for decision support systems (Keen and Morton 1978) for policy-making under complexity. It is also a personal statement whose meaning is conveyed, in part, in personal terms. It is *evolutionary systems design*.

REFERENCES

Ackoff, R. L. 1974. *Redesigning the Future*. Wiley, New York.

______. 1979a. *A Guide to Corporate Planning*. Wiley, New York.

______. 1979b. "Resurrection the Future of Operational Research." *Journal of the Operational Research Society*, **30**, No. 3, March.

______. 1981. "The Art and Science of Mess Management." *Interfaces*, **11**, No. 1, February.

Ackoff, R. L., and Emery, F. E. 1972. *On Purposeful Systems*. Aldine-Atherton, Chicago.

Adams, J. L. 1974. *Conceptual Blockbusting*. Stanford Alumni Association, Stanford, CA.

Ashby, W. R. 1952. *Design for a Brain*. Wiley, New York.

Atkins, R. H. 1974. *Mathematical Structure in Human Affairs*. Heinemann Educational Books Ltd., London.

*Frankl (1962, 1969) suggests such intermediate values as creative values (e.g., work), experiential values (e.g., love), and attitudinal values, i.e., attitudes to pain (suffering), guilt, and death.

Axelrod, R., and Hamilton, W. D. 1981. "The Evolution of Cooperation." *Science*, **211**, No. 4489, March 27.

Beer, S. 1972. *Brain of the Firm*. Allen Lane, Penguin Press, London.

Bensoussan, A., Hurst, E. G., and Maslund, B. 1974. *Management Applications of Modern Control Theory*. North-Holland, Amsterdam.

Bensoussan, A., Kleindorfer, P. R., and Tapiero, C. S. (eds.), 1978. *Applied Optimal Control*. North-Holland, Amsterdam.

Bentov, I. 1977. *Stalking the Wild Pendulum: On the Mechanics of Consciousness*. E. P. Dutton, New York.

Blaquiere, A. 1973. *Topics in Differential Games*. North-Holland, Amsterdam.

Casti, J. L. 1977. *Dynamical Systems and Their Applications Linear Theory*. Academic Press, New York.

_____. 1979. *Connectivity, Complexity and Catastrophe in Large-Scale Systems*. Wiley, London.

Checkland, P. B. 1972. "Towards a System-Based Methodology for Real-World Problem Solving." *Journal of Systems Engineering*, **3**, No. 2, pp. 87–116.

_____. 1977. "Science and the Systems Paradigm." *International Journal of General Systems*, **3**, No. 2, pp. 127–128.

_____. 1980. "The Systems Movement and the 'Failure' of Management Science." *Cybernetics and Systems*, **11**, pp. 317–324.

Churchman, C. W. 1971. *The Design of Inquiring Systems*. Basic Books, New York.

Dando, M. R., and Bennett, P. G. 1981. "A Kuhnian Crisis in Management Science?" *Operational Research Society*, **32**, pp. 91–103.

Deal, K. R. 1979. "Optimizing Advertising Expenditures in a Dynamic Duopoly." *Operations Research*, **27**, No. 4, July–August.

Deutsch, K. W. 1966. *The Nerves of Government*. Free Press, New York.

Feichtinger, G., and Jorgesen, S. 1983. "Differential Game Models in Management." *European Journal of Operational Research*, **14**, No. 2, October.

Fogelman-Soulie, F., Munier, B., and Shakun, M. F. 1980. "Bivariate Negotiations as a Problem of Stochastic Terminal Control Management Science." **29** no. 7, July.

Frankl, V. E. 1962. *Man's Search for Meaning: An Introduction to Logotherapy*, rev. Ed. Beacon Press, Boston.

_____. 1969. *The Will to Meaning: Foundations and Applications of Logotherapy*. World Publishing Company, New York.

Friedman, A. 1971. *Differential Games*. Wiley-Interscience, New York.

Ghosal, A., and Shakun, M. F. 1980. "Discontinuous Change in a Planning Process." A. Ghosal (ed.), *Applied Cybernetics and Planning*. South Asian Publishers, New Delhi.

Hamilton, W. F., and Moses, M. A. 1973. "An Optimization Model for Corporate Financial Planning." *Operations Research*, **21**, No. 2, May–June.

_____. 1974. "A Computer Based Corporate Planning System." *Management Science*, **21**, No. 2, October.

Hermann, P., and Krener, A. 1977. "Nonlinear Controllability and Observability." *IEEE Transactions ON Automatic Control*, **AC-22**, pp. 728–740.

Hesse, H. 1970. *Klein and Wagner*. In *Klingsor's Last Summer*. Winston R., and Winston, C. (tr.). Farrar, Straus and Giroux, New York.

Holling, C. S. 1976. "Resilience and Stability of Ecosystems." In E. Jantsch, et al., (eds.). *Evolution and Consciousness*. Addison-Wesley, Reading, MA.

Jantsch, E. 1975. *Design for Evolution*. George Braziller, New York.

______. 1978. Personal Letter, October 29.

______. 1979. "The Unifying Paradigm Behind Dissipative Structures, Autopoiesis, Hypercycles and Ultracycles." Center for Research in Management Science, University of California, Berkeley.

______. 1980. *The Self-Organizing Universe*, Pergamon Press, New York.

______. 1981. "Unifying Principles of Evolution." In Jantsch, E. (ed.). *The Evolutionary Vision*. Wetview Press, Boulder, CO.

Jantsch, E., and Waddington, C. H. 1976. *Evolution and Consciousness*. Addison-Wesley, Reading, MA.

Jaynes, J. 1976. *The Origin of Consciousness in the Breakdown of the Bicameral Mind*. Houghton Mifflin, Boston.

Keen, P. G., and Scott Morton, M. S. 1978. *Decision Support Systems: An Organizational Perspective*. Addison-Wesley, Reading, MA.

Kenney, R. L., and Raiffa, H. 1976. *Decisions with Multiple Objectives: Preferences and Value Tradeoffs*. Wiley, New York.

Krishnamurti, J. 1979. *Meditations*. Harper and Row, New York.

Kuhn, H. W., and Szego, G. P. (eds.). 1971. *Differential Games and Related Topics*. North-Holland, Amsterdam.

Landry, M. 1979. *Qu'Est-Ce Qu'Un Probleme?*. Laval University, Quebec, Canada.

______. 1980. *Doit-On Concevoir Ou Analyser Les Problemes Complexes*. Laval University, Quebec, Canada.

Lao Tzu. 1961. *Lao Tzu, Tao Te Ching*. Lau, D. C. (tr.) Penguin Books, Harmondsworth, England.

Layzer, D. 1975. "The Arrow of Time." *Scientific American*, **233**, No. 6, December, pp. 56–68.

Lewin, A. Y., and Shakun, M. F. 1976. *Policy Sciences: Methodologies and Cases*. Pergamon Press, New York.

Lilly, J. C. 1972. *The Center of the Cyclone*. Julian Press, New York.

Mason, R. O. 1969. "A Dialectical Approach to Strategic Planning." *Management Science*, **15**, B403–B414.

Merton, T. 1969. *The Way of Chuang Tzu*. New Directions, New York.

Michael, D. N. 1983. "Competence and Compassion in an Age of Uncertainty." *World Future Society Bulletin*, **17**, No. 1, January/February.

Mitroff, I. I., and Mason, R. O. 1980a. *On the Structure of Dialectical Reasoning in the Social and Policy Sciences*. University of Pittsburgh and University of Southern California.

______. 1980b. *Policy As Argument: A Logic for Ill-Structured Decision Problems*. University of Pittsburgh and University of Southern California.

Moses, M. A. 1975. "Implementation of Analytical Planning Systems." *Management Science*, **2**, No. 10, June.

Negoita, C. V. 1979. *Management Applications of Systems Theory*. Birkäuser Verlag, Basel.

Nicolis, G., and Prigogine, I. 1977. *Self Organization in Nonequilibrium Systems from Dissipative Structures to Order Through Fluctuation*. Wiley-Interscience, New York.

Nystrom, H. 1974. "Uncertainty, Information, and Organizational Decision-Making: A Cognitive Approach." *Swedish Journal of Economics*, **76**.

Ornstein, R. E. 1977. *The Psychology of Consciousness*, 2nd ed. Harcourt Brace Jovanovich, New York.

Prigogine, I. 1980. *From Being to Becoming*. W. H. Freeman, San Francisco.

Rao, A. G., and Shakun, M. F. 1974. "A Normative Model for Negotiations." *Management Science*, **20**, No. 10, June.

Ring, K. 1980. *Life at Death: A Scientific Investigation of the Near Death Experience*. Coward, McCann & Geohegan, New York.

______. 1982. "Near Death Studies: A New Area of Consciousness Research." *Institute of Noetic Sciences Newsletter*, **10**, No. 2, Fall, pp. 17–19.

Rosen, R. 1974. "Planning, Management, Policies and Strategies: Four Fuzzy Concepts." *International Journal of General Systems*, **1**, pp. 234–252.

Rufus, I. 1965. *Differential Games*. Wiley, New York.

Sagan, C. 1977. *The Dragons of Eden*. Random House, New York.

Shakun, M. F. 1975. "Policy Making Under Discontinuous Change: The Situational Normativism Approach." *Management Science*, **22**, No. 2, October.

______. 1976. "Design, Values, Negotiations, and Cognition in Adaptation of Purposeful Systems." *Computer Sciences and Management*, Institut de Recherche d'Informatique et d'Automatique, Le Chesnay, France.

______. 1977. "Cybernetics and Yom Kippur." *Cybernetics Forum*, Spring/Summer.

______. 1978. "Conflict Resolution as Design of Purposeful Systems." *Proceedings of the International Workshop on Conflict Resolution*, University of Haifa, June.

______. 1979. "Metaphors for General Systems: Religion, Evolution, Cybernetics, and Policy Making." In *General Systems Research: A Science, a Methodology, a Technology*. Proceedings of the 1979 North American Meeting, Society for General Systems Research, Louisville, KY.

______. 1981a. "Formalizing Conflict Resolution in Policy Making." *International Journal of General Systems*, **7**, No. 3.

______. 1981b. "Policy Making and Meaning as Design of Purposeful Systems." *International Journal of General Systems*, **7**, No. 4.

Shakun, M. F., and Sudit, E. F. 1978a. "Applying Conflict Resolution as Design of Purposeful Systems to the Arab-Israel Conflict." *Proceedings of the International Workshop on Conflict Resolution*. University of Haifa, June.

______. 1978b. "A Goals/Values Solution to Mandatory Retirement Age for Tenured Faculty: An Application of Situational Normativism." *Management Science*, **24**, No. 14, October.

Tapiero, C. 1977. *Management Planning: An Optimum & Stochastic Control Approach*, vols. 1 and 2. Gordon & Breach, New York.

Tart, C. T. 1975. *States of Consciousness*. E. P. Dutton, New York.

Thom, R. 1975. *Structural Stability and Morphogenesis*. W. A. Benjamin, Reading, MA.

Vickers, G., 1965. *The Art of Judgment*. Basic Books, New York.

Wilson, E. O. 1975. *Sociobiology: The New Synthesis*. Harvard University Press, Cambridge, MA.

Woodhouse, M. 1980. "Holistic Theories of Time." *Revision*, **3**, No. 2, Fall.

Zeeman, E. C. 1976. "Catastrophe Theory." *Scientific American*. April, pp. 65–83.

______. 1977. *Catastrophe Theory: Selected Papers 1972–1977*. Addison-Wesley, Reading, MA.

Zeleny, M. 1977. "Adaptive Displacement of Preferences in Decision Making." *Multiple Criteria Decision Making*. North-Holland, Amsterdam.

CHAPTER 3

Conflict Resolution as Evolutionary Systems Design: The Arab-Israeli Conflict*

3.1 Resolving the Conflict: June 1978, Before Camp David

Chapters 1 and 2 developed evolutionary systems design as a methodology for conflict resolution. Solutions to conflict within that methodology deliver values in the form of operational goals as defined by participants. Values and operational goals are not fixed but may be redefined by a goals/values referral process. At any stage in the process there is a relation of controls to goals to values (Figure 1.2). Table 3.1 developed by Shakun and Sudit (1978) is such a stage in the resolution of the Arab-Israeli conflict as of June 1978, prior to the Camp David accords.

In Table 3.1 we present a multicomponent (14 components) feasible solution X (controls u in Chapters 1 and 2) that satisfies the stated values and goals of the participants listed. The solution is a synthesis of ideas from various sources as well as our own. It is not necessarily a final proposed solution. Rather, it is a starting point (in June 1978) for an iterative search for a solution to be undertaken by the participants themselves.

The solution X will no doubt result in some other values not delivered to the listed participants. There can also be additional participants. These additional values and participants should be listed along with associated operational goals and the X solution modified to attempt delivery of all values identified. As part of the search process, operational goals might be changed, i.e., different operational goals found that deliver identified values, or new values can be identified along with associated operational goals and corre-

*This chapter is based on Shakun and Sudit (1978) and Casti and Shakun (1978).

Table 3.1 A Goals/Values Approach to the Arab-Israeli Conflict

Feasible Solution $X = (x_1, x_2, x_3, x_4, x_5;\ x_6, x_7, x_8;\ x_9;\ x_{10}, x_{11}, x_{12}, x_{13}, x_{14})$

x_1, x_2, x_3, x_4, x_5: Israeli-Palestinian-Jordanian relations

x_6, x_7, x_8: Israeli-Egyptian relations

x_9: Israeli-Syrian relations

$x_{10}, x_{11}, x_{12}, x_{13}, x_{14}$: Israel and all other participants

Feasible solution, X	*Operational goals satisfied*	*Values satisfied*
x_1: Autonomous Palestinian state in the West Bank and Gaza that is part of a Jordanian-Palestinian confederation having joint foreign and military affairs–otherwise separate.	Palestinians: End to Israeli occupation, self-rule Arab government. Jordan: Return of territories lost in 1967 war. Israel: Removal of major hurdle for peace agreement; end to occupation problem. Egypt, Syria, Saudi Arabia: Removal of major hurdle to peace agreement. Egypt, Saudi Arabia: Reducing military costs. Egypt, Saudi Arabia: Reducing potential for Russian influence.	Palestinians: Recognition of national entity, self-determination. Jordan: Territorial integrity. Israel: Achieving lasting peace treaties with neighbors. Egypt, Syria, Saudi Arabia: Securing legitimate rights of Palestinians. Egypt, Saudi Arabia: Economic development. Egypt, Saudi Arabia: Political self-preservation.
x_2: Long-term leasing on the part of Israel of military outposts along the Jordan River in exchange for long-term leasing by Palestinians of a corridor connecting the West Bank with Gaza. x_3: Demilitarization of West Bank and Gaza. The Palestinian Jordanian Confederation will maintain only police forces there. Their joint army will be stationed on the East Bank of the Jordan.	Israel: Maintenance of military strategic positions. Palestinians, Jordanians: not giving up territory. Territorial contact between West Bank and Gaza. Access to sea.	Israel: Military security. Palestinians, Jordanians: Territorial sovereignty, economic development.
x_4: Granting Vatican-type status (autonomous rule) to Muslim and Christian religious shrines in Jerusalem in exchange for secure status for Jewish religious shrines on West Bank. x_5: Jerusalem remaining under Israeli administration with the Arab citizens of the old city given dual citizenship in Israel and in the Jordanian-Palestinian confederation.	Arab countries: Control of and free access to Muslim religious centers in Jerusalem. Israel: Administering an undivided Jerusalem as capital of Israel. Free access to and control of Jewish religious centers on West Bank. Arab citizens of Jerusalem: Opportunity to choose between living under Israeli rule or living under Jordanian-Palestinian rule.	Muslims: Religious self-actualization. Jews: National and religious self-actualization. Arab citizens of Jerusalem: National identification and sense of belonging.

Table 3.1 *(continued).*

Feasible solution, X	*Operational goals satisfied*	*Values satisfied*
x_6: Return of Sinai to Egypt. Demilitarization of the area.	Egypt: Return of lost territory.	Egypt: Territorial sovereignty and integrity.
x_7: Long-term leasing by Israel from Egypt of Sharm El Sheik as an Israeli military post. x_8: Territorial exchange between Israel and Egypt Northern Sinai Israeli settlements for central Negev territory.	Israel: Control of militarily strategic areas of Northern Sinai and safeguarding navigation through the straits of Tiran.	Israel: Security.
x_9: Return of Golan Heights to Syria. Long-term Israeli leasing of military outposts on the Heights. Limited Syrian police forces only.	Syria: Return of lost territory. Israel: Control of militarily strategic areas on heights safeguarding settlement below.	Syria: Territorial sovereignty and integrity. Israel: Security.
x_{10}: Full-fledged diplomatic relations between Israel and all neighboring Arab states. Ambassadorial exchanges, trade agreements, and free travel. x_{11}: Elimination of the Arab economic boycott against Israel. x_{12}: A comprehensive antiterrorist treaty signed by Israel and all Arab states including binding agreement to extradite terrorists.	Israel: Recognition by neighbors secured and institutionalized. Palestinians: Recognition of statehood and national rights secured and institutionalized. All countries: New trade and cultural opportunities. Stable foundation of peace. All countries: Preventing bloodshed and terror against their citizens.	Israel: Political acceptance, security. Palestinians: Political legitimization, security. All countries: Peace, stability, economic development, well-being.
x_{13}: International agreement on limitations on arm supplies to all Middle East countries by all major weapon producers.	All Middle East states: Limitations on arms race and associated costs, fewer external alliances.	All Middle East states: Security, economic development, reduction of potential for outside intervention by foreign powers.
x_{14}: Agreement by Arab countries to grant citizenship to Palestinians who choose to remain living within their boundaries.	Palestinians: Increased choice, elimination of "people without country" status, alleviation of population density in Palestinian state.	Palestinians: Security, status, well-being, economic development.

sponding modified X solutions. Considering the participants involved in the Middle East conflict, appeal to common religious values in tune with evolution would appear to be useful in the goals/values referral process.

From the point of view of methodology we note that one often begins with identification of the participants, their values, and associated operational goals and then proceeds to search for a feasible solution X. However in cases such as the Arab-Israeli conflict, where there are many participants and many goals/values, it may be easier to begin the iterative search process with a proposed solution X and then systematically identify the operational goals and values satisfied by it as a starting point for the next iteration in the search process.

3.2 A Postscript After Camp David

The above starting solution was outlined in June 1978, before the Camp David talks in September 1978. This postscript was written three weeks after Camp David.

The Egyptian-Israeli accord at Camp David may be considered with respect to components x_6, x_7, x_8, x_{10}, and x_{11} of our feasible solution. Components x_6, x_{10}, and x_{11} were literally agreed upon and were scheduled to be operationally implemented between Israel and Egypt in the framework of the peace treaty to be signed by the two countries before the end of 1978. Components x_7 and x_8 are absent from the agreement. The strategic importance to Israel of retaining its settlements in Northern Sinai was probably overestimated. Israel presumably preferred to use them as bargaining cards, thereby facilitating the Egyptian concession to formulate the two Camp David accords (the Israeli-Egyptian and the West Bank documents) separately from each other. (However the Egyptians are under pressure from the other Arab participants to link the two agreements.) Therefore the territorial exchange deal proposed in x_8 was probably deemed too complex relative to its benefits. The same was probably true for the x_7 proposal. Israeli presence in Sharm El Sheik does not by itself guarantee Israel free access to the Indian Ocean because the Bab-El-Mandeb straits at the entrance to the Red Sea are under Arab control.

The five-year autonomy plan for the West Bank agreed upon in Camp David is by its nature a temporary interim solution. In any case no specific solutions could be negotiated without the participation of Jordan and the Palestinians. Agreement on implementation of durable solutions was therefore effectively deferred for five years. Our suggestions for a feasible starting solution for a West Bank and Gaza agreement thus remain valid as a focal point for future negotiations. In fact the five-year autonomy plan, if smoothly implemented, may well evolve toward this type of solution. The same prognosis seems to hold for the suggested Syrian-Israeli solution, provided that the peace movement momentum will be strong and durable enough to lure Syria to the negotiating table.

3.3 *Q*-Analysis of the Conflict

We employ a simplified form of Table 3.1 to motivate the use of q-analysis—a mathematical method for studying the structure of a relation (see Atkin 1974 and Casti 1982)—for analyzing the λ relation (goals/values matrix, see Chapters 1 and 2) in the Arab-Israeli conflict. We consider one general value, "well-being" (superordinate to the values listed in Table 3.1) and treat the solution X (in effect the controls u) as a set of issues or goals rather than what is labeled "operational goals" in Table 3.1 because the controls (solution X) are more specific. Hierarchically, the controls at one level can be considered as goals with respect to a level below (see Chapter 14). We then focus on the problem of goal dimension agreement among six players (Middle Eastern countries).

For X we use 10 issues (goals) derived from the 14-component solution X given in Table 3.1. With one value (well-being) and 10 issues the relation is represented by a (1×10) matrix. The entry at each row-column intersection is a 6-tuple of 1s or 0s corresponding to $N = 6$ Middle Eastern countries being "for" or "neutral" (1) or "against" (0) the value being delivered by the particular goal. The questions here are goal dimension agreement, and if not achieved at this stage, whether there is a subset of goal dimensions on which all or nearly all players agree, thus providing a basis for further iterations, and whether in the $N = 6$ player conflict situation a subset of the N players can be identified that shows promise for successfully negotiating to form a coalition C. Without going into the theory (for that see Atkin 1974, Casti 1982), we interpret the results of a q-analysis to help with these questions. For this purpose, because there is only one value, it is simpler to represent the λ relation as a matrix in which the columns are goals, as before, but now the rows represent the six players. This matrix, as presented below, has only one entry (a 0 or 1) at each row-column intersection so that the previous 6-tuple of 1s and 0s is now a column.

Casti and Shakun (1978) interpret the q-analysis as follows: The set X of issues (goals) are $X = (x_1, x_2, \ldots, x_{10})$ where

x_1 = autonomous Palestinian state in the West Bank and Gaza
x_2 = return of the West Bank and Gaza to Arab rule
x_3 = Israeli military outposts along the Jordan River
x_4 = Israel retains East Jerusalem
x_5 = free access to all religious centers
x_6 = return of Sinai to Egypt
x_7 = dismantle Israeli Sinai settlements
x_8 = return of Golan Heights to Syria
x_9 = Israeli military outposts on Golan Heights
x_{10} = Arab countries grant citizenship to Palestinians choosing to remain within their borders

The set of countries (players) is

$$Y = (y_1, y_2, \ldots, y_6)$$
$$= (\text{Israel, Egypt, Palestinians, Jordan, Syria, Saudi Arabia}).$$

We choose $\lambda = 1$ to mean that the country is favorable ("for") or neutral toward the goal and $\lambda = 0$ to mean the country is against it. The matrix for λ is:

	x_1	x_2	x_3	x_4	x_5	x_6	x_7	x_8	x_9	x_{10}
y_1	0	1	1	1	1	1	0	0	1	1
y_2	1	1	1	0	1	1	1	1	1	0
y_3	1	1	0	0	1	1	1	1	1	1
y_4	1	1	0	0	1	1	1	1	1	0
y_5	1	1	0	0	1	1	1	1	0	0
y_6	1	1	1	0	1	1	1	1	1	1

Focusing on controls, the q-analysis (Casti and Shakun 1978) shows that the most likely negotiating partner for Israel is Saudi Arabia, which is neutral or favorable on all issues except one. However both Egypt and the Palestinians are nearly as likely candidates because they are simplices of dimension only one less than Saudi Arabia. As the Camp David talks demonstrated, Egypt is indeed a favored negotiating partner due also to psychological and other factors not incorporated into the above relation.

Focusing upon issues, the q-analysis identifies the high-dimensional objects x_2 = return of the West Bank and Gaza to Arab rule, x_5 = free access to religious centers, and x_6 = return of the Sinai to Egypt. All six players view these goals as favorable or neutral. Therefore they provide a good basis for further negotiating toward settlement of the conflict. This has been borne out by the Camp David talks and subsequent developments.

In addition to the above relation λ, three other cases were considered: favorable only, against only, and neutral only. The results of these studies confirmed that (1) Israel is highly disconnected from the other parties in the dispute, (2) Saudi Arabia is the most moderate of the Arab states, (3) Syria is by far the most rigid and inflexible, and (4) the single issue that tends to bring all the parties together is free access to all religious centers.

REFERENCES

Atkin, R. H. 1974. *Mathematical Structure in Human Affairs*. Heineman, London.

Casti, J. 1982. "Topological Methods for Social and Behavioral Systems." *International Journal of General Systems*, 8, pp. 187–210.

Casti, J., and Shakun, M. F. 1978. "Goals/Values Geometry and Conflict Resolution." New York University.

Shakun, M. F., and Sudit, E. F. 1978. "Applying Conflict Resolution as Design of Purposeful Systems to the Arab-Israeli Conflict." *Informal Proceedings of the International Workshop on Conflict Resolution*, University of Haifa, June 1978.

CHAPTER 4

A Goals / Values Solution to Mandatory Retirement Age for Tenured Faculty: An Application of Situational Normativism*

Melvin F. Shakun[†] and Ephraim F. Sudit[††]

4.1 Introduction

On September 15, 1977, the House of Representatives of the U.S. Congress passed the Age Discrimination in Employment Act, which would prohibit mandatory retirement of employees before the age of 70.** The Senate version of the same bill contained the Chafee amendment, which would exclude tenured university faculty members and certain high-salaried executives from the retirement extension provisions of the act. Consequently, the Chafee amendment would permit universities to require mandatory retirement of tenured faculty at the age of 65.

The Chafee amendment was introduced in the Senate at the urging of the administrators of many well-known universities. Their primary concern was that the extra costs entailed in carrying high-salaried senior faculty on payroll for several additional years would impose additional financial burdens on universities in times of crisis and restrict the number of openings for young prospective entries into the academic ranks. This amendment generated opposition from the American Association of University Professors and many

*Minor additions have been made to this chapter, which first appeared in *Management Science*, Vol. 24, No. 14, October 1978.

[†] New York University.

[††] Rutgers—The State University of New Jersey.

**The previous existing law allowed mandatory retirement at age 65.

scholarly associations on the grounds that the exclusion of tenured faculty from extended retirement is discriminatory in nature and at the same time would pose a threat to the academic tenure system. These opposing positions, which led to a spirited public debate, are anchored in a clear-cut conflict among operational goals.

4.2 Nature of the Conflict: Goals / Values Analysis

Following situational normativism (Lewin and Shakun 1976; Shakun 1975), we identify the participants, their operational goals, and their underlying values. For tenured faculty a key goal is economic. The ability of a university administration to deny a tenured professor the possibility of staying in his position beyond 65 could result in diminished salaried income and pension. As a second goal many professors would like to hold an active appointment until age 70. The values underlying this goal include desires for continuing contributions to research, teaching, and university affairs; self-actualization; status; and influence. Also, because only tenured academic personnel were singled out for exception, some fears were aroused that it could be the beginning of a wider attack on academic tenure and the values of academic freedom and job security it guarantees.

Many university administrators saw some of their more important goals threatened if legally compelled to extend mandatory retirement age to 70. In view of the shaky financial status of many institutions of higher education and the relatively gloomy forecast for the 1980s due to demographics, administrators felt that mandatory retirement at age 70 is likely to impose extra financial burdens on their institutions at a particularly difficult period. Further, in the opinion of many administrators, raising the retirement age from 65 to 70 would severely limit the number of openings for new prospective entrants to professorial ranks, as well as deny tenure to eligible nontenured faculty members, thereby forcing many of them to leave academia. Such newly created de facto entry barriers can reduce overall quality of research and decelerate the advance of knowledge. In many scientific disciplines practitioners are generally most creative and most capable of original and important contributions during their relatively younger years. Severely limited opportunities may also discourage bright young people contemplating academic careers, thus further reducing the pool of talent for years to come. Deterioration of the quality of research may have adverse spillovers on the rate of advancement in productivity and technological innovation, thereby aggravating the condition of the national economy and impeding its growth. The latter is a concern of society as a whole.

Table 4.1 lists the key participants in the mandatory retirement problem along with the operational goals and underlying values. In terms of the solutions proposed by the House of Representatives and the Senate, namely, retirement at ages 65 and 70, respectively, there is no feasible solution, i.e., neither 65 nor 70 as a retirement age is able to satisfy all the operational goals

Table 4.1. **Participant Goals / Values Relevant to Mandatory Retirement of Tenured Faculty**

Participants	*Operational goals*	*Underlying values*
Tenured faculty	Economic goals (full salary to age 70 and associated increased pension benefits)	Economic well-being and security: equal treatment of all work categories (nondiscrimination)
	Active appointment until age 70	Continued contributions to research, teaching, and university affairs, self-actualization, status, and influence; equal treatment of all work categories (nondiscrimination)
	Integrity of tenure system	Academic freedom and job security
University administrators	Financial goals	Financial viability of university
	Academic entry job opportunities for new Ph.D.s	Maintain quality of research; fairness of employment opportunities
	Tenure opportunities for nontenured faculty	Maintain quality of research; fairness of employment opportunities
Nontenured faculty	Tenure opportunities	Academic freedom and job security
New Ph.D.s	Academic entry job opportunities	Pursue academic interests; economic benefits
Society as represented by Congress	Same retirement age for all	Nondiscrimination
	Integrity of tenure system	Academic freedom
	Academic entry job opportunities for new Ph.D.s	Maintain quality of research and socioeconomic well-being; fairness of employment opportunities

of the participants. In order to find a feasible solution, following situational normativism, we refer to the values (Table 4.1) underlying these operational goals. We seek modifications of the latter, which, while delivering the values (or other values now identified), permit a feasible solution.

First, we discuss the notion of "discrimination." The term "discrimination" has general negative connotations and overtones. However it is obviously impossible to set policy criteria on almost any issue of importance without some form of "discrimination." The issue therefore does not reside in "discrimination" itself but on the negative impact on other values or operational goals aspired to by certain participants in society. A case in point is the tenure system. In the U.S. the tenure system can be regarded as discriminatory in the sense that it guarantees almost absolute job security to one profession only—university professors. However this type of discrimination is socially acceptable because it is believed to secure cherished values of freedom of research, teaching, and the advancement of knowledge. In the context of our discussion, the Chafee amendment was not opposed by tenured faculty and many members of Congress just because it is discriminatory in an abstract sense, but because of negative impacts of this discriminatory provision on tenured faculty. Namely, it denies tenured faculty the values listed in Table 4.1 for this group.

4.3 Proposed Solution

Consider therefore the following proposed solution: University administrators would be required by law to extend to tenured faculty the option of retaining between 65 and 70 their tenure privileges and their full rank and professional status. These professors would be employed half-time and receive 50 percent of their regular salary.

Such a solution we feel would deliver the operational goods and values. The part-time tenured status after 65 of these professors would free their full-time tenure slots, so that hiring of new Ph.D.s and bestowing of tenure could continue. Note that half a senior professor's salary approximately covers an entry-level salary. Such an arrangement would eliminate most, if not all, the negative discriminatory effects of the Chafee amendment. The academic tenure system would not be threatened because university professors would be guaranteed the right to retain their tenure until 70. Further, professors would have an active appointment until age 70. As to the financial aspects, half-pay would be supplemented by pension, so that differences in actual income during half-time employment should be small. The erosion in future pension benefits beyond age 70 because of earlier drawing of pension would be offset by the extra leisure afforded by half-time compared to full time employment between 65 and 70. Alternatively, if so desired, the half-time made available could be used to seek supplemental income in this period.

Our proposed solution is empirically supported by the Ladd and Lipset (1977) faculty survey, which finds that "48% of all faculty members and 30% of those aged 55 to 62 would consider retiring sooner if assured part time employment with a proportionate decrease in salary."

In conclusion, we believe that our proposed solution takes care of the essential concerns of the various participants as revealed through the goals/values analysis.

4.4 Implementation

An important criterion for implementation of management science is that models, or at least their results, be understandable to policymakers. The goals/values analysis, as presented in Table 4.1 and associated discussion, is compact and understandable, thus facilitating its usefulness as an input to policymakers.

The previous analysis was submitted as a memorandum to members of the House-Senate Conference Committee considering resolution of the House and Senate versions of the bill. Note that the analysis was intended to deal with the policy problem before the committee. Such questions as abolishing or modifying the present tenure system were not considered.

It is clear from the responses received from individual senators and representatives on the committee that the analysis provided a framework and solution useful in rethinking the issues before the committee. Representative Paul Findley (Illinois) wrote us on February 21, 1978, as follows:

> Your proposal for half-time basis for tenured faculty members between the age of 65 and 70 is certainly intriguing and I feel holds promise. It does not look as though the House/Senate Conference Committee, however, will move in that direction to resolve the exemption for college professors dilemma. Rather, the exemption will probably be allowed for a few years and then terminated.
>
> I hope that in actual practice many colleges and universities will work out with tenured senior faculty arrangements similar to the one that you suggest on a voluntary basis. Such arrangements would solve the problems of creating opportunity for young professors and providing meaningful activity for senior professors.

The solution actually adopted by the conference committee, then passed by the House and Senate and signed into law by President Jimmy Carter, establishes mandatory retirement age for tenured professors at age 70 but makes this effective in 1982—three years after the 70-year retirement for others in the population goes into effect. This solution does not deliver the operational goals of tenure opportunities for nontenured faculty, academic entry job opportunities for new Ph.D.s, and related university administrators' financial goals because it does not free up funds for accomplishing these, as with our proposed solution. Evidently, Congress was not concerned with delivering these particular goals.* The act as passed (as well as our own solution) does deliver the remaining goals. The three-year implementation delay in the act does give the participants additional time to adjust—a tacit recognition of other operational goals that will not be delivered in the future by the act as passed. This was noted in Senator John Chafee's (Rhode Island) letter of March 8, 1978, in response to our proposal: "It is anticipated that during the period from now until July 1, 1982, universities and colleges will be developing alternative retirement policies and innovative programs to meet the challenges presented by this legislation." The price of this implementation

*There are a number of possible explanations for this lack of concern. Constraints implemented in the future with regard to tenure opportunities and academic entry job opportunities do not have the same force as constraints implemented now. The future is often heavily discounted by politicians. Further, the bargaining power of new Ph.D. entrants and nontenured faculty is relatively weak within such organizations as the American Association of University Professors and relative to other participants. Also, the time horizon of university administrators is often short-term.

delay is that tenured faculty reaching the applicable retirement age 65 within this three-year period will not be guaranteed an opportunity to continue to 70.*

REFERENCES

Ladd, G. C., Jr., and Lipset, M. 1977. "Many Professors Would Postpone Retirement If Law Were Changed." *Chronicle of Higher Education*, November 7.

Lewin, A. Y., and Shakun, M. F. 1976. *Policy Sciences: Methodologies and Cases*. Pergamon Press, New York.

Shakun, M. F. 1975. "Policy Making Under Discontinuous Change: The Situational Normativism Approach." *Management Science*, 22, No. 2 (October).

*Although not undertaken here, we note that q-analysis could be applied using a λ matrix whose rows are the participants and whose columns are the operational goals in Table 4.1, similar to Section 3.3. Situational normativism (Lewin and Shakun 1976; Shakun 1975) is a forerunner of evolutionary systems design. Goals/values methodology is discussed in Chapters 1 and 2.

CHAPTER 5

Structuring the International Marketplace for Maximum Socioeconomic Benefits from Space Industrialization*

William A. Good[†], **George S. Robinson**[††], **Melvin F. Shakun**[†††], **and Ephraim F. Sudit**[††††]

5.1 Introduction

The purpose of this chapter is to determine the present status of space industrialization in order to develop strategic options and policy recommendations.

STAGES OF DEVELOPMENT FOR COMMERCIAL TECHNOLOGY

In broad terms the commercial viability of a new technology may involve a natural evolution through the following five stages:

1. concept dissemination via mythology, fiction, and/or science fiction
2. scientific progress leading to technology of demonstrable practicality
3. entrepreneurial process of bringing technology to the marketplace

*This chapter is a slightly changed version of the one that appeared in the *Proceedings of the Ninth International Symposium on Space Economics and Benefits*, International Academy of Astronautics, Munich, Germany, September 1979.

[†] Earth Space Transport System Corporation.
[††] The Smithsonian Institution.
[†††] New York University.
[††††] Rutgers—The State University of New Jersey.

4. governmental process of structuring the marketplace via regulation
5. maturation of the technology in a competitive international marketplace

The successful introduction and development of new technology seems to require an orderly progress through each of the five stages in the given sequence, although it is recognized that there is considerable overlap between some of the stages. For instance, if the concept has not been broadly disseminated throughout the population, the entrepreneurial process will have to include substantial resource expenditures on advertising for the purpose of educating potential customers in addition to marketing the product. This educational process is a requirement that acts as a negative motivational factor facing potential entrepreneurs because competitors may take advantage by offering a similar product at a lower price before the market leader can recover total market development costs.

HISTORY OF AIR AND SPACE COMMERCE

With the recent passage in the United States of the 1978 Airline Deregulation Act, the development of commercial aviation technology appears to be entering the fifth stage, which is maturation in a competitive international marketplace. The legislative history of the Airline Deregulation Act indicates that the air transport industry is viewed as having completed the first four stages in order (Senate Committee on Commerce 1978). However, a review of the writings and speeches of the early aviation entrepreneurs indicates that their vision extended beyond subsonic jumbo jets, including both supersonic flight and interplanetary travel (Lindbergh 1976).

It may be necessary to consider the space shuttle as a *sui generis* mode of transportation (Sloup 1978). This is not due to the unique characteristics of the space shuttle orbiter, which is technically just an *aircraft* capable of flying higher and faster. Rather, it is due to the unique characteristics of the space transport market (Barsh 1978).

For the first time the transport of people and goods will become routinely possible from Earth to space, from space to Earth, *and* between different locations in outer space (Committee on Science and Technology 1977). In terms of these capabilities space transport is on the verge of transitioning from the second to the third stage of development.

In order to deal with the complexity of numerous participants having different operational goals and values, the methodology of evolutionary systems design has been utilized to develop strategic options and policy recommendations that will accelerate the evolution of space transport technology into the third and fourth stages of development.* We identify participants, values, goals, technologies (market structures), and controls.

*See Chapters 1 and 2 for details of the evolutionary systems design methodology.

PARTICIPANTS IN SPACE INDUSTRIALIZATION

Space industrialization is a global phenomenon. The participants in this process can be classified according to their general goals and values as follows:

1. U.S. private sector
2. U.S. public sector
3. U.S. domestic associations
4. multinational corporations and other private international organizations
5. U.N. and other public international organizations
6. other industrialized nations (OECD)
7. USSR and its allies
8. China and its allies
9. less developed countries

The goals and values of all these groups of participants must be accounted for during the goals/values referral process, but this investigation is done from the perspective of the U.S. private and public sectors (Laszlo 1978), the two players forming coalition C.

VALUES OF THE U.S. PRIVATE SECTOR

In general terms the values of the U.S. private sector can be stated as follows:

1. technological and marketing advantage over international competitors
2. strong market position with adequate return on stockholders' equity
3. decreased reliance on foreign energy resources and improved energy technology
4. improved reliability of resource supplies through diversification of sources
5. improved public transportation and communications systems
6. stability in the international marketplace and security of foreign investments
7. economic survival and avoidance of unnecessary risk
8. prestige and growth, as well as freedom for entrepreneurship in outer space
9. opportunity for development of projects with long-term corporate economic benefits but negative net present value for private investment (Benjamin and Hydrean, 1979)

VALUES OF THE U.S. PUBLIC SECTOR

In general terms the values of the U.S. public sector can be stated as follows:

1. maintenance of national security through improved defense capabilities
2. social stability with strong family ties and family planning

3. improved education, communication, information, and transportation systems
4. clean and safe environment with improved health care services
5. adequate food supply, improved standard of living and quality of life
6. industrial innovation, improved productivity and employment opportunities
7. advancement of science and growth in the technology base
8. stable and growing international economy with decreasing international unrest
9. global human rights and justice, along with equitable and reliable resource distribution to the factors of production

These values are manifest in various programs and activities within the U.S. private and public sectors. There may be others that have been overlooked. The potential commercial applications of space technology and the process of space industrialization are analyzed in terms of the goals and values of the global participants. The economic concepts discussed in Section 5.2 are also taken into account. An analysis based on the theory of individual needs and behavior will be left for further investigation (Cheston 1979; Maslow 1970).

5.2 Economic Regulation

The market structure for public transportation can generally be described using a continuum that extends from pure competition through oligopoly and regulated oligopoly to a monopoly controlled by government. In a competitive or slightly oligopolistic market structure the participants' emphasis is generally on service, including its frequency. The result of such an open structure is a mobility and dispersion of capital resources, as well as diminished leverage for labor when considered on a career basis. At the other extreme a monopoly structure emphasizes economies of scale, centralized planning and control, and intensive capital investment. This results in an emphasis on cost/price considerations, often at the expense of service. There is generally more economic leverage for organized labor due to a greater concentration of economic resources.

In between these two extremes exists the regulated oligopoly represented by the U.S. air transport industry prior to the recent Airline Deregulation Act. The 1938 Civil Aeronautics Act established the concept of certificates of public convenience and necessity for the airline industry in order to enable aviation entrepreneurs to attract capital and to make technological progress in air transport. The concept was justified using an infant industry argument. This led to a relatively stable environment at the expense of airline management flexibility. Trade-offs existed between frequency and economies of scale. Competition took place in the offering of so-called frills. When the supply of air transport services exceeded demand, institutionalized price inflexibilities led to the parking or selling of aircraft rather than the reduction of prices.

Prior to the 1938 Civil Aeronautics Act and the 1934 Air Mail Act, the infant airline industry was operated under the 1923 Kelly Act. This established a competitive contractual system supporting the development of air mail service. It was in the operation of the Chicago–St. Louis air mail contract that Charles A. Lindbergh received his most valuable flight experience following U.S. Army pilot training.

Politically inspired accusations of antitrust violations led to the cancellation of all air mail contracts by President Roosevelt in 1934. The U.S. Army then flew the air mail routes for the post office. The U.S. air transport industry had not been a government monopoly for more than a few months when it was replaced by more competitive contracts and the subsequent regulated oligopoly structure (Dade and Vecsey 1979; Frankum 1971; Gray 1948).

PRIVATE GOODS, PUBLIC GOODS, AND EXTERNALITIES

In contrast to the segmentation of the infant air transport market by the granting of route awards, there is no obvious geographical basis for segmenting the space transport market. In fact there is still not even a clear legal definition of where the Earth's atmosphere ends and outer space begins. Therefore a normative analysis of space transport market structure should be done on the basis of the economic nature of the goods provided by the application of technology in space. For this purpose a clear understanding of three simple economic concepts is required.

In economic terms a pure private good has two fundamental characteristics. First, the total cost of production is a function of the total number of users and/or the total number of units produced. Second, any particular user can be excluded from enjoying the economic benefits of production even after the good has been produced. Transportation services and consumer durables are examples of private goods.

A pure public good has the opposite fundamental characteristics. First, the marginal cost of serving an additional user is zero. In other words there is a capability for serving a varying number of users at the same total cost. Second, it is impossible to exclude any particular user from enjoying the economic benefits once the good is produced. Nuclear defense, clean air, and a lighthouse are examples of public goods. Most economic goods would fall somewhere on a continuum between a pure private good and a pure public good.

Externalities are economic activities that affect the welfare of parties who are not directly involved either as producers or consumers. An example would be the mutual benefit attained in the colocation of a bee hive and an apple orchard. Externalities may be either beneficial (technology transfer from the public sector to the private sector) or detrimental (automobile emissions). Externalities may be either private goods (military pilot training) or public goods (airport noise pollution). Also, externalities may arise from either production (radioactive waste) or consumption (tobacco smoke).

Because externalities are often public goods or semipublic goods, there is a lack of market motivation to generate the optimal quantities of positive and negative externalities. The consequence of this market failure is that there is a need for

1. government regulation and control of externalities
2. participation of the public sector in the provision of public goods
3. nonmarket government participation in the production of certain goods and services where such production yields substantial externalities that are public goods

Examples of such situations include aircraft certification, air traffic control services, and defense contracts to aerospace manufacturers, respectively.

5.3 Feasible Solutions for Technology (Market Structure)

Using the perspective of the U.S. public and private sectors, the following feasible solutions are examined for alternative aerospace transport system market structures (technologies). This process will lead to a structuring of the international marketplace for maximum socioeconomic benefits from space industrialization. The subsequent normative analysis is based on the economic nature of the goods produced and the five stages of commercial technological development given previously.

1. NASA ownership and management of the space transportation system (STS)
2. NASA ownership of the space transportation system (STS) with management contracts to private firms
3. joint venture between NASA and private firms for ownership and operation of the aerospace transport system (ASTS), but with domestic participants only
4. market organization of the aerospace transport system (ASTS) with the public sector excluded from both ownership and management
5. market organization of the aerospace transport system (ASTS) with the public and private sectors included, but with domestic participants only
6. market organization of the aerospace transport system (ASTS) with the public and private sectors included, as well as both domestic and international participants
7. market organization of the earth space transport system (ESTS) with the public and private sectors included, as well as both domestic and international participants with at least a nonmarket involvement of less developed countries

The name of the solution system is varied in the list in order to emphasize the sometimes radical change in concept among the various alternatives. The first option is essentially the status quo as expressed in the October 11, 1978, White

House Fact Sheet on U.S. Civil Space Policy. The second option for the NASA space transportation system is the concept proposed in the STS Operations Management Study completed for NASA Kennedy Space Center (Booz Allen & Hamilton 1977). The NASA Johnson Space Center has also initiated a study to define their own organizational structure for the operation of a mature space transportation system (McKinsey 1980). It appears doubtful, however, that either NASA or any NASA contractor will go beyond the first two options at this time due to their underlying goals and values.

The aerospace transport system (ASTS) indicates potential involvement of common carriers from the air transport industry in the development of commercial space transport services. This implies a close examination and refinement of existing antitrust constraints that might impede collaboration among one or more air transport and/or aerospace manufacturing firms (either domestic or international).

Earth space transport systems (ESTS) refers to the vertical integration of multinational and multimodal transport services in a deregulated environment. Firms in the communications industry and firms in the energy resources industry could also be integrated into such a multinational corporation. The minimum cash flow for the resulting organization to enter the space transport market would be $10 billion per year in 1975 U.S. dollars. Additional requirements include assets of $10 billion and equity of $5 billion. The ratio of debt to equity must be in line with the average for the parent industries. The criteria need to be met by June 30, 1985.

The above figures assume a risk exposure on entering the space transportation market of less than 10 percent of total cash flow. Also assumed is a discrepancy of less than 10 percent on the NASA space transportation system performance in terms of its impact on the published NASA STS pricing policy. The availability of an additional space shuttle orbiter (and supporting equipment) every six months would be sought at an orbiter cost of less than $250 million. A total production run of at least 25 (including modified and improved versions) would be guaranteed for Rockwell International. Various guarantees would be sought from the U.S. government in return for fixed cost performance in providing space transport services (Good 1979).

5.4 Goal Delivery Classification Analysis: Controls

The economic criteria for discrimination between private and public participation in the delivery of operational goals (economic goods) may be summarized as follows:

1. Private goods should be provided primarily by the private sector unless positive or negative externalities outweigh this consideration.
2. Public goods should be provided primarily by the public sector unless positive externalities justify subsidized private sector participation.
3. Market failures caused by externalities may require public sector participation via public financing and/or subcontracts.

***Table 5.1.* Controls For Goal Delivery Classification Analysis**

Control	Description
A	private good to be delivered by the private sector
B	public good to be delivered by the public sector
A C	private good requiring comprehensive international agreement
A D	private good requiring partial international agreement
A C E	private good requiring comprehensive international agreement due to positive or negative externalities
AB E	joint private and public participation due to market failure or infant industry and substantial positive or negative externalities
A DE	private good requiring partial international agreement due to negative externalities in production having international impact
AB DE	joint private and public participation requiring partial international agreement due to substantial positive or negative externalities
B DE	public good requiring partial international agreement due to substantial positive or negative externalities

This methodology is applied to all the potential commercial applications of technology in outer space that were identified in the space industrialization studies (Rockwell International 1978; Science Applications, Inc. 1978). Each application (such as pocket telephone, electronic mail, teleconferencing, materials processing, satellite power system, etc.) is classified according to one of the controls in Table 5.1, which gives the goal (commercial application) using the technology (market structure, which we argue in Section 5.6 should be feasible solution 7 from Section 5.3):

In addition to the commercial application goals, transport service goals (cost, reliability, etc.) are outputs of the technology (market structure).

5.5 Results

The analysis and classification of well over 100 potential applications of technology in outer space indicates that the space transport market could be divided up into no more than six but at least three individual markets. Potential operators could bid on contracts or rights to develop these markets. Monopoly rights could be granted for fixed time periods using an infant industry argument. Competition could then be gradually increased as the developing markets mature. Mergers could be allowed in existing related industries contingent on the merged corporation entering new space transport markets. Airlines could be encouraged to develop space transport capabilities by allowing mergers and acquisitions among any transport modes for the purpose of achieving earth space transport systems balance sheet constraints for entering space transport markets.

The goals/values referral process for conflict resolution is required because NASA aspirations are presently being adequately met by the first feasible solution of NASA ownership and management of the space transportation system (STS). However, it is apparent that NASA lacks the legal status of a "common carrier," and NASA authority beyond research and development is open to question (Good and Robinson 1979). If the process of space industrialization is in fact full of potential commercial possibilities, NASA could become a highly visible target for merger strategies among existing firms in the private sector, especially those firms that might be under pressure from antitrust constraints. Some of these firms may also be potential consumers of space transport services.

There is also the problem of extreme turbulence in the international air transport industry, much of which is viewed by industry leaders as having been caused by U.S. government policies. The planning horizon of air transport executives has therefore become shorter, but a new generation of industry leaders will emerge during the 1980s due to the retirement of many chief executive officers. Pressure exists for many U.S. air carriers to merge for survival and growth, but recent experience indicates that antitrust sentiments may impede mergers based on simple efficiency arguments ("Merger Hopes Boosted" 1979). It is difficult to imagine that the NASA monopoly in high technology transportation will remain completely unnoticed by the air transport industry after the completion of the first orbital test flights of the space shuttle. The flight characteristics of the space shuttle orbiter might also serve as a catalyst for the development of more carefully defined regulations for supersonic flight based upon the likelihood of actual noise impact at ground level ("AIAA Lone Voice" 1977).

U.S. public sector values would best be served by maintaining NASA budgets for the purpose of improving the commercial space technology base while restricting NASA from commercial operations. Applications should be transferred to the private sector if they are defined as private goods. The policy goal of the public sector should be to find suitable participants in the private sector who are willing and able to develop the commercial markets in conjunction with the space transport operator under a suitably protected market structure. The guaranteed presence of the public sector as a consumer of space technology should be used as an incentive for new space transport firms, just as it was by the post office in the infancy of commercial airlines.

5.6 Conclusions

Many policy questions must be answered prior to completing the space shuttle orbiter flight tests. No presently proposed U.S. legislation would move the NASA space transportation system directly to the private sector. The ongoing assessment of applications of technology in space being conducted by the Office of Technology Assessment for the U.S. Congress will examine "the utilization, pricing policy, and future development of the Space Transportation System" as a broad issue, but not necessarily as the primary issue in the process of space industrialization (Office of Technology Assessment 1979). Also, a study by the National Transportation Policy Study Commission failed even to consider space transport as a mode of transportation or the evolution of air transport into aerospace transport (National Transportation Policy Study Commission 1979).

The analysis undertaken here in order to structure the international marketplace for maximum socioeconomic benefits from space industrialization indicates that the *key issue* is the market organization for space transport services. The optimal structuring of the space transport market will yield the maximum benefits from space technology. The nature of the economic goods provided by space industrialization indicates that the private sector must play a key role in developing space industry, but not just as consumers of space transport services. The nature of space transportation as a private good indicates that it should be provided by the private sector. The principal government role after developing space transport technology (a public good) should be that of the primary consumer of space transport services while the market is in its infancy. This is already the case for the NASA space transportation system.

The magnitude of the required investment and the risk involved indicate that size and profitability constraints will have to be met by potential private providers of space transport services. The international nature of space industrialization indicates that multinational corporations will have an advantage in determining and meeting the needs of international consumers. Most important, the intermingling of public and private aspects of space technology requires a closely integrated effort by governments and industry, with each participant's roles and responsibilities clearly defined. Thus we are recommending feasible solution 7 from Section 5.3—market organization of the earth space transport system (ESTS).

(By way of acknowledgment, the authors would like to express their gratitude to many individuals for their encouragement and their concern about the future of commercial aerospace transport services. In particular, these include George C. Dade (director, Long Island Cradle of Aviation Museum), William K. Kaiser (curator, Long Island Cradle of Aviation Museum), Alexander R. Ogston (chairman, Historical and Educational Committee, The Wings Club), and James T. Pyle (retired FAA administrator), as well as many others. The conclusions reached, however, are solely the responsibility of the authors.)

REFERENCES

"AIAA Lone Voice for Concorde." 1977. *Astronautics & Aeronautics*, 15, No. 11, p. 13.

Barsh, R. L. 1978. "The Prospect for Space Industrialization: A Closer Look." *Journal of Contemporary Business*, 7, pp. 1–5.

Benjamin, J. S., and Hydrean, P. P. 1979. "The Effect of the Time Value of Money on R & D." Unpublished report for the Applications of Technology in Space Work Group, Office of Technology Assessment, U.S. Congress, Washington, DC (rough draft).

Booz Allen & Hamilton. 1977. *STS Operations Management Study*. NAS10-9098.

Cheston, T. S. 1979. "Space Social Science: Suggested Paths to an Emerging Discipline." *The Space Humanization Series*, 1, pp. 1–14.

Committee on Science and Technology. 1977. *Space Transportation System*. U.S. House of Representatives, Washington, DC.

Dade, G. C., and Vecsey, G. 1979. *Getting off the Ground: Pioneers of Aviation Speak for Themselves*. E. P. Dutton, New York.

Frankum, J. E. 1971. *Legacy of Leadership: A Pictorial History of Trans World Airlines*. Walsworth Publishing, Marceline, MO.

Good, W. A. 1979. "Strategy, Structure, and Environment in Multinational, Multimodal Transportation Under Deregulation in the Age of Space Industrialization." An outline of a proposed dissertation for the Ph.D. degree submitted for approval by the reading committee. New York University, New York.

—, and Robinson, G. S. 1979. "The Shuttle: Uncommon, Common Carrier." *Astronautics & Aeronautics*, 17, No. 2, p. 8.

Gray, G. W. 1948. *Frontiers of Flight*. Knopf, New York.

Laszlo, E. 1978. *Goals for Mankind: A Report to the Club of Rome on the New Horizons of Global Community*. New American Book, New York.

Lindbergh, C. A. 1976. *Autobiography of Values*. Harcourt Brace Jovanovich, New York.

Maslow, A. H. 1970. *Motivation and Personality*. Harper & Row, New York.

McKinsey and Co. 1980. *STS Operations Management Study*. NAS 9-15781.

"Merger Hopes Boosted for Continental/Western." 1979. *Aviation Week & Space Technology*, 110, No. 21, p. 45, May 21.

National Transportation Policy Study Commission. 1979. *National Transportation Policies Through the Year 2000*. Final Report. Washington, DC.

Office of Technology Assessment. 1979. *Proposal for an Assessment of Applications of Technology in Space*. U.S. Congress, Washington, DC.

Rockwell International. 1978. *Space Industrialization*. NAS 8-32198.

Science Applications, Inc. 1978. *Space Industrialization*. NAS 8-32197.

Senate Committee on Commerce, Science, and Transportation. 1978. *Amending the Federal Aviation Act of 1958*. U.S. Senate, Washington, DC, pp. 1–5.

Sloup, G. P. 1978. "The NASA Space Shuttle and Other Aerospace Vehicles: A Primer for Lawyers on Legal Characterization." *California Western International Law Journal*, 8, pp. 403–453.

CHAPTER 6

A Normative Model for Negotiations*

Ambar G. Rao[†] **and Melvin F. Shakun**[†]

6.1 Introduction

The relationship between a multinational corporation (MNC) and a host government (HG) typically involves elements of conflict and cooperation. Not infrequently differences in goals between the two parties, unless resolved through negotiations, can prevent realization of potential gains for both inherent in mutual cooperation. For background on MNC-HG relationships see Kapoor and Grub (1972).

The general subject of negotiations has been widely studied in the literature. A review has been undertaken by Patchen (1970). Recent mathematical models of negotiation have included descriptive models by Cross (1969) and Coddington (1968) and normative models by Saraydar (1971) and Harsanyi and Selten (1972). From a management science viewpoint the problem is one of developing a normative negotiation model that uses appropriate behavioral assumptions and is rich enough to be relevant to real negotiations while maintaining feasibility with respect to input data and computability.

The research described in this chapter is oriented in this direction. Various negotiation behavior concepts are discussed. These are used in a dynamic programming model of negotiations yielding normative recommenda-

*This chapter originally appeared in *Management Science*, Vol. 20, No. 10, June 1974.

[†] New York University

tions as to concession making. The negotiation situation (described below) involves a possible joint venture between a MNC and a HG.

In addition to the above references, our work has been influenced by Tsaklanganos (1971) and Tsaklanganos and Rao (1976), who develop a mathematical model for considering the effect of incentives on foreign private investment and HG economic goals but do not consider negotiations as such; Shakun (1970), who explores a game theory approach to international joint venture negotiations under incomplete information that provides game theory orientation to the present work; Rao and Shakun (1972), whose work provides behavior concepts orientation; Harsanyi (1961; 1966) who develops Zeuthen's (1930) principle of concession making involving the concept of risk limit, the maximum probability of a break off in negotiations that a player would be willing to face rather than to make a concession; Kapoor (1970), who qualitatively studies the joint venture negotiation between an American consortium and the government of India that provides the basic negotiation situation for our mathematical model.

Briefly, the negotiation situation involves fertilizer production in India. A group of large, private oil, chemical, and engineering companies based in the United States form a consortium (to be referred to here as the MNC), that presents to the government of India (HG) a general proposal for collaboration on a massive fertilizer program. The project involves the establishment of five fertilizer factories in India. Negotiations involve such issues as percent equity of the two parties, percent of crude oil supply right (i.e., the right to supply crude oil, from which naphtha, used in fertilizer manufacture, is obtained), percent return on the investment of the consortium, management control, and the like. (For a more detailed description see Kapoor 1970.)

We consider initially a single negotiation variable, such as the percent of crude oil supply right going to the MNC. The latter would ideally like this to be 100 percent, i.e., to have the right to import crude oil sufficient to produce all the naphtha needed in the fertilizer production. (Naphtha is obtained from the refining of crude oil as are, of course, other petroleum products. The right of the consortium to import this crude oil is also viewed as a lever to open the way for further participation by consortium members in oil refining in India.) Since surplus naphtha is already produced in India, the HG would ideally like the MNC's percent of crude oil supply right to be zero.

Our model assumes that a total of $2n$ negotiation sessions are available to the two players ($i = 1$ will represent the MNC and $i = 2$, the HG). It is assumed that the players have made initial proposals as to the percentage of crude oil supply right going to the MNC, P_1^{2n+1} and P_2^{2n+1} say, $P_1^{2n+1} > P_2^{2n+1}$. The superscript indicates that these proposals were obtained in a preliminary negotiation session where there were $2n + 1$ sessions available (including the preliminary one). In other words, with $2n$ sessions remaining the proposals of the MNC and the HG are P_1^{2n+1} and P_2^{2n+1} respectively. Now the MNC must make a new offer conceding some amount $C_1 \geq 0$. After the HG receives the offer, it must make a counteroffer conceding some amount $C_2 \geq 0$. Then the MNC must make a new offer, etc. Thus concessions are made sequentially. We let $t = 2n, 2n - 1, \ldots, 1$ representing negotiation sessions remaining with the understanding that player 1 (MNC) makes concessions with an even number of time periods remaining and player 2 (HG), with an odd number of periods remaining. Therefore, considering points in time with $t + 1, t, t - 1, t - 2$ negotiation sessions remaining where t is even, player 2 must decide on the magnitude of his concession in periods $t + 1$ and $t - 1$ and player 1, in t and $t - 2$. The concession made by player i when he must make a new offer and t negotiating sessions remain will be denoted by C_i^t so that a typical sequence of concessions will be represented by $C_2^{t+1}, C_1^t, C_2^{t-1}, c_1^{t-2}$.*

Now consider that t sessions remain, and player i must choose C_i^t. Suppose the choice of C_i^t is based on the last concession made by player j ($j = 1, 2$; $j \neq i$), namely C_j^{t+1}, and the current proposals by the two parties, P_1^{t+1} and P_2^{t+1}. (Note: $C_i^t = P_i^{t+1} - P_i^t$.) Thus C_j^{t+1}, P_1^{t+1} and P_2^{t+1} are the state variables based upon which C_i^t must be chosen. Let ϕ_{it} $(C_j^{t+1}, P_1^{t+1}, P_2^{t+1}, C_i^t, C_i^{t-2} \ldots, C_j^{t-1}, C_j^{t-3} \ldots)$ be the utility to player i for the last t periods (stages) given that the negotiations are in state $C_j^{t+1}, P_1^{t+1}, P_2^{t+1}$ and player i selects $C_i^t, C_i^{t-2} \ldots$ etc. Player j selects C_j^{t-1}, C_j^{t-3} etc. Thus the utility that player i gets depends not only on his own concessions C_i^t, C_i^{t-2} etc., but also on player j's concessions C_j^{t-1}, C_j^{t-3} etc., which in turn can be influenced by player i's concessions (as discussed in detail below). Let $F_{it}(C_j^{t+1}, P_1^{t+1}, P_2^{t+1}, C_i^t)$ and $G_{it}(C_j^{t+1}, P_1^{t+1}, P_2^{t+1}, C_i^t)$ be respectively the expected utility and the minimum utility of the best overall policy for player i for the last t periods given that the negotiations are in state $C_j^{t+1}, P_1^{t+1}, P_2^{t+1}$ and player i selects C_i^t (the assumption being that player i's future concessions C_i^{t-2}, C_i^{t-4}, etc. will be chosen optimally). Thus if E stands for expected value, we may write $F_{it} = \max_{C_i^{t-2}, C_i^{t-4}, \ldots} E(\phi_{it})$, the expectation being taken with respect to player j's concessions C_j^{t-1}, C_j^{t-3}, etc., using player i's subjective probabilities of player j's concessions. Let $\underline{C}_i^{t-2}, \underline{C}_i^{t-4}, \ldots$ represent the optimum values of player i's concessions; then $G_{it} = \min_{C_j^{t-1}, C_j^{t-3}, \ldots} \phi_{it}(C_j^{t+1}, P_1^{t+1}, P_2^{t+1}, C_i^t, \underline{C}_i^{t-2}, \underline{C}_i^{t-4}, \ldots C_j^{t-1}, C_j^{t-3}, \ldots)$.

We note that the only way negotiations can break off and players receive the conflict payoffs is through successive nonconcessions by each of the players. Therefore player j can give player i the conflict payoff only with the latter's concurrence, i.e., with player i also not conceding once player j has not conceded. Thus, in general, $G_{it} \neq S_i$.

*We require that $2(P_1^{2n+1} - P_2^{2n+1})/\delta \leq 2n$ where δ is the smallest unit of concession. This is needed to avoid the possibility of the player making the next-to-last concession decision giving an ultimatum to the other, that is, forcing the other in the last session to concede the remaining amount separating the two players or be faced with "irrationally" taking a smaller conflict pay-off.

***Table 6.1.* Negotiation behavior concepts for player i in time period t.**

1. $\text{Max}_{C_i^t} F_{it}(C_j^{t+1}, P_1^{t+1}, P_2^{t+1}, C_i^t)$
2. $\text{Max}_{C_i^t} G_{it}(C_j^{t+1}, P_1^{t+1}, P_2^{t+1}, C_i^t)$
3. If $r_{it} \leq r_{jt}$ choose $C_i^t = \delta$ the smallest unit of negotiation; otherwise choose $C_i^t = 0$.

6.2 Negotiation Behavior Concepts

Before developing the negotiation model further, some discussion of the types of negotiation behavior to be employed is in order. The approach to negotiations taken here visualizes conceptually the bargaining game in extensive form but in effect deals with only part of the game tree. The approach uses negotiation behavior concepts to guide concession making and considers plausible dynamics of adjustment to a solution. Table 6.1 lists the behavior concepts to be considered here (others are, of course, possible).

Under behavior concept 1, player i is choosing C_i^t so as to try to maximize his expected utility F. The maximized value of F will be denoted by f, that is, $\max_{C_i^t} F_{it}(C_j^{t+1}, P_1^{t+1}, P_2^{t+}, C_i^t) \equiv f_{it}(C_j^{t+1}, P_1^{t+1}, P_2^{t+1})$.

Under behavior concept 2, player i is choosing C_i^t to maximize his minimum utility—a conservative type of behavior. We define $\max_{C_i^t} G_{it} \equiv h_{it}$.

Behavior concept 3 is an extension of Harsanyi's (1961, 1966) concept of risk limit. Under behavior concept 3, player i concedes with t sessions remaining if r_{it}, his minimax risk (minimum risk limit)—the minimum value of the maximum probability of player j breaking off negotiations that player i would be willing to face rather than make a concession—is less than or equal to his estimate of player j's minimax risk, $\hat{r}_{jt}$. Player i computes his minimax risk as follows: Letting δ = the smallest unit of negotiation, we set the minimum utility under concession, $C_i^t = \delta$, namely, $G_{it}(C_j^{t+1}, P_1^{t+1}, P_2^{t+1}, \delta)$, equal to the expected utility under no concession, $C_i^t = 0$, namely, $F_{it}(C_j^{t+1}, P_1^{t+1}, P_2^{t+1}, 0)$. That is, we set

$$F_{it}\left(C_j^{t+1}, P_1^{t+1}, P_2^{t+1}, 0\right) = G_{it}\left(C_j^{t+1}, P_1^{t+1}, P_2^{t+1}, \delta\right). \quad (1)$$

We use the following notation:

p_{it} = player i's subjective probability, with t sessions remaining, of player j breaking off negotiations, with $t-1$ sessions remaining, if player i does not now concede = $\text{Prob}(C_j^{t-1} = 0 | C_j^{t+1} > 0, C_i^t = 0)$. (Note two successive nonconcessions are considered to mean a break off in negotiations.)

S_i = player i's conflict payoff if negotiations are broken off.

$f_{i,t-2}(\delta, P_i^{t+1}, P_j^{t+1} \pm \delta)$ = expected utility to player i if he does not concede and player j concedes $C_j^{t-1} = \delta$ with $t-1$ sessions remaining ($\pm$ means + is used if $j = 2$ and − if $j = 1$). Using this additional notation we may write (1) as

$$F_{it}\left(C_j^{t+1}, P_1^{t+1}, P_2^{t+1}, 0\right) = (1 - p_{it}) f_{i,t-2}\left(\delta, P_i^{t+1}, P_j^{t+1} \pm \delta\right) + p_{it} S_i$$

$$= G_{it}\left(C_j^{t+1}, P_1^{t+1}, P_2^{t+1}, \delta\right). \tag{2}$$

The value of p_{it} that satisfies (2) is player i's minimax risk r_{it}, i.e.,

$$r_{it} = \frac{f_{i,t-2}\left(\delta, P_i^{t+1}, P_j^{t+1} \pm \delta\right) - G_{it}\left(C_j^{t+1}, P_1^{t+1}, P_2^{t+1}, \delta\right)}{f_{i,t-2}\left(\delta, P_i^{t+1}, P_j^{t+1} \pm \delta\right) - S_i} \tag{3}$$

Player i's estimate of player j's minimax risk, $\hat{r}_{ji}$ is based on player i, reasoning as follows: If in fact player j had to make the concession decision with t stages remaining rather than me (player i), then I estimate that player j's minimax risk would be

$$\hat{r}_{jt} = \frac{\hat{f}_{j,t-2}\left(\delta, P_j^{t+1}, P_i^{t+1} \pm \delta\right) - G_{jt}\left(C_i^{t+1}, P_1^{t+1}, P_2^{t+1}, \delta\right)}{\hat{f}_{j,t-2}\left(\delta, P_j^{t+1}, P_i^{t+1} \pm \delta\right)} \tag{4}$$

and he would concede if his $r_{jt} \leq \hat{r}_{it}$, the estimate he would have to make of my r_{it}. (In (4), $\pm$ means $+$ is used if $i = 2$ and $-$ if $i = 1$.)

The symbol $\hat{}$ means that quantities for player j are being estimated by player i. In writing (4) we have assumed that player i knows player j's conflict payoff S_j, his utility function ϕ_{jt} and hence G_{jt}, but not $f_{j,t-2}$. This is because $f_{j,t-2}$ is an expected value requiring player j's subjective probabilities of player i's concessions C_i^{t-3}, C_i^{t-5}, etc., which subjective probabilities player i does not know and will have to estimate. (If player i does not know S_j, ϕ_{jt} and hence G_{jt}, he will, of course, have to estimate these also.)

Behavior concept 3 can be viewed as a kind of platinum rule of behavior —"Do unto others as you believe they would do unto you were they in your place" observed by Emshoff (1971)—with regard to the concession-making procedure. Thus player i may reason that if player j were in his place, player j would concede if his $r_{jt} \leq \hat{r}_{it}$, the estimate player j would make of player i's r_{it}. Consequently, following the platinum rule, player i concedes if his $r_{it} \leq \hat{r}_{jt}$.

A behavior concept combination is a $2n$-tuple $(b_{2n}, b_{2n-1}, \ldots, b_t, \ldots b_1)$ of behavior concepts subscripted on the number of sessions remaining. The symbol b_t (where $b_t = 1, 2, 3$) represents the type of behavior followed by player i when t negotiation sessions remain and he must make a new offer and is one of the three behavior concepts in Table 6.1. (Remember that player 1 makes concessions with an even number of periods remaining and player 2 with an odd number remaining. Also, the applicability of behavior concept b_t is conditional on negotiations continuing, i.e., that agreement has not been reached nor negotiations broken off.) For example, combination $(1, 1, \ldots, 1)$ represents expected utility maximizing type of behavior 1 by both players throughout the negotiations. Combination $(2, 2, \ldots, 2)$ represents maximin behavior by both players throughout. Combination $(2, 1, 2, 1, \ldots, 2, 1)$ represents maximin behavior by player 1 and expected utility maximizing behavior by player 2. Of course, combinations like $(1, 2, 1, 2, 2, 2, \ldots, 2, 2)$ are possible where player 1 switches his behavior concept from 1 to 2 during the course of negotiations.

In the analysis that follows, the effects of various behavior concept combinations are explored by using them in a dynamic programing model of negotiations, which we now develop in further detail.

6.3 Mathematical Model

In the previous sections the problem of concession making has been formulated as a sequential decision process. Let us consider this process from the viewpoint of one of the players, say the MNC. (A similar analysis can be made from the viewpoint of the HG.) Say there are t negotiation sessions remaining, and the MNC must decide on the magnitude of its concession. The state variables that the MNC will face when it next must make a decision (when only $t-2$ sessions will remain) will depend on the current state, the magnitude of the concession made, and the response of the HG. Because this response is not completely predictable, the state variables that the MNC will face have random components.

In general the response by player i in a given session may depend on the sequence of moves by both players up to that time. We shall assume, however, that the magnitude of player i's concession depends on the last concession made by player j and the current proposals by the two parties. Suppose, for the sake of definiteness, that there are $K+1$ possible amounts of concession at any negotiation session of magnitude $0, 1, 2, \ldots k, \ldots K$ units respectively. Let p^t_{mkl} define the conditional probability that the HG will next concede l units given that its previous concession was m units and the NMC concedes k units, $0 \leq m \leq K$, $0 \leq k$, $0 \leq l \leq K$.

Clearly if the NMC follows behavior concept 1, the maximization of expected value, then it must have subjective or objective estimates of the p^t_{mkl}, for each t and for all m, k, l combinations. Ideally, one would like to develop some normative rule or psychological mechanism generating a plausible probability distribution for each player over the other player's possible concession making. These probabilities are likely to be influenced by the behavior concept employed by each player. An attempt to specify a mechanism at the present state of understanding of negotiation behavior is difficult.* For example,

*One possibility is for a player to specify his subjective probability distribution over the possible behavior concepts (see Table 6.1) employed by the other. In principle this would allow the computation of subjective probabilities of various concessions by the other player. However it is not clear whether this approach is preferable to a player specifying directly his subjective probabilities of concession making by the other. In many cases a negotiator may have a clearer idea of the subjective probabilities of concession making than of the subjective probabilities of the behavior concepts employed by the other players.

realism would require that the probability estimates change with t as new information, not explicitly introduced in the model, becomes available to the players. One way of obtaining estimates of the p_{mkl}^t is through a Bayesian scheme; unfortunately, such a scheme cannot comprehend the many informal bits of information that are exchanged in negotiations or the "feeling" that the players have about the discussions as they proceed.

At the present time the best way of using such "feelings" and informal information is to incorporate them via *subjective* probability estimates supplied by a player. These may be revised at each stage as necessary. Such an approach is commonly used when the underlying mechanism is not clear. A good example is afforded by the use of the Delphi technique for estimating subjective probabilities of future events. Such a technique could be used by the members of each negotiating team (in this chapter we use the term "player" to represent what is in reality a negotiating team). In the light of the above, we must limit ourselves to examination of the consequences of the use of a particular behavior concept rather than attempt to find an "optimal" behavior concept. We therefore develop the model on the basis that subjective estimates of the p_{mkl}^t will be supplied at each stage and that at each stage the decision will be evaluated under the assumption that these estimates will hold until the end of the negotiations, that is, $p_{mkl}^t = p_{mkl}^{t-2} = \cdots = p_{mkl}^2$ for each mkl combination. As such we shall drop the superscript where possible without ambiguity. Of course, the MNC may use a Bayesian scheme to give it plausible estimates of these probabilities at each stage and then modify these estimates on the basis of subjective considerations. The estimates will be revised if necessary from session to session, but the new estimates will be assumed to hold constant to the end of the negotiations in each case.

We can characterize the various states in which the MNC may find itself as either "terminal states," in which a payoff is obtained, or nonterminal states, where negotiations continue. A terminal state can be reached if (1) a 0 concession by the HG is followed by a 0 concession by the MNC, or vice versa, leading to a conflict payoff; (2) the two players do not reach agreement in $2n$ negotiation sessions, again resulting in the conflict payoff being received; and (3) agreement is reached, that is, $P_1 = P_2$ at some negotiation session, in which case the payoff depends on P_1. In addition we assume that no state where $P_1 < P_2$ can be reached. These characteristics supply the boundary conditions of the problem.

The above considerations lead naturally to the modeling of the concession-making decisions as one of stochastic terminal control, which can be solved using dynamic programming (See Bellman and Dreyfuss, 1962).

Let $f_t(m, P_1, P_2)$ be the maximum expected return to the MNC, given that there are t negotiating sessions remaining, the HG has just conceded an amount m, the MNC's previous demand is for P_1 percent of the crude oil supply right, the HG's current offer is P_2 percent, and an optimal policy is followed from here on. With these definitions we no longer need superscripts for the arguments in $f_t(,,)$. Then

$$f_t(m, P_1, P_2)$$

$$= \begin{cases} \text{Max}\left[\text{Max}_{0<k\leq K} \sum_{l=0}^{K} p_{mkl} f_{t-2}(l, P_1 - k, P_2 + 1), \right. & m > 0 \\ \left. \qquad \sum_{l=1}^{K} p_{m0l} f_{t-2}(l, P_1, P_2 + l) + p_{m00} S_1 \right] & \\ \text{Max}\left[S_1, \text{Max}_{0<k\leq K} \sum_{l=0}^{K} p_{0kl} f_{t-2}(l, P_1 - k, P_2 + l)\right] & m = 0 \end{cases} \tag{5}$$

with the boundary conditions:

$$f_t(m, P_1, P_1) = \phi_1(P_1)$$

$$f_2(m, P_1, P_2) = \text{Max}_k \left\{ S_1 \sum_{k+l<P_1-P_2} p_{mkl} + \sum_{k+1\geq P_1-P_2} \phi_1(P_1 - K) p_{mkl} \right\} \qquad m > 0$$

$$\text{and Max}\left[S_1, \text{Max}_{k>0}\left\{ S_1 \sum_{k+l<P_1-P_2} p_{0kl} \right.\right.$$

$$\left.\left. + \Sigma_{k+l\geq P_1-P_2} \phi_1(P_1 - k) p_{0kl} \right\}\right] \qquad m = 0$$

If the MNC employs behavior concept 2, that of maximizing the minimum gain, we can again use the dynamic programing approach, with similar boundary conditions but a somewhat different recursive relationship defined as follows.

Let $h_t(m, P_1, P_2)$ be the maximin return to the MNC, where m, P_1 and P_2 have their previously defined meanings and an optimal policy is followed for the remaining sessions. Then

$$h_t(m, P_1, P_2)$$

$$= \begin{cases} \text{Max}_k \text{Min}_{0\leq l\leq K} h_{t-2}(l, P_1 - k, P_2 + l) & m > 0 \\ \qquad \text{where } h_{t-2}(0, P_1, P_2) = S_1 & \\ \text{Max}[S_1, \text{Max}_{0<k\leq K} \text{Min}_{0\leq l\leq K} h_{t-2}(l, P_1 - k, P_2 + l)] & m = 0 \end{cases} \tag{6}$$

with the boundary conditions: $h_t(m, P_1, P_1) = \phi_1(P_1)$ and
$h_2(m, P_1, P_2)$

$$= \begin{cases} \mathrm{Max}_k \left[S_{1k=0}, \min_{0<k<P_1-P_2} \{ \phi_1(P_1 - k), S_1 \}, \phi_1(P_2)_{k=P_1-P_2} \right], & P_1 - P_2 \leq K \\ \mathrm{Max}_k \left[S_{1k=0,1,\ldots,P_1-P_2-K-1}, \min_{P_1-P_2-K \leq k < K} \{ \phi_1(P_1 - k), S_1 \} \right], & K < P_1 - P_2 \leq 2K \\ S_1 & P_1 - P_2 > 2K \end{cases}$$

6.4 Numerical Example

In order to illustrate the models developed we present several numerical examples. Suppose in the preliminary negotiations the MNC has demanded a 75 percent share of the crude oil supply rights and the HG has offered a 50 percent share. The models are employed from this point on. Assume for simplicity that at each session the decision maker (MNC or HG) can either make no concessions or concede 5 percent and that a maximum of ten negotiation sessions are available ($2n = 10$), with the MNC having to make the first decision. Let the conflict payoffs be $\$4.4 \times 10^6$ for the MNC and $\$1.2 \times 10^6$ for the HG, implying that the MNC has a more profitable alternative investment than the HG in case negotiations break off. Let the utility be linear in share for both players, the value of 100 percent supply rights being $\$8.0 \times 10^6$ to the MNC and $\$4.0 \times 10^6$ for the HG. Suppose the MNC has made the following initial subjective estimates of the HG's response to MNC concessions:

$$p_{555} = 0.6 \qquad p_{550} = 0.4$$
$$p_{055} = 0.3 \qquad p_{050} = 0.7$$
$$p_{505} = 0.7 \qquad p_{500} = 0.3.$$

Note that two successive "no concessions" lead to negotiations breaking off. Using these specifications, we can rewrite equations (5) and (6) as follows, from the viewpoint of the MNC.

$$f_t(0, P_1, P_2) = \mathrm{Max}\,[\underbrace{4.4}_{\text{(MNC does not concede)}}, \underbrace{0.3 f_{t-2}(5, P_1 - 5, P_2 + 5) + 0.7 f_{t-2}(0, P_1 - 5, P_2)}_{\text{(MNC concedes)}}] \quad (7.1)$$

$$f_t(5, P_1, P_2) = \mathrm{Max}\,[\underbrace{0.7 f_{t-2}(5, P_1, P_2 + 5) + 0.3 \times 4.4}_{\text{(MNC does not concede)}}, \underbrace{0.6 f_{t-2}(5, P_1 - 5, P_2 + 5) + 0.4 f_{t-2}(0, P_1 - 5, P_2)}_{\text{(MNC concedes)}}] \quad (7.2)$$

with the boundary conditions:

$$f_t(0, P_1, P_1) = f_t(5, P_1, P_1) = 0.08P_1$$

$$f_2(5, P_1, P_2) = f_2(0, P_1, P_2) = S_1 = 4.4 \text{ if } P_1 - P_2 > 10$$

$$f_2(5, P_1, P_1 - 10) = \text{Max}\,[\underset{\text{(MNC concedes)}}{0.4S_1 + 0.6 \times 0.08(P_1 - 5)}, \quad \underset{\text{(MNC does not concede)}}{S_1}]$$

$$f_2(0, P_1, P_1 - 10) = \text{Max}\,[\overset{\text{(MNC concedes)}}{0.7S_1 + 0.3 \times 0.08(P_1 - 5)}, \quad \overset{\text{(MNC does not concede)}}{S_1}]$$

$$f_2(5, P_1, P_1 - 5) = \text{Max}[\overset{\text{(MNC concedes)}}{0.08(P_1 - 5)}, \quad \overset{\text{(MNC does not concede)}}{0.3S_1 + 0.7 \times 0.08(P_1)}]$$

$$f_2(0, P_1, P_1 - 5) = \text{Max}[\overset{\text{(MNC concedes)}}{0.08(P_1 - 5)}, \quad \overset{\text{(MNC does not concede)}}{S_1}]$$

$$h_t(0, P_1, P_2) = \text{Max}[\underset{\text{(MNC does not concede)}}{4.4}, \quad \underset{\text{(MNC concedes)}}{\text{Min}\{h_{t-2}(0, P_1 - 5, P_2), h_{t-2}(5, P_1 - 5, P_2 + 5)\}}] \qquad (8.1)$$

$$h_t(5, P_1, P_2) = \text{Max}[\underset{\text{(MNC does not concede)}}{\text{Min}\{h_{t-2}(5, P_1, P_2 + 5), 4.4\}}, \quad \underset{\text{(MNC concedes)}}{\text{Min}\{h_{t-2}(5, P_1 - 5, P_2 + 5), h_{t-2}(0, P_1 - 5, P_2)\}}] \qquad (8.2)$$

with the boundary conditions:

$$h_t(0, P_1, P_1) = h_t(5, P_1, P_1) = 0.08P_1$$

$$h_2(5, P_1, P_2) = h_2(0, P_1, P_2) = 4.4 \quad \text{if} \quad P_1 - P_2 > 10$$

$$h_2(5, P_1, P_1 - 10) = \text{Max}[\text{Min}\{\underset{\text{(MNC concedes)}}{S_1, 0.08(P_1 - 5)}, \quad \underset{\text{(MNC does not concede)}}{S_1}] = h_2(0, P_1, P_1 - 10)\}$$

$$h_2(5, P_1, P_1 - 5) = \text{Max}[\underset{\text{(MNC concedes)}}{0.08(P_1 - 5)}, \quad \underset{\text{(MNC does not concede)}}{\text{Min}\{0.08P_1, S_1\}}]$$

$$h_2(0, P_1, P_1 - 5) = \text{Max}[\underset{\text{(MNC concedes)}}{0.08(P_1 - 5)}, \quad \underset{\text{(MNC does not concede)}}{S_1}].$$

Note that in equations (7.1), (7.2) and (8.1), (8.2) the only admissible state spaces have $P_1 \geq P_2$.

Using these equations, we can compute the optimal strategies for the MNC when it employs behavior concepts 1 and 2. In order to compute the

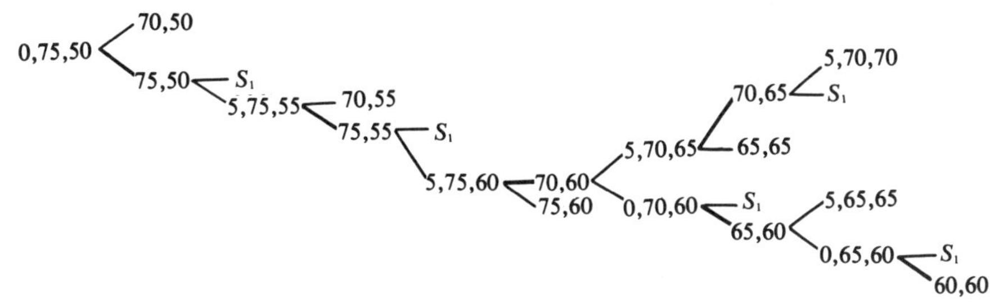

Figure 6.1. **Optimal concession (heavy lines) by MNC for any response by HG (expected value maximization).**

strategy under behavior concept 3, first note that (8.1) and (8.2) can be rewritten in terms of notations introduced previously as

$$h_t(0, P_1, P_2) = \mathrm{Max}[4.4G_{it}(0, P_1, P_2, 5)] \tag{8.3}$$

$$h_t(5, P_1, P_2) = \mathrm{Max}[\mathrm{Min}\{h_{t-2}(5, P_1, P_2 + 5), 4.4\}, G_{it}(5, P_1, P_2, 5)]. \tag{8.4}$$

Recursive relationships similar to (7.1)–(8.4) can be written from the viewpoint of the HG. Using all these results, we can compute the optimal concessions for the MNC when it employs behavior concept 3.

Figure 6.1 displays the optimal policies for the MNC under behavior concept 1 in the form of a tree diagram.

It is of interest to compute the actual course of negotiations when each player employs a particular behavior concept, using the numerical values above. We have computed the policies for the HG assuming that its estimates of the conditional probabilities of the MNC conceding or not conceding given the two previous concessions are the same as those of the MNC. Table 6.2 illustrates behavior concept combination $(1, 1, 1, 1, \ldots)$ with both players employing expected utility-maximizing behavior. Table 6.3 illustrates combination $(2, 1, 2, 1, \ldots)$ with the MNC using maximin behavior and the HG maximizing expected utility. As may be expected, the MNC does better in the first case (unless because of indifference the HG breaks off negotiations at the end). In the second case we have illustrated two extreme policies for the MNC

Table 6.2. **Concession-making sequence with both sides maximizing expected value. Bargaining terminates with MNC receiving 70 percent or with conflict payoff.**

Sessions remaining	10	9	8	7	6	5	4	3	2	1
Decision maker	MNC	HG	MNC	HG	MNC	HG	MNC	HG	MNC	HG
Concession amount	0	5	0	5	5	5	0	5 or 0	(negotiations breakoff) or	
State at which decision is made	0, 75, 50	0, 75, 50	5, 75, 55	0, 75, 55	5, 75, 60	5, 70, 60	5, 70, 65	0, 70, 65	5, 70, 70	

Table 6.3. **Concession making when the MNC uses maximin and HG maximizes expected value. The first row of concessions reflects the situation when if indifferent between conceding and not conceding, MNC never concedes. The second row shows the results when the opposite is true. In the first case the final agreement is at 65 percent and in the second, at 60 percent.**

Sessions remaining	10	9	8	7	6	5	4	3	2	
Decision maker	MNC	HG	MNC	HG	MNC	HG	MNC	HG	MNC	H
Concession	0	5	0	5	5	5	5			
amount	5		5		5					
State at which	0, 75, 50	0, 75, 50	5, 75, 55	0, 75, 55	5, 75, 60	5, 70, 60	5, 70, 65	5, 65, 65		
decision is made	0, 75, 50	5, 70, 50	5, 70, 55	5, 65, 55	5, 65, 60	5, 60, 60				

Table 6.4. **Concession making when both MNC and HG use maximin behavior. Note the numerous possibilities for negotiations to break off. If negotiations do not break off, agreement is at 65 percent.**

Sessions remaining	10	9	8	7	6	5	4	3	2	1
Decision maker	MNC	HG	MNC	HG	MNC	HG	MNC	HG	MNC	HG
Concession amount	0	5 or 0 (breakoff)	0	5 or 0 (breakoff)	5	5 or 0 (breakoff)	5			
State at which decision is made	0, 75, 50	0, 75, 50	5, 75, 55	0, 75, 55	5, 75, 60	5, 70, 60	5, 70, 65	5, 65, 65		

—the first being one where, when it is indifferent between conceding and not conceding, it never concedes, and the second being the opposite. Again, as one expects, the second variation is less successful than the first. Table 6.4 illustrates the play when both parties employ maximin behavior, that is, the behavior concept combination (2, 2, 2, 2, . . .). Note that because of the conservative outlook of both parties, there are numerous opportunities for negotiations to break off. In fact the probability that the negotiations end successfully is rather small in this case. Finally, Table 6.5 illustrates behavior concept combination (3, 1, 3, 1, . . .) with the MNC using the minimax risk limit and the HG maximizing expected value.

Table 6.5. **Concession making when MNC uses minimax risk comparison behavior and HG maximizes expected value. Agreement is reached at 60 percent.**

Session remaining	10	9	8	7	6	5	4	3	2	
Decision maker	MNC	HG	MNC	HG	MNC	HG	MNC	HG	MNC	H
Concession amount	5	5	5	5	5					
State at which decision is made	0, 75, 50	5, 70, 50	5, 70, 55	5, 65, 55	5, 65, 60	5, 60, 60				

Most of the literature referred to previously considers the concession-making problem in the framework of game theory, with players making concessions simultaneously rather than sequentially. The Nash equilibrium solution is considered to be a "fair" solution in such cases. In our case the Nash solution is for the MNC to receive 55 percent of the crude oil supply rights, equal in utility to its conflict payoff. This is considerably less than what the MNC obtains in the computations shown above. On the basis of this limited numerical work, however, it is not possible to make any general comparisons among the various behavior concept combinations and the Nash solution.

6.5 Extension to Multivariate Case

In many real conflict situations more than one variable is subject to negotiation and concession making. If the trade-offs between the variables are known explicitly, then the formulation presented in this chapter can be generalized to this multivariate situation. The state variables m, P_1, and P_2 would now be vectors each with R elements where R is the number of variables subject to negotiation. For instance, if the MNC and HG were negotiating both the percent crude oil supply right (variable 1) and the percent equity (variable 2), then the state at any negotiation facing the MNC could be represented by (m_1, m_2), (P_{11}, P_{12}), and (P_{21}, P_{22}) where m_1 and m_2 are the latest concessions by the HG on variables 1 and 2 respectively, P_{11} and P_{12} are the previous demands by the MNC for the supply rights and equity respectively, and P_{21} and P_{22} are the latest offers by the HG. Similarly, the utility function ϕ for each player will be dependent on both variables.

Although there is no conceptual problem in such a formulation, two practical difficulties arise. First, the trade-offs between the various variables subject to negotiations may not be available. Second, the computational burden will become prohibitive as the number of variables increases. Section 6.6 offers some suggestions for further work to solve these difficulties.

6.6 Concluding Remarks

The present work may be viewed as a special case of a general policy-making system of relationships such as discussed by Lewin and Shakun (1971) and Shakun (1972). For example, if $x_1, x_2, x_3, \ldots, x_N$, are decision variables, the MNC-HG system of relationships might be represented by

$$g_s(x_1, x_2, \ldots, x_N) \geq B_s, \qquad s = 1, \ldots, M \tag{9}$$

where some of the constraints are aspiration level constraints (i.e., the function $g_s[x_1, x_2, \ldots, x_N]$ must be greater than or equal to aspiration level B_s), and the other constraints involve technological, resource, etc. limitations. We consider the MNC-HG system goals (including the goals of both the MNC and the HG) to be contained in the whole set of constraints. (9) to be satisfied.

The left-hand side of (9) corresponds to the means available to achieve the right-hand side, the ends. Thus goals are viewed in terms of the whole set of means-ends relationships that solutions must satisfy, i.e., in terms of the whole set of constraints. In general the MNC-HG system of relationships (9) as seen by the MNC may be different from that seen by the HG. The MNC model (9) may be incompatible with the HG model (9) in the sense that no solution $(x_1, x_2, \ldots, x_N)$ may satisfy both. However we do not view the MNC and HG models as fixed. The solution (negotiation and organization design) process involves a search for change in the constraints (goals, means-ends) represented by the MNC and HG models (9) as well as a search for solutions satisfying the changed constraints. In other words we are searching for solutions satisfying a changing joint solution space (which initially may be empty, i.e., when MNC and HG models (9) are incompatible). Thus we have a problem in negotiations and organization design in which the structures (9) as seen by the MNC and HC change.

Using appropriate behavioral concepts and mathematical structures, research can develop the search process involving negotiation and organization design by which the MNC and HG can arrive at solutions (reach agreements) in (9).

This chapter deals with a special case of the above general MNC-HG system of relationships. Here we have considered the constraints (aspiration levels, resource limitations, etc.) to be fixed over the course of negotiations and that these define a nonempty joint solution space. The behavior concepts guide the search by the MNC and HG for a solution—an agreed-upon point—in this joint solution space. As presently formulated, if there are several negotiation variables, the behavior concepts are applied to multivariate utility functions to guide concession making. Future research should explore formulations in which behavior concepts are applied to one dimension at a time where it becomes difficult to determine multivariate utilities. Future research should also develop behavior concepts that guide changes in constraints (e.g., in cases in which the joint solution space is empty) as well as guide the search for solutions satisfying the constraints.

REFERENCES

Bellman, R., and Dreyfus, S. 1962. *Applied Dynamic Programming*. Princeton University Press, Princeton, NJ.

Coddington, A. 1968. *Theories of the Bargaining Process*. Aldine Publishing Company, Chicago.

Cross, J. G. 1969. *The Economics of Bargaining*. Basic Books, New York.

Emshoff, J. R. 1971. *Analysis of Behavioral Systems*. MacMillian, New York.

Harsanyi, J. C. 1961. "On the Rationality Postulates Underlying the Theory of Cooperative Games." *Journal of Conflict Resolution*, 5, No. 2.

______. 1966. "A General Theory of Rational Behavior in Game Situations." *Econometrica*, 34, No. 3, July.

______, and Selten, R. 1972. "A Generalized Nash Solution for Two Person Bargaining Games with Incomplete Information." *Management Science*, 18, No. 5, January, Part II.

Kapoor, A. 1970. *International Business Negotiations, A Study in India*. New York University Press, New York.

______, and Grub, P. D. (eds.) 1972. *The Multinational Enterprise in Transition*. Darwin Press, Princeton, NJ.

Lewin, A. Y., and Shakun, M. F. 1971. "Situational Normativism: A Descriptive-Normative Approach to Decision Making and Policy Sciences." Paper presented at the Cost Effectiveness Conference of the International Federation of Operational Research Societies, Washington, DC, April 12–15.

______. 1976. *Policy Sciences: Methodologies and Cases*. Pergamon Press, New York.

Patchen, M. 1970. "Models of Cooperation and Conflict: A Critical Review." *Journal of Conflict Resolution*, 14, No. 3, September.

Rao, A. G., and Shakun, M. F. 1972. "A Quasi-game Theory Approach to Pricing." *Management Science*, 18, No. 5, January, Part II.

Saraydar, E. 1971. "A Certainty Equivalent Model of Bargaining." *Journal of Conflict Resolution*, 15, No. 3, September.

Shakun, M. F. 1970. "International Joint Ventures: A Decision Analysis–Game Theory Approach." In Lawrence, J. R. (ed.) *Proceedings of the Fifth International Conference of Operational Research*. Tavistock Publications, London.

______. 1972. "Management Science and Management: Implementing Management Science Via Situational Normativism." *Management Science*, 18, No. 8, April.

Tsaklanganos, A. A. 1971. "National Strategies Concerning Foreign Private Investment: An OR Approach." Doctoral dissertation, Graduate School of Business Administration, New York University.

______, and Rao, A. G. 1976. "National Policies Towards Foreign Private Investment." In Lewin, A. Y., and Shakun, M. F. (1976) *Policy Sciences: Methodologies and Cases*. Pergamon Press, New York.

Zeuthen, F. 1930. *Problems of Monopoly and Economic Warfare*. G. Routledge and Sons, London.

CHAPTER 7

Bivariate Negotiations as a Problem of Stochastic Terminal Control*

Francoise Fogelman-Soulie[†], Bertrand R. Munier[††], and Melvin F. Shakun[†††]

7.1 Introduction

Rao and Shakun (1974) developed a mathematical model for negotiations between two players involving concession making on a single negotiation variable. (Each player has a univariate utility function defined on this variable.) The model treats concession making as a problem of stochastic terminal control (Bellman and Dreyfus 1962). Rao and Shakun note that, if there are several negotiation variables and if multivariate utility functions for the players can be determined, e.g., perhaps using multiattribute utility theory (MAUT) (Kenney and Raiffa 1976), then a straightforward generalization of the model follows. In this paper we assume that multivariate utility functions are difficult to determine, but at any negotiation stage local preference information can be obtained interactively so that a concession decision can be made. We do not specify a particular interactive multicriteria method. There are several possibilities, e.g., see Jacquet-Lagreze and Siskos 1982; Kempf, Duckstein, and Casti 1979; Roy 1977; Starr and Zeleny 1977; Zeleny 1982, which include some comparisons, as well as a suggested direct visual assessment that uses the gestalt capabilities of players in the case of bivariate probability distributions (see Section 7.2).

*This chapter first appeared in *Management Science*, Vol. 29, No. 7, July 1983.

[†]CREA, Ecole Polytechnic, France

[††]University of Aix-Marseille

[†††]New York University

The approach to negotiations taken here (stochastic terminal control) is different from those in the standard game theory literature as summarized by Harsanyi (1977) in developing his own game theoretic approach. Harsanyi, noting there are a variety of solution concepts (some determinate, some not) in the game theory literature, develops a unifying theory yielding determinate solutions for all classes of "classical" games: two-person and n-person games, zero-sum and non-zero-sum, games with and without transferable utility, cooperative and noncooperative games, etc. Harsanyi defines a "classical" game as one that satisfies the following conditions: (1) be a game with complete information where players know fully their own and other players' utility functions and strategies; (2) be either fully cooperative or noncooperative and not mixed or intermediate, i.e., players are either permitted to make binding and enforceable agreements before playing the game (cooperative) or not (noncooperative); (3) be a game representable in normal form, which eliminates games with delayed commitment where commitment to specific strategies may occur after onc or more chance and/or personal moves have been made.

For classical games Harsanyi obtains determinate solutions related to the Nash-Zeuthen solution, to the modified Shapley value, and to their generalizations. This is based on the principle of mutually expected rationality, i.e., if a player follows certain rationality postulates, then, if he is a rational player, he must expect and act on the expectation that other rational players will do so likewise. Harsanyi shows this restricts the subjective probabilities that a rational player can entertain about another's behavior and leads to Zeuthen's principle. This says that at any given stage of bargaining between two rational players the next concession must always come from the player less willing to risk a conflict as measured by the highest probability of conflict a player would be willing to face rather than accept the terms offered by his opponent. Zeuthen's principle in turn leads to the Nash solution; the modified Shapley value is also obtained.

Harsanyi's (1977) discussion is restricted to classical games even though elsewhere (1967–1978; Harsanyi and Selten 1972) he has shown the analysis can be extended to certain classes of nonclassical games, e.g., games with incomplete information and games with delayed commitment.

Thus Harsanyi has developed an important unifying theory for classical games with a possibility of generalization to nonclassical games. If we accept his rationality postulates, he has shown us that the Nash-Zeuthen/modified Shapley value results apply at least in classical games. However in real negotiations his key principle of mutually expected rationality may apply only imperfectly because real players have cognitive limits or bounds to their rationality (Simon 1976). Consequently, subjective probabilities may be far less restricted than mutually expected rationality indicates, and so Zeuthen's principle may not in fact be realized. Under bounded rationality, useable information on the nature of these restrictions constitutes an area for behavioral research (Kadane and Larkey 1982). Also Harsanyi (1977, p. 176) does not discuss in detail the computational problem of finding solutions to specific games, simply saying in brief that iterative methods are required.

For these reasons—(1) limitation of existing general theories mainly to classical games, (2) imperfections in the applicability of rationality postulates (such as mutually expected rationality) to real negotiations due to bounded rationality, and (3) computational difficulties in computing solutions following existing game theory solution concepts—alternative approaches to negotiations are of interest. This paper provides an alternative modeling approach. For one thing our model is prescriptive from the point of view of a player who currently must decide on concessions; game theory in general is jointly prescriptive of solutions rather than singly prescriptive of the dynamics of concession making. Our approach is richer than classical game theory in that it: (1) allows incomplete information; (2) accommodates mixed cooperative and noncooperative games; (3) allows strategies to change as the game progresses and permits delayed commitments (Harsanyi 1977) so we are not limited to games representable in normal form—in our model the strategy vector is revised after each negotiation session, and only the concession for that session is actually implemented; (4) permits a player to concede consistent with what he knows (information) about his changing implicit utilities and changing subjective probabilities. Our approach does not require use of rationality postulates, i.e., it uses instead subjective probabilities; and although the computational burden can become heavy, there is a definite computational method—dynamic programming. Thus the advantage of our stochastic terminal control approach compared to standard game theory is that negotiations are more realistically and flexibly modeled. In return for this with the stochastic terminal control approach we must obtain revised subjective probabilities from players after each concession round, and (except for games with complete information and fixed subjective probabilities) we necessarily cannot predict how negotiations are going to end, i.e., predict the solution to the game.

The earlier work of Rao and Shakun (1974) on negotiations uses the stochastic terminal control approach but differs from this chapter in that it assumes that utility functions are known, i.e., it assumes complete information; however, it retains the other advantages over classical game theory.

To illustrate the above remarks, consider an actual case of negotiations reported by Kapoor (1970)—a joint venture negotiation between an American consortium and the government of India. Briefly, the negotiation situation involves fertilizer production in India. A group of large, private oil, chemical, and engineering companies based in the United States forms a consortium that presents to the government of India a general proposal for collaboration on a massive fertilizer program. The project involves the establishment of five fertilizer factories in India. Negotiations involve such issues as percent equity, percent of crude oil supply right going to the consortium (i.e., the right of the consortium to supply crude oil from which naptha used in fertilizer manufacturing is obtained. Naptha is obtained from refining crude oil, as are, of course, other petroleum products. The right of the consortium to import this crude oil is also viewed as a lever to open the way for further participation by consortium members in oil refining in India), percent return on the investment of the consortium, management control, and the like. For a detailed description of the case, see Kapoor (1970).

The negotiation situation in this case is complex, the cognitive demand on players large, so that the applicability of rationality postulates used in standard game theory is open to question. We have a multivariate negotiation in which the utility functions are not known, i.e., we have incomplete information. We have a game intermediate between cooperative and noncooperative in that players can make binding and enforceable agreements after the game has started. We have a game in which delayed strategic commitments can occur. Clearly for this negotiation and other complex real-world cases like it (e.g., see Fayerweather and Kapoor 1974; Kapoor 1975) classical game theory is not sufficient. We suggest the stochastic terminal control approach developed in this chapter as a more realistic and flexible modeling alternative.

In the chapter we treat a two-player negotiation. We consider a situation at first involving negotiations on a single variable that have become deadlocked but that would normally result in conflict payoffs. Then in an attempt to avoid the conflict payoffs, a second negotiation variable is introduced. Thus a new situation, a two-dimensional negotiation, arises. We wish to model concession making prescriptively (normatively) from this point on without bivariate utilities.

Redefining the negotiation variable (operational goal) space—here increasing the dimensionality from one to two—is a method of conflict resolution discussed by Shakun (1981a, 1981b) and in Chapter 2. He illustrates conceptually the negotiation between a large bank and a women's rights group. We shall model this negotiation mathematically. We note our approach applies where a negotiation is initially two-dimensional as well as the situation here where a second dimension is introduced after deadlock on a first one. We consider the two dimensions jointly so that concessions on either one or both of the dimensions at a time are possible.

7.2 A Two-Player Bivariate Negotiation Model

Consider we have two players ($i = 1, 2$) with $i = 1$ representing the women's group and $i = 2$ the bank. Suppose the players have been negotiating (making concessions) on a single goal dimension, e.g., the number of women the bank should promote as vice-presidents. Concession making here could have been proceeding using the Rao-Shakun univariate model. Now suppose negotiations have become deadlocked, with the women's group demanding nine women vice-presidents and the bank offering five. At this point the bank—to avoid a conflict payoff and show it supports the value "women's rights"—offers to give $100,000 toward women's scholarships at a local graduate school of business administration, i.e., the bank introduces a second negotiation dimension.* Thus the bank's current offer is (5, 100). The women's group replies that $100,000 is too small to affect its concession making on women vice-presidents and asks $400,000 to reduce its demand for vice-presidents to eight. The current demand of the women's group is therefore (8, 400).

The negotiations have become two-dimensional with current (opening) positions $P_1 = (8, 400)$ and $P_2 = (5, 100)$ for the women's group and bank, respectively. The bank, which now must decide on further concessions, considers the state of the negotiations facing it to be represented by state vectors $x = (x_1, x_2)$, $P_1 = (P_{11}, P_{12})$, and $P_2 = (P_{21}, P_{22})$, where x_1, x_2 are the latest concessions by the women's group on vice-presidents and scholarships, respectively; P_{11} and P_{12} are the latest demands by the women's group for vice-presidents and scholarships, respectively; P_{21} and P_{22} are the previous offers of the bank. At this point the negotiation facing the bank has $x = (1, 0)$, where $x_1 = 1$ is associated with the women's group concession from nine to eight vice-presidents and by convention $x_2 = 0$ is associated here with the women's group opening figure of 400 on scholarships. In addition $P_1 = (8, 400)$ and $P_2 = (5, 100)$.

We let P_1^* and P_2^* be numbers of vice-presidents and amounts of scholarships, respectively. We wish to bypass the need for determining bivariate utility functions $U_1(P_1^*, P_2^*)$ and $U_2(P_1^*, P_2^*)$ for the women's group and bank, respectively. Toward this end we shall consider the probability of a player obtaining payoff $P^* = (P_1^*, P_2^*)$. In terms of the negotiation now facing the bank the set of current possible negotiated payoff vectors P^* is defined by the intersections and respective probabilities of attainment associated with the grid shown in Figure 7.1

In Figure 7.1 we have taken $\delta_1 = 1$ and $\delta_2 = 50$ where $\delta = (\delta_1, \delta_2) > 0$ specifies minimum meaningful concession steps δ_1 and δ_2 for negotiation variables 1 (vice-presidents) and 2 (scholarships), respectively. Behaviorally, we assume concession making involves a choice of 0 (no concession) or δ_1 on variable 1 and 0 or δ_2 on variable 2.**

We assume, as in the Rao-Shakun model, a known or estimated finite number t of negotiation sessions remaining, after which, if the two players do not reach agreement, they receive conflict payoffs: $S_1 = (S_{11}, S_{12})$ and $S_2 = (S_{21}, S_{22})$, also shown in Figure 7.1: $S_1 = (5, 100)$, $S_2 = (8, 400)$.*** Here S_{11} and S_{12} are conflict payoffs for the women's group in vice-presidents and

*New negotiation dimensions (operational goals) may be defined using the goals/values referral process discussed by Shakun 1981a.

**Concessions $0, \delta_1, 2\delta_1, \ldots,$ and $0, \delta_2, 2\delta_2, \ldots,$ are also possible but require considerably more subjective probability estimates and computations. Also, negative concessions can be accommodated in the model. See Section 7.5.

***In case of conflict the bank is willing to give (5, 100), so that the conflict payoff for women $S_1 = (5, 100)$. In case of such conflict the women will picket the bank with consequences for it assumed worse than the payoff (8, 400); therefore we may use as the bank's conflict payoff $S_2 = (8, 400)$. Of course, in any real situation other appropriate conflict payoffs assumptions can be made.

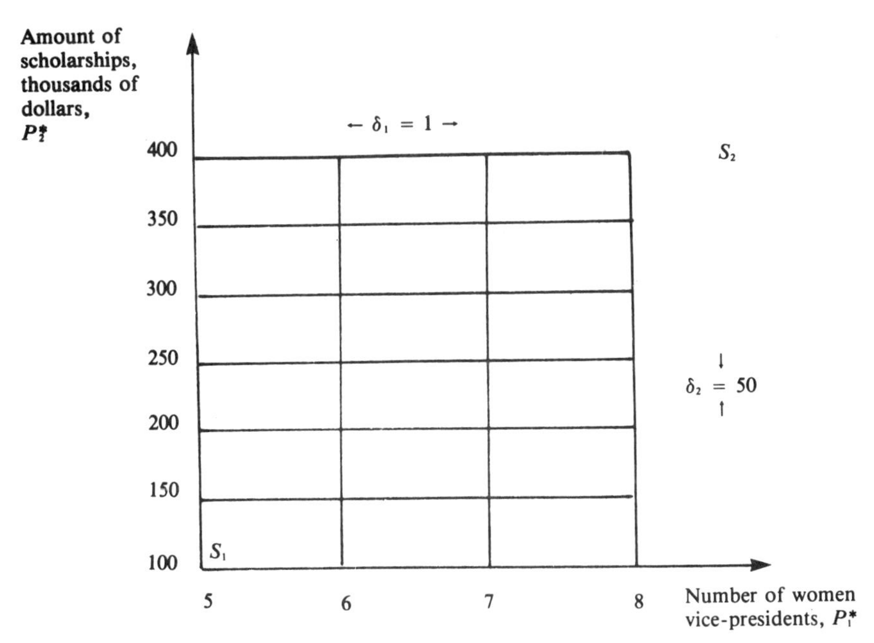

Figure 7.1. **Payoff grid**

scholarships, respectively; S_{21} and S_{22} are corresponding conflict payoffs for the bank. Two successive zero concession vectors, one player after the other, also result in conflict payoffs. These two situations and that of reaching agreement characterize the terminal states of the negotiation. We analyze concession making from the viewpoint of the bank. The analysis for the women's group is similar. With t periods remaining, the bank (who, we assume by convention, decides when t is even) must decide on a concession.

Under the model the bank considers $p^t(x, y, z)$, the conditional probability that the women's group will next concede $z = (z_1, z_2)$, given that its previous concession was $x = (x_1, x_2)$ and the bank now concedes $y = (y_1, y_2)$, that is:*

$$p^t(x, y, x) = \text{Prob}(z/x, y, t, b) \quad \text{where} \quad \begin{matrix} x_1, y_1, z_1 = 0 \text{ or } \delta_1 \\ x_2, y_2, z_2 = 0 \text{ or } \delta_2 \end{matrix}. \qquad (1)$$

In the same fashion we denote $q^{t-1}(x, y, z)$, with t even, the conditional probability that the bank will next concede $z = (z_1, z_2)$ given that its previous concession was $x = (x_1, x_2)$ and the women's group now concedes $y = (y_1, y_2)$, that is:

$$q^{t-1}(x, y, x) = \text{Prob}(z/x, y, t-1, w) \qquad \text{(same conditions on } x, y, z). \qquad (2)$$

*Note that here the probability is taken under the assumption that the bank (b) is to decide. We often write Prob($z/x, y, t$) when there is no ambiguity.

As in the Rao-Shakun approach, we assume that needed subjective estimates of the p^t and q^{t-1} will be supplied at each negotiation stage and that at each stage the concession decision will be evaluated under the assumption that these estimates will hold until the end of negotiations, that is:*

$$p^t(x, y, z) = p^{t-2}(x, y, z) = \cdots = p^2(x, y, z),$$

$$q^{t-1}(x, y, z) = q^{t-3}(x, y, z) = \cdots = q^1(x, y, z)$$

for each x, y, z.

Thus we shall drop the superscript where possible without ambiguity. Considering (1) at session t, we note that x is known to the bank, which therefore must specify $4 \times 4 = 16$ subjective probabilities $p^t(x, y, z)$ subject to the constraint that:

$$\sum_z p^t(x, y, z) = 1 \tag{3}$$

so that, in effect, $16 - 4 = 12$ subjective probabilities must be specified. However some of these may be estimated to be 0: For example, if the additional behavioral assumption is made that a player will never simultaneously concede a positive amount on both negotiation dimensions, then:

$$p^t(x, y, z) = 0 \quad \text{if} \quad x,\ y \text{ or } z = (\delta_1, \delta_2).$$

This would reduce the number of subjective probabilities to be specified to $3 \times 3 - 3 = 6$ (because of constraint (3)).

In the same way we have:

$$\sum_z q(x, y, z) = 1. \tag{4}$$

Let $F_t(P^*, P_1, P_2, x, y)$ be the probability that:

if t is even, the bank
if t is odd, the women's group } will receive a payoff P^* when:

- there are t negotiating sessions remaining
- the opponent has just conceded x
- the women's current demand is P_1
- the bank's current offer is P_2
- the negotiator decides to concede y at time t

$$F_t(P^*, P_1, P_2, x, y) = \text{Prob}(P^*/P_1, P_2, t, x, y). \tag{5}$$

Note that $F_t(P^*, P_1, P_2, x, y) \neq 0$ if the following 3 conditions hold:

- $P_2 \leq P_1$: bank's current offer has both components less than women's group offer; offers do not "cross over"

*This is a simplifying assumption. In principle $p^t, \ldots, p^2$ and $q^{t-1}, \ldots, q^1$ can be individually specified. Of course subjective probabilities p^t and q^{t-1} can be revised at any time t so as to include information recently gained from the opponent's previous concession.

• $P_2 \leq P^* \leq P_1$: only points lying between current offers can be finally reached
• $P_2 + y \leq P_1$ if t is even $\}$ offers do not "cross over": negotiators
• $P_2 \leq P_1 - y$ if t is odd $\}$ would never concede more than necessary

Then, for $t > 2$:

$$F_t(P^*, P_1, P_2, x, y) = \sum_z \text{Prob}(P^*, z/P_1, P_2, t, x, y)$$

$$= \sum_z \text{Prob}(P^*/P_1, P_2, t, x, y, z)\text{Prob}(z/P_1, P_2, t, x, y).$$

When t is even:

$$F_t(P^*, P_1, P_2, x, y) = \sum_z p(x, y, z)\text{Prob}(P^*/P_1, P_2, t, x, y, z) \quad (5A)$$

where, because two successive zero concession vectors result in conflict payoffs (break off), we have:

$$\text{Prob}(P^*/P_1, P_2, t, x, 0, 0) = \begin{cases} 1 & \text{if} \quad P^* = S_2, \\ 0 & \text{if} \quad P^* \neq S_2, \end{cases} \quad (5B)$$

$$\text{Prob}(P^*/P_1, P_2, t, 0, 0, z) = \begin{cases} 1 & \text{if} \quad P^* = S_2, \\ 0 & \text{if} \quad P^* \neq S_2. \end{cases} \quad (5C)$$

Except if negotiations break off in periods t or $(t - 1)$, the probability of obtaining payoff P^* starting from state (P_1, P_2, x) with t sessions remaining, the bank conceding y and the women z is exactly the same as obtaining P^* from state $(P_1 - z, P_2 + y, z)$* with $(t - 2)$ sessions remaining. Let $y_{t-2}^{\text{opt}}(s)$ denote the optimum concession with $(t - 2)$ sessions remaining when the state of the system is (s). Choice of the optimum (meaning preferred) concession is discussed below following equation (9). Then, except if negotiations break off in periods t or $(t - 1)$, we have

$$\text{Prob}(P^*/P_1, P_2, t, x, y, z)$$

$$= \text{Prob}\big(P^*/P_1 - z, P_2 + y, t - 2, z, y_{t-2}^{\text{opt}}(P_1 - z, P_2 + y, z)\big)$$

$$= F_{t-2}\big(P^*, P_1 - z, P_2 + y, z, y_{t-2}^{\text{opt}}(P_1 - z, P_2 + y, z)\big)$$

where the definition of F_{t-2} follows from (5). Thus this F_{t-2} may replace $\text{Prob}(P^*/P_1, P_2, t, x, y, z)$ in (5A) except when (5B) or (5C) apply due to break off. Hence we may write (5A) as (6) to obtain the recursive relation of dynamic programming:

$$F_t(P^*, P_1, P_2, x, y)$$

$$= \sum_z p(x, y, z) F_{t-2}\big(P^*, P_1 - z, P_2 + y, z, y_{t-2}^{\text{opt}}(P_1 - z, P_2 + y, z)\big) \quad (6)$$

*Note that if concessions z or y are not "feasible" (points $P_1 - z$ or $P_2 + y$ outside payoff grid), the corresponding probabilities are 0.

except, noting (5B) and (5C), if $x > 0$, $y = z = 0$, or if $x = y = 0$ we replace F_{t-2} by 1 if $P^* = S_2$ and by 0 if $P^* \neq S_2$.

When t is odd, using the same reasoning, we get:

$$F_t(P^*, P_1, P_2, x, y)$$
$$= \sum_z q(x, y, z) F_{t-2}(P^*, P_1 - y, P_2 + z, z, y_{t-2}^{\text{opt}}(P_1 - y, P_2 + z, z)) \quad (7)$$

again employing the exception indicated for (6).

If $t = 2$:

$$F_2(P^*, P_1, P_2, x, y) = \sum_z p(x, y, z) \text{Prob}(P^*/P_1, P_2, 2, x, y, z).$$

$$\cdot \text{If } x > 0 \quad \text{Prob}(P^*/P_1, P_2, 2, x, y, z) = \begin{cases} 1 & \text{if } P^* = P_1 - z = P_2 + y, \\ & \text{or } P^* = S_2 \text{ and } P_1 - z \neq P_2 + y, \\ 0 & \text{otherwise.} \end{cases}$$

(Note that the $y + z$ should never be strictly larger than $P_1 - P_2$ because women would never concede more than just necessary.)

$$\cdot \text{If } x = 0 \quad \text{Prob}(P^*, P_1, P_2, 2, x, y, z) = \begin{cases} 1 & \text{if } P^* = P_1 - z = P_2 + y \text{ and } y \neq 0 \\ & \text{or } P^* = S_2 \text{ and } y = 0 \\ & \text{or } P^* = S_2,\ y \neq 0 \\ & \text{and } P_1 - z \neq P_2 + y, \\ 0 & \text{otherwise.} \end{cases}$$

Hence

$$x > 0 \quad F_2(P^*, P_1, P_2, x, y) = \begin{cases} p(x, y, P_1 - P^*) & \text{if } P^* = P_2 + y \\ & \text{and } P_1 - P^* \geq 0, \\ 1 - p(x, y, P_1 - P_2 - y) & \text{if } P^* = S_2,\ P^* \neq P_2 + y \\ & \text{and } P_1 - P_2 - y \geq 0, \\ 0 & \text{otherwise.} \end{cases} \quad (8)$$

$$x = 0 \quad F_2(P^*, P_1, P_2, 0, y) = \begin{cases} p(0, y, P_1 - P^*) & \text{if } P^* = P_2 + y,\ y \neq 0 \\ & \text{and } P_1 - P^* \geq 0, \\ 1 & \text{if } P^* = S_2 \text{ and } y = 0, \\ 1 - p(0, y, P_1 - P_2 - y) & \text{if } P^* = S_2,\ y \neq 0, \\ & P^* \neq P_2 + y \\ & \text{and } P_1 - P_2 - y \geq 0, \\ 0 & \text{otherwise} \end{cases}$$

If $t = 1$:

$$F_1(P^*, P_1, P_2, x, y) = \begin{cases} 1 & \text{if } P^* = P_1 - y = P_2 \\ & \text{or } P^* = S_1 \text{ and } P_1 - y = P_2 \text{ is untrue,} \\ 0 & \text{otherwise.} \end{cases} \quad (9)$$

Equations (6) and (8) for the bank, (7) and (9) for the women's group define the basic relationship between successive periods that will allow us to design our dynamic programming process.

At each step—time t—the bank (for t even) or women's group (for t odd) knows:

- the state of the system (x, P_1, P_2)
- the subjective—possibly revised at the end of preceding session—probabilities $p^t(x, y, z)$ or $q^t(x, y, z)$ (assumed to hold constant through the negotiating session when making calculations at time t)
- the optimal concessions $y_{t-2}^{\text{opt}}, y_{t-4}^{\text{opt}} \ldots y_2^{\text{opt}}$ (or y_1^{opt}) computed previously, defined for each possible state of the system, i.e., function $y_j^{\text{opt}}(x^j, P_1^j, P_2^j)$ for any state (x^j, P_1^j, P_2^j)

It can then build a probability distribution for *each* y on the payoff grid (with t sessions remaining). See Figure 7.2.

Let us denote μ_y^t this probability distribution on the payoff grid (for a given state x, P_1, P_2):

$$\mu_y^t(P^*, x, P_1, P_2) = F_t(P^*, P_1, P_2, x, y). \quad (10)$$

When there is no ambiguity, we shall drop the state (x, P_1, P_2) and write $\mu_y^t(P^*) = \mu_y^t(P^*, x, P_1, P_2)$.

To decide which concession it has to make—y_t^{opt}—the bank or women's group will choose among the μ_y^t the one it prefers. We do not specify a particular multicriteria method or specific criteria for choosing among these probabilities distributions—criteria are related implicitly to the negotiator's unspecified bivariate preference field. One suggestion is to use direct visual assessment that uses gestalt capabilities of players. Various visual displays of the probability distributions may be interactively generated by computer. These could include three-dimensional visual perspectives, use of dots on the payoff grid whose blackness or size could be graduated according to probability, and various two-dimensional displays (univariate marginal or conditional distributions). One could also compute means, standard deviations, probabilities of ruin, etc., which are all possible choice criteria. In addition to direct visual assessment, various other multicriteria methods employing such criteria as means, standard deviations, probabilities of ruin, etc., can be used. See Jacquet-Lagreze and Siskos 1982; Kempf, Duckstein, and Casti 1979; Roy 1977; Starr and Zeleny 1977; Zeleny 1982, which include some comparisons among methods.

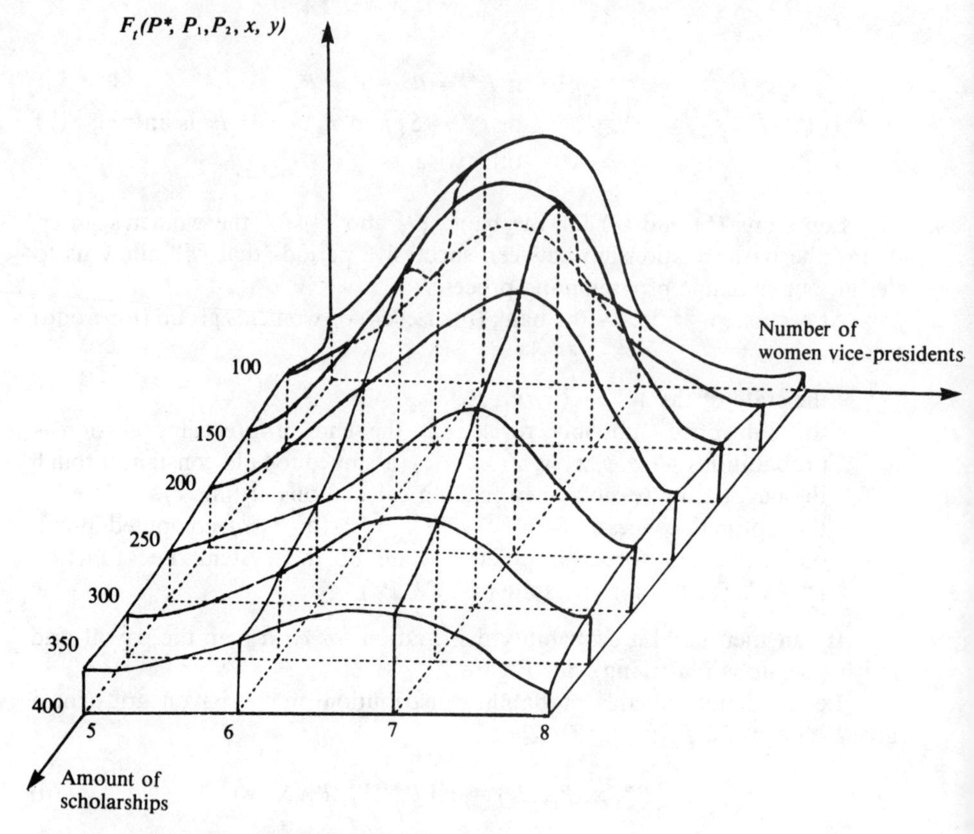

Figure 7.2. **Probability distribution on the payoff grid**

In any case in period t, t even, the bank concedes the amount y_t^{opt} just determined, after which the women's group computes its dynamic programming problem and concedes its calculated amount. Then the bank computes a new t-stage problem (which now has two stages less) using revised subjective probabilities and the then current state of the system, and so forth.

The choice of negotiator is thus based upon the preferred distribution, which we call θ_t:

$$\theta_t(P^*, x, P_1, P_2) = \underset{y}{\text{opt}}\, \mu_y^t(P^*, x, P_1, P_2)$$

$$= \underset{y}{\text{opt}}\, F_t(P^*, P_1, P_2, x, y) \qquad (11)$$

where "opt" means preferred. Then $y_t^{\text{opt}}(P_1, P_2, x)$ is a solution of this program.

7.3 Numerical Example

In order to illustrate the model we present a numerical example for the negotiation discussed above in which we suppose:

- The negotiations will take place in $n = 10$, 8 or 6 sessions.
- At the beginning the bank faces a state of the system such that, $x = (1, 0)$, $P_1 = (8, 400)$, $P_2 = (5, 100)$.
- The conflict payoffs are $S_1 = (5, 100)$, $S_2 = (8, 400)$ as discussed above.
- $x^t = (x_1^t, x_2^t)$ will denote the previous concessions (with $t + 1$ sessions remaining) by the opponent on women vice-presidents and amounts of scholarships; $\Rightarrow x^n = (1, 0)$.
- $P_1^t = (P_{11}^t, P_{12}^t)$ the present offer by women; $P_2^t = (P_{21}^t, P_{22}^t)$ the present offer by the bank; $\Rightarrow P_1^n = (8, 400)$, $P_2^n = (5, 100)$.

The calculations are made by repeating the following sequence for $t = 2, 4, \ldots, n - 2, n$.

With $t - 1$ sessions remaining, for each state x^{t-1}, P_1^{t-1}, P_2^{t-1}:

1. Possibly revise $q(x, y, z)$. If these $q(x, y, z)$ are changed, then recalculate $y_1^{\text{opt}}, y_3^{\text{opt}}, \ldots, y_{t-3}^{\text{opt}}$ for the new q's.
2. Compute $F_{t-1}(P^*, P_1^{t-1}, P_2^{t-1}, x^{t-1}, y)$ for each y and each P^* on women's payoff grid.
3. Compute $\theta_{t-1}(P^*, x^{t-1}, P_1^{t-1}, P_2^{t-1},)$ and hence concede $y_{t-1}^{\text{opt}}(P_1^{t-1}, P_2^{t-1}, x^{t-1})$.
4. Then compute x^{t-2}, P_1^{t-2}, P_2^{t-2} using the following:

$$x^{t-2} = y_{t-1}^{\text{opt}}\left(P_1^{t-1}, P_2^{t-1}, x^{t-1}\right),$$

$$P_1^{t-2} = P_1^{t-1} - y_{t-1}^{\text{opt}}\left(P_1^{t-1}, P_2^{t-1}, x^{t-1}\right),$$

$$P_2^{t-2} = P_2^{t-1}.$$

With t sessions remaining, for each state x^t, P_1^t, P_2^t:*

5. Possibly revise $p(x, y, z)$ and recalculate $y_2^{\text{opt}}, \ldots, y_{t-2}^{\text{opt}}$ if necessary.
6. Compute $F_t(P^*, P_1^t, P_2^t, x^t, y)$ for each y and P^* on bank's payoff grid.
7. Compute $\theta_t(P^*, x^t, P_1^t, P_2^t)$ and hence concede $y_t^{\text{opt}}(P_1^t, P_2^t, x^t)$.
8. Then compute x^{t-1}, P_1^{t-1}, P_2^{t-1}) using the following:

$$x^{t-1} = y_t^{\text{opt}}\left(P_1^t, P_2^t, x^t\right),$$

$$P_1^{t-1} = P_1^t,$$

$$P_2^{t-1} = P_2^t + y_t^{\text{opt}}\left(P_1^t, P_2^t, x^t\right).$$

*For $t = n$ make the calculations for only one state of the system (m^n, P_1^n, P_2^n) that is known.

It is clear that in parts 2 and 6 one will have to work in backwards fashion using the recursive relations (6) through (9). Choices in 3 and 7 can be made as discussed in Section 7.2. In this numerical example we have used the following criteria:

Let $m^t(y) = (m_1^t(y), m_2^t(y))$ denote the mean-vector and $V^t(y) = (V_1^t(y), V_2^t(y))$ the variance-vector of distribution μ_y^t (for a given state P_1^t, P_2^t, x^t).

Assuming that the number of vice presidents is of paramount importance for the bank and the women, we shall take y_t^{opt}—for t even—as the solution of the following:

—Let Y^t denote the set of all possible concessions:

$$Y^t = \left\{ y \in \{(0,0),(1,0),(0,50),(1,50)\} : \begin{array}{ll} P_2^t + y \leq P_1^t & \text{if } t \text{ is even} \\ P_2^t \leq P_1^t - y & \text{if } t \text{ is odd} \end{array} \right\}.$$

$$- \ Y_1 = \left\{ y^* \in Y^t : m_1^t(y^*) = \underset{y \in Y^t}{\text{Min}}\, m_1^t(y) \right\},$$

$$Y_2 = \left\{ y^* \in Y_1 : m_2^t(y^*) = \underset{y \in Y_1}{\text{Min}}\, m_2^t(y) \right\},$$

$$Y_3 = \left\{ y^* \in Y_2 : V_1^t(y^*) = \underset{y \in Y_2}{\text{Min}}\, V_1^t(y) \right\},$$

$$Y_4 = \left\{ y^* \in Y_3 : V_2^t(y^*) = \underset{y \in Y_3}{\text{Min}}\, V_2^t(y) \right\},$$

y_t^{opt} is chosen at random in Y_4.

The same choice criteria are assumed to apply for the women's group (except that they are maximizing the means).

The optimal concession path for the bank and the women's group will then be given by:

$$x_n = y_n^{\text{opt}}(P_1^n, P_2^n, x^n).$$

$$x_{n-1} = y_{n-1}^{\text{opt}}(P_1^n, P_2^n + x_n, x_n),$$

$$x_{n-2} = y_{n-2}^{\text{opt}}(P_1^n - x_{n-1}, P_2^n + x_n, x_{n-1}),$$

$$\cdots$$

$$x_2 = y_2^{\text{opt}}(P_1^n - x_{n-1} - x_{n-3} - \cdots - x_3, P_2^n + x_n + x_{n-2} + \cdots + x_4, x_3),$$

$$x_1 = y_1^{\text{opt}}(P_1^n - x_{n-1} - x_{n-3} - \cdots - x_3, P_2^n + x_n + x_{n-2} + \cdots + x_4 + x_2, x_2).$$

Now, if

$$P_1^n - x_{n-1} - x_{n-3} - \cdots - x_3 - x_1 = P_2^n + x_n + x_{n-2} + \cdots + x_4 + x_2,$$

this will be the final payoff. Otherwise bank and women's group will get their conflict payoffs $= S_2$ and S_1, respectively.

Table 7.1 **Subjective Probability Distribution.**

x		*(0, 0)*				*(0, 50)*				*(1, 0)*				*(1, 50)*			
z	y	*0, 0*	*0, 50*	*1, 0*	*1, 50*	*0, 0*	*0, 50*	*1, 0*	*1, 50*	*0, 0*	*0, 50*	*1, 0*	*1, 50*	*0, 0*	*0, 50*	*1, 0*	*1, 50*
(0, 0)		1	0.20	0.40	0.20	0.25	0.10	0.20	0.10	0.25	0.20	0.10	0.10	0.70	0.40	0.10	0.05
(0, 50)		0	0.30	0.20	0.40	0.30	0.35	0.30	0.20	0.30	0.30	0.20	0.30	0.10	0.25	0.30	0.45
(1, 0)		0	0.30	0.20	0.30	0.30	0.35	0.30	0.60	0.30	0.30	0.50	0.30	0.10	0.25	0.40	0.45
(1, 50)		0	0.20	0.20	0.10	0.15	0.20	0.20	0.10	0.15	0.20	0.20	0.30	0.10	0.10	0.20	0.05

In this numerical example, we use for both negotiators the subjective probability distribution shown in Table 7.1.

From Table 7.1 we see that:

$$\text{Prob}(y = z = (0,0)) = \begin{matrix} 0.25 & \text{if } x = (0,50) \text{ or } (1,0), \\ 0.70 & \text{if } x = (1,50), \end{matrix}$$

Hence the subjective probability of negotiations break off is relatively high. This feature explains why, in this numerical example, it is difficult to get an agreement. Appropriate choices of subjective probabilities should lead to more frequent agreement. Results are shown in Figures 7.3, 7.4, and 7.5 for $n = 10$, 8, and 6.

- In the figures concessions (0, 0), (0, 50), (1, 0), (1, 50) are denoted 1, 2, 3, 4. Decision makers B and W stand for bank and women's group.

These figures show the influence of n on the possibility of an agreement: With the present opening positions, at least six concessions are needed to fill the gap. In fact it appears here that, with the optimum behavior used by the negotiators, they are not able to reach agreement in six, eight, or ten sessions.

Figure 7.3. **Numerical Example for $n = 10$. Negotiators get their conflict payoffs after two successive (0, 0) concessions at times 5 and 4. But the choice of y_5^{opt} and y_4^{opt} is made at random in Y_4. If Y_4 contains more than one element, a different choice could have led to an agreement or to final positions closer than here. In the computer program, we did not take this possibility into account.**

Sessions remaining		10	9	8	7	6	5	4	3	2	1
Decision maker		*B*	*W*	*B*	*W*	*B*	*W*	*B*	*W*		
Concession amount		4	3	2	1	2	1	1	Break off		
State at which decision is made											
Nb of women V.P.	*W*	8	8	7	7	7	7	7			
	B	5	6	6	6	6	6	6			
$ Scholarship	*W*	400	400	400	400	400	400	400			
	B	100	150	150	200	200	250	250			
Last concession *X*		3	4	3	2	1	2	1			

Figure 7.4. **Numerical Example for $n = 8$. Negotiators get their conflict payoffs because they fail to agree at the end of negotiating time. Bank should expect women to accept its offer if one or more sessions were added. Note that for $n = 8$ final positions stand closer than for $n = 10$. It could have been different with a different choice of Y_4 (see Figure 7.3). This shows the sensitivity of final results to the optimality criteria used.**

		8	7	6	5	4	3	2	1	
Sessions remaining		8	7	6	5	4	3	2	1	
Decision maker		*B*	*W*	*B*	*W*	*B*	*W*	*B*	*W*	
Concession amount		4	2	1	2	1	4	1	4	Break off
State at which decision is made										
Nb of women V.P.	*W*	8	8	8	8	8	8	7	7	
	B	5	6	6	6	6	6	6	6	
$ Scholarship	*W*	400	400	350	350	300	300	250	250	
	B	100	150	150	150	150	150	150	150	
Last concession *X*		3	4	2	1	2	1	4	1	

Figure 7.5. **Numerical Example for $n = 6$. Negotiators get their conflict payoffs because of two successive (0, 0) concessions and end of negotiating time. Note that final positions stand far away. Distance between initial positions *B* and *W* is such that only three paths can lead to an agreement, all of which begin with women conceding (0, 50)**

Sessions remaining		6	5	4	3	2	1	
Decision maker		*B*	*W*	*B*	*W*	*B*	*W*	
Concession amount		4	3	2	3	1	1	Break off
State at which decision is made								
Nb of women V.P.	*W*	8	8	7	7	6	6	
	B	5	6	6	6	6	6	
$ Scholarship	*W*	400	400	400	400	400	400	
	B	100	150	150	200	200	200	
Last concession *X*		3	4	3	2	3	1	

Nevertheless the final positions get closer when n increases, so there is some hope that, for $n = 12$ or 14, they would have succeeded to avoid conflict payoffs.

See Section 7.5 for the use negotiators could make of these figures.

The computer can be used throughout the negotiating process as a conversational tool: It can accommodate revision of the subjective probabilities p and q. We do not specify a precise underlying mechanism. However, clearly, the revised probabilities incorporate each negotiator's learning about his opponent's behavior. A negotiator tries to influence his opponent's probability revisions by his own concession choice. Additional research is needed to make more precise underlying mechanism on how to revise subjective probabilities. (For partial use and limitation of Bayesian schemes, see discussion in Rao and Shakun 1974. Also see Kadane and Larkey 1982.)

7.4 Empirical Results

We have not yet tried using the model in experimental situations to any great extent. However, in response to a suggestion, we did run ten experimental negotiations simply to check the assumption that a negotiator is capable of specifying and changing his subjective probabilities $p^t(x, y, z)$ in response to concession decisions by the other player. Ten different students represented the bank in 10 negotiations, with the women's group represented by one of the authors. The prior concession of the women's group was $x = (1, 0)$ in each case. The bank (student) in each case specified 16 subjective probabilities $p^t(x, y, z)$, of which 12 are independent because of constraint equation (3). Then the bank selected 1 of 4 concessions $y = (0, 0), (0, 50), (1, 0)$, or $(1, 50)$. Following this the women selected a concession; in each case it deliberately was $(0, 0)$. The bank then revised its subjective probabilities and decided on a concession y. Then the women announced a concession; in each case again it was $(0, 0)$. The bank again revised its subjective probabilities and decided on concession y. The women then changed their concession from $(0, 0)$.

Past observation of negotiations tells us that a number of behavioral principles can be operative. One principle operates as follows: Given that the women concede $x = (0, 0)$, the bank thinks that if it concedes $y = (0, 50)$, $(1, 0)$, or $(1, 50)$, i.e., chooses a nonzero concession vector y, then the women will take this as a sign of weakness and will tend to choose $z = (0, 0)$ more.

Table 7.2. **Bank's Change in Subjective Probability. The direction (positive, negative, no change) of six bank probability respecifications in ten experimental negotiations is indicated. A positive change means that the bank's subjective probability increases from one time period to the next. The women's concession vector remains at (0, 0).**

Student (bank) number	*Number of changes in subjective probability*		
	Positive change	*Negative change*	*No change*
1	Immediate	Negotiations	Breakoff
2	2	0	4
3	2	4	0
4	4	0	2
5	3	1	2
6	3	0	3
7	5	0	1
8	3	2	1
9	3	3	0
10	6	0	0
Totals	31	10	13

Hence with continued operation of this principle of the bank's subjective probabilities $p^t[x = (0,0);\ y = (0,50),\ (1,0)$ or $(1,50);\ z = (0,0)]$ will increase (or at least not decrease) over time, i.e., as the number of sessions remaining, t, decreases. In our experiments this behavior is observable for five students (students 2, 4, 6, 7, 10) as shown in Table 7.2. These students changed their subjective probabilities in a manner completely consistent with the above behavior principle. As shown in Table 7.2, the number of positive or no changes in subjective probability is 6 out of 6 probability respecifications for each of these students. The number of positive to no change for these students as a group is 20 to 10 obtained by adding the entries for them in the respective columns.

Following this, when the women changed their concession vector from (0, 0) to (0, 50), these students entered negative changes in their subjective probabilities (the number of negative to positive changes is 8 to 1).

We do not formally test hypotheses statistically here. We simply point out that the data appear to support the assumption that a negotiator is capable of meaningfully specifying and changing his subjective probabilities in response to concession decisions by the other player.

7.5 Concluding Remarks

Bivariate negotiations have been modeled as a problem of stochastic terminal control in which both of two players attempt through concession making to steer (control) the negotiations to a preferred terminal state characterized at any stage by a preferred (optimum) payoff probability distribution. The latter (target) is not fixed but changes in the course of negotiations. The problem can be formulated and solved by dynamic programming. The research models mathematically for the case studied, aspects of a general approach to conflict resolution and design of purposeful systems, are discussed by Shakun (1981a, 1981b) and in Chapter 2.

Although the model has been formulated for zero or positive concessions, it can accommodate negative concessions as well. For example, suppose concession making can involve a choice of 0 (no concession), δ_1 (moving closer to the other player's current position on variable 1), and $-\delta_1$ (moving further away from the other player's current position on variable 1). Similarly, for variable 2, suppose concessions 0, δ_2, and $-\delta_2$ are considered. The combinations $(-\delta_1, -\delta_2)$, $(-\delta_1, 0)$, $(0, -\delta_2)$ could be ruled out by agreement between the players; the other six combinations would be legitimate. When deciding on concessions, given he knows x, a player must now specify $6 \times 6 = 36$ subjective probabilities $p(x, y, z)$ (compared to 16 when concessions are limited to zero or single positive amounts), again subject to constraint (3), which reduces the number of probabilities to be specified to $36 - 6 = 30$ (compared to 12). Negative concessions also increase the number of possible states of the system.

We conclude with some comments on the practical use of the model:*

1. The model could be used "on line," with the computer providing at each step of the negotiations the optimum concession given the present state of the system.
   "On line" use would require further study of computer processing time because of the large amounts of information to be processed.
2. Using the computer the model can be used before negotiations begin to provide various simulations. From these, the negotiator could gain insight beforehand into the negotiation process and the consequences of certain choices:
   a. The opening positions. This can be of significant importance, e.g., small variations in opening positions could produce drastically different final payoffs (as in Thom's Catastrophe Theory).
      Simulations can help in choosing the opening position because the final consequences of such a choice are demonstrated.
   b. The allowed or estimated number n of negotiating sessions available at the beginning of negotiations. The influence of n on final results can be studied by computer negotiations. If institutionally the negotiator can influence the choice of n, he can make use of these simulations. We also note in practice that negotiators sometimes allow additional sessions as the end of negotiations approaches, hoping to avoid the conflict payoff, i.e., t is reestimated. The effect of this can also be studied in the simulations.
   c. The subjective probabilities. Computer simulations can reveal the sensitivity of final results to changes in subjective probabilities.
3. A combination of 1 and 2 above would be desirable to give negotiators insight into the model and negotiating process beforehand and to aid them "on line" in their actual concession choices.
4. In principle the bivariate negotiation model can be extended to the multivariate case using multicriteria decision-making methods as noted (Jacquet-Lagreze and Siskos 1982; Kempf, Duckstein, and Casti 1979; Roy 1977; Starr and Zeleny 1977; Zeleny 1982) to choose a preferred multivariate payoff probability distribution. However as a practical matter the subjective probability requirements and computational burden become heavy, suggesting that simplifying behavior assumptions in the multivariate case may be necessary.

REFERENCES

Bellman, R., and Dreyfus, S. 1962. *Applied Dynamic Programming*. Princeton University Press, Princeton, NJ.

*The authors are indebted to participants in a colloquium sponsored by Compagnie Française des Pétroles and GRASCE, where CFP negotiators discussed the practical use of this model.

Fayerweather, J., and Kapoor, A. 1974. *Strategy and Negotiations for the International Corporation*. Ballinger Publishing Company, Cambridge, MA.

Harsanyi, J. C. 1967–1968. "Games with Incomplete Information Played by 'Bayesian' Players, I–III." *Management Science*, 14, Nos. 3, 5, 7.

______. 1977. *Rational Behavior and Bargaining Equilibrium in Games and Social Situations*. Cambridge University Press, Cambridge.

______, and Selten, R. 1972. "A Generalized Nash Solution for Two-Person Bargaining Games with Incomplete Information." *Management Science*, 18, No. 5, Part II (January).

Jacquet-Lagreze, E., and Siskos, J. 1982. "Assessing a Set of Additive Utility Functions for Multicriteria Decision Making, The UTA Method." *European Journal of Operational Research*, 10, No. 2 (June), pp. 151–164.

Kadane, J. B., and Larkey, P. D. 1982. "Subjective Probability and the Theory of Games." *Management Science*, 28, No. 2 (February). Also comments by J. C. Harsanyi appearing after this article in the same journal.

Kapoor, A. 1970. *International Business Negotiations: A Study in India*. New York University Press, New York.

______. 1975. *Planning for International Business Negotiations*. Ballinger Publishing Company, Cambridge, MA.

Kenney, R. L., and Raiffa, H. 1976. *Decisions with Multiple Objectives: Preferences and Value Tradeoffs*. Wiley, New York.

Kempf, J., Duckstein, L., and Casti, J. 1979. "Polyhedral Dynamics and Fuzzy Sets as a Multicriteria Decision Making Aid." Department of Industrial Engineering, University of Arizona, Tucson.

Rao, A. G., and Shakun, M. F. 1974. "A Normative Model for Negotiations." *Management Science*, 20, No. 10 (June). (See also Chapter 6 of the present volume.)

Roy, B. 1977. "Partial Preference Analysis and Decision-Aid: The Fuzzy Outranking Relation Concept." In Bell, E., Keeney, R. C., and Raiffa, H. (Eds.). *Conflicting Objectives in Decisions*. Wiley, New York.

Shakun, M. F. 1981a. "Formalizing Conflict Resolution in Policy Making." *International Journal of General Systems*, 7, No. 3. (See also Chapter 2 of the present volume.)

______. 1981b. "Policy Making and Meaning as Design of Purposeful Systems." *International Journal of General Systems*, 7, No. 4. (See also Chapter 2 of the present volume.)

Simon, H. A. 1976. *Administrative Behavior*, 3rd ed. Free Press, New York.

Starr, M. K., and Zeleny, M. 1977. "MCDM—State and Future of the Arts." In Starr, M. K., and Zeleny, M. (eds.). *Multiple Criteria Decision Making*. North-Holland, New York.

Zeleny, M. 1982. *Multiple Criteria Decision Making*. McGraw-Hill, New York.

CHAPTER 8

Decision Support Systems for Semistructured Buying Decisions*

Eric Jacquet-Lagreze† **and Melvin F. Shakun**††

8.1 Introduction

The decision to purchase a product is often difficult for an individual or an organization. Even if (1) considerable data on the available alternatives and their characteristics exist, (2) information is known of the behavior of previous decision makers (the criteria used, the biases and pitfalls encountered), and (3) there are experts who know a lot about the product, the buying decision can remain difficult. It seems that society as a whole knows a lot about this type of decision, but a given decision maker may know relatively little or not enough about the particular problem before him. Thus the decision problem can be almost perfectly structured from the point of view of society as a whole and relatively unstructured from the point of view of the decision maker. In this chapter this is what we refer to as a *semistructured buying decision problem*. Typical examples of such decision problems for the individual or family consumer would be:

- buying a car
- buying or renting a dwelling (house, apartment)
- buying a long distance travel ticket or vacation package
- buying heating equipment for a house (e.g., solar heating)
- buying a microcomputer

*This chapter first appeared in *European Journal of Operational Research*, Vol. 16, No. 1, April 1984.

† LAMSADE, University of Paris—Dauphine
†† New York University

Some examples for organizations include:

- choosing a type of car for a given use
- buying a new production machine tool
- buying or leasing a new computer
- buying a new copying machine

We may list the features of the semistructured buying decision problem as follows:

1. There usually exists a large number of alternative products; the decision maker (DM) knows a few of them.
2. Considerable data exist on the characteristics and performance warranties of these alternatives, but only some of this information is of interest to a particular decision maker.
3. There is certainty regarding the characteristics, although there might well be some uncertainty on the needs of the decision maker.
4. Despite the large amount of information on each alternative, some information—subjective in nature—may be missing. The decision maker must be able to introduce such subjective criteria as taste, liking or disliking the shape of a car, etc.
5. The criteria are conflicting. Thus the DM has to learn about his own preferences, considering his values, tastes, and needs on one hand and the available alternatives and their characteristics on the other hand.

Research on consumer behavior (see for instance Assael 1981; Chan Park, and Yu 1980; Engell, Kollat, and Blackwell 1978) shows that the decision-making process for purchasing a new product in which there is a high stake can be described as an individual and sometimes a multiparticipant conflict resolution process. Shakun (1981a, 1981b and in Chapters 1 and 2) has developed a general methodology—called evolutionary systems design (ESD)—for such processes. These same studies and others such as that of Janis and Mann (1977) suggest that, because of time pressure, limited cognitive capacities, and the existence of conflicting criteria, the buying decision process could be much more efficient if it were supported by some decision support system (DSS). Keen and Scott Morton (1978), Sprague and Carlson (1982), Bonczek, Holsapple, and Whinston (1981) suggest that semistructured decision problems are a fruitful area for DSS.

We shall design a DSS for the semistructured buying decision that will enable the decision maker to learn in an efficient way what his preferences and goals are in relation to what is available on the market and thus to make a buying decision.

To achieve this, we shall use an updated data base on the existing alternatives and their characteristics. We shall build into the DSS a set of goals (criteria) that have been commonly used by previous decision makers and those suggested by experts in the field. Thus we include society's collective memory (knowledge) on the subject. There is of course not a unique answer to the buying decision—different decision makers will choose different products. Thus we must be able to assess the preference function of the particular decision maker involved. The DSS should be user-oriented in the sense that the decision maker is cognitively comfortable in using the system.

In Section 8.2 we present the main theoretical framework used in the DSS. Section 8.3 is a detailed presentation of the DSS. Section 8.4 is an illustrative example of the decision to purchase a car. Section 8.5 presents concluding remarks.

8.2 Decision Making as Conflict Resolution

THE GENERAL FRAMEWORK: EVOLUTIONARY SYSTEMS DESIGN

We may view decision making as a process of evolutionary systems design involving conflict resolution in which decision makers (players) define and attain goals as operational expressions of underlying values. In the methodology developed by Shakun (1981a, 1981b, and in Chapters 1 and 2) N players are viewed as playing a dynamical (difference) game in which a coalition C of the set of N players can form provided it can deliver to itself (and hence to its members) a set of agreed-upon goals. Formally this means that for each time period the intersection of the coalition goal target $Y^C(t)$ and its technologically feasible performance $y^C(t)$ is nonempty. Thus the geometry of conflict and conflict resolution is shown in Figures 8.1 and 8.2.

For a given operational goal space, if the intersection of $Y^C(t)$ and $y^C(t)$ in Figure 8.1 is empty, then one or both of these sets may expand to give a nonempty intersection to resolve conflict as shown in Figure 8.2. By expansion we mean that new points are added to the sets $Y^C(t)$ and $y^C(t)$; this process of expansion does not preclude the dropping of some other points of these sets. However the dimensions of the operational goal space itself may be redefined using a goals/values referral process. Within the new goal space—either originally or after goal target and/or technological feasible

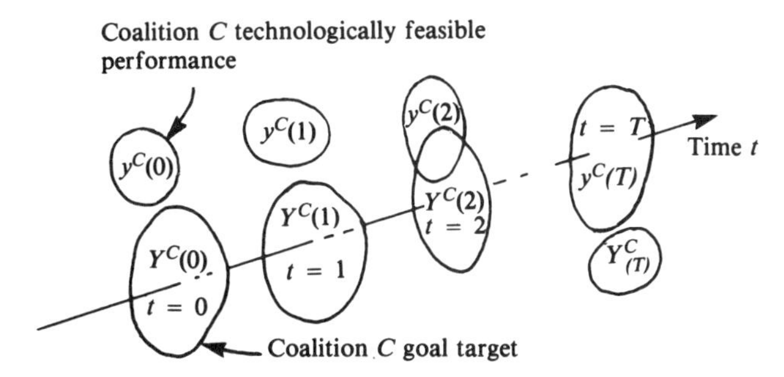

Figure 8.1. **The Geometry of Conflict.**

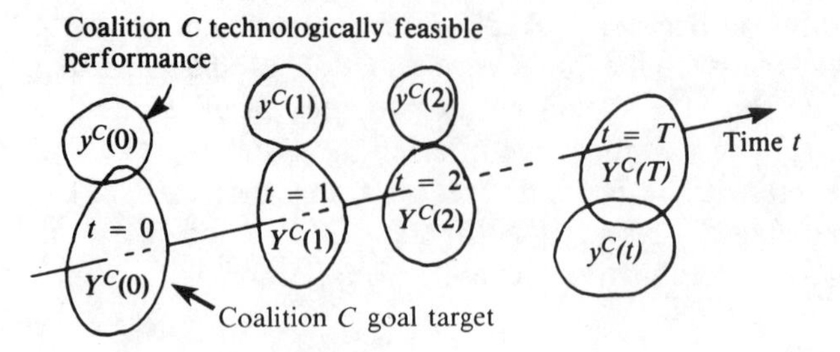

Figure 8.2. **Conflict Resolution.**

performance expansion—the target-performance intersection may be non-empty.

If the target-technological performance intersection has one point, then the point is the solution in output goal space. If the intersection contains more than one point, then $Y^C(t)$ and/or $y^C(t)$ may contract (negative expansion) to give a unique solution. It is also possible that the dimensions of the operational goal space could be redefined. The controls (inputs) that give this unique solution point in output goal space represent the decision to be taken.

The above methodology—called evolutionary systems design—is developed in Shakun (1981a, 1981b and in Chapters 1 and 2) and discussed for hierarchical systems in Shakun and Sudit (1982 and Chapter 14). It constitutes an approach to problem definition and solution in complex contexts involving multiparticipant, multicriteria, ill-structured, dynamic problems and provides a basis for decision support systems.

EVOLUTIONARY SYSTEM DESIGN IN SEMISTRUCTURED BUYING DECISIONS

For DSS for semistructured buying decisions a simpler version of the general evolutionary systems design methodology may be employed. Frequently in these buying decisions we may consider only one decision maker—the buyer. (We shall comment on the case of more than one decision maker in Section 8.5.) He has multiple criteria, and his problem is semistructured in the sense discussed above. For many buying decisions the general dynamical framework can be reduced to the present time period for purposes of analysis. For example, purchase costs or other costs (e.g., maintenance costs) paid in the future may be discounted to the present time. With these simplifications the evolutionary systems design methodology suggests the following framework for DSS for semistructured buying decisions.

Consider that we have a set A of products (purchase choices, controls, or inputs), e.g., all the different types of cars we may buy. Let g be a function from A into $\mathbf{R}^p$ (the goal space, the p-dimensional real vector space); then $y = g(a)$ for $a \in A$ is the vector of $\mathbf{R}^p$ representing the outputs of a particular

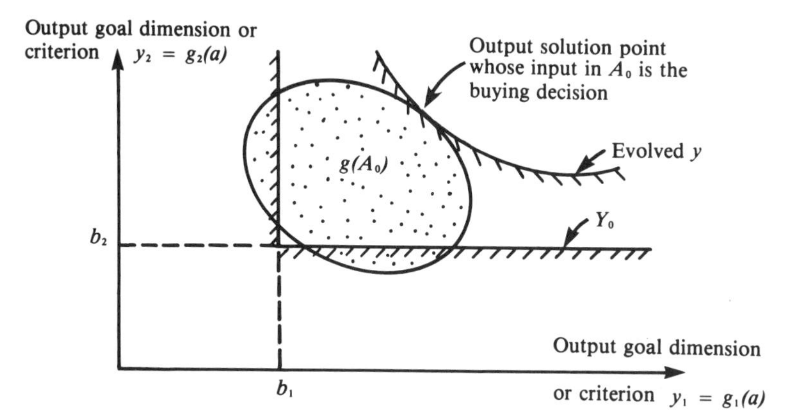

Figure 8.3. **Output Goal or Criterion Space.**

purchase choice and $y = g(A)$ represents technologically feasible performance, i.e., a set of possible outputs (characteristics) that may be obtained from the inputs (purchase choices) A. These outputs are generally constrained a priori by preliminary goal target Y_0 information. For example, these constraints could be $Y_0 = \{y \in \mathbf{R}^p\colon y_i \geq b_i, i = 1, \ldots, p\}$. The intersection of Y_0 and $g(A)$ is called $g(A_0)$, a set of a priori admissible outputs; $A_0 = \{a \in A\colon g(a) \in Y_0\}$ is the corresponding set of a priori admissible inputs. Typically in buying (e.g., buying cars) the set $g(A_0)$ has many points. By learning how to operationally contract Y_0 (see below), the decision maker reaches a point where the intersection of the evolved goal target Y and $g(A_0)$ is a single output point whose input in A_0 is the buying decision. In this learning process the output goal space itself may be redefined (e.g., a new criterion added). Figure 8.3 illustrates the output goal (criterion) space.

EVOLUTION OF THE GOAL TARGET AS A PREFERENCE LEARNING PROCESS

As discussed above from an operational point of view, we need a methodology that enables the decision maker to assess and contract goal target Y in order to obtain a solution leading to a buying decision. This can be achieved by using some recent research on multicriteria decision making. Once an initial goal space $g_1 \ldots g_p$ and an initial target set Y_0 have been defined we can use a utility function $u(y)$ in order to contract Y_0. If we define u^* as a particular level of utility, evolving goal target Y is defined by:

$$Y = \{y | u(y) \geq u^*\}.$$

In the evolution of Y, for a given goal space $g_1 \ldots g_p$, the aspiration level u^* will change (increase), and the utility function $u(y)$ may also change. In addition the goals $g_1 \ldots g_p$ themselves may change. The finally evolved Y (see

Figure 8.3) corresponds to a maximum level of unity

$$U_{\max} = \max_{a \in A_0} u(g(a)).$$

In this section we discuss general elements of the learning process by which the DM defines $u(g(a))$. We then detail these general concepts as they are used in the DSS in Section 8.3.

Usually, assuming a nonlinear additive utility function $u(y) = \Sigma_{i=1}^{p} u_i(y_i)$ is a quite reasonable assumption that we make here. (The nonadditive case is discussed later—see step 2.4 in Section 8.3 and Section 8.6). To assess the parameters of the utility function, it is generally understood (see Fishburn 1967) that one can use either a direct or an indirect procedure:

A *direct procedure* consists in asking the DM to make judgments on each criterion in order to estimate separately each marginal utility $v_i(y_i)$ and then asking for judgments on weights and/or trade-offs in order to assess the relative weights w_i of $v_i(y_i)$ such that $u(y) = \Sigma_i w_i v_i(y_i) = \Sigma_i u_i(y_i)$. When u is normalized, the weight w_i equals the utility $u_i(y_i)$ at the most preferred value of y_i.

An *indirect procedure* consists in asking for wholistic judgments based on several criteria at a time in order to estimate the weights using multiple linear regression or even nonlinear marginal utilities using an ordinal regression method such as UTA (see Jacquet-Lagrèze and Siskos 1982).

If we consider the process of assessing preference as a *learning process* (see Jacquet-Lagrèze 1979, 1982), then a more general methodology consists of using both procedures interactively. A highly analytical decision maker is more likely to use the direct procedure (aggregation of the criteria), and a highly intuitive decision maker is more likely to use the indirect procedure (disaggregation of a wholistic preference). But many decision makers will feel cognitively comfortable in using both procedures involving *aggregation* phases and *disaggregation* phases, see Figure 8.4. Furthermore, using such a procedure interactively enables the decision maker to find some hidden criteria (see

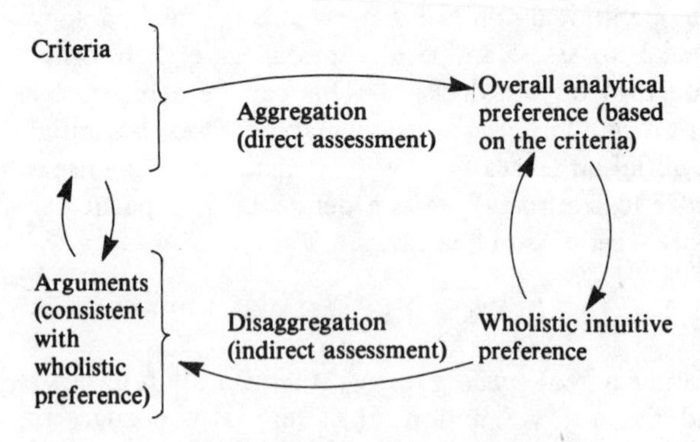

Figure 8.4. **Aggression-disaggregation of Preferences.**

Jacquet-Lagrèze 1982). Thus a wholistic preference order may be inconsistent with a given set of criteria. Revealing such inconsistencies might help the decision maker to find hidden criteria or operational goals y_i that better express underlying values, tastes, or needs of the decision maker. Thus the aggregation-disaggregation procedure is well suited as a learning process in which the DM defines $u(y)$. We shall incorporate it in the DSS in Section 8.3.

In the practical use of the aggregation-disaggregation procedure (see Jacquet-Lagrèze 1982) the DM will express wholistic preference on a subset A_1 of A_0, the set of a priori admissible outputs as defined previously. A_1 generally consists of some alternative products the consumer already knows. The wholistic preference needed as an input to UTA that is used in the DSS (Section 8.3) consists of rank-order $R(A_1)$ of the alternatives in A_1. A_1 should represent a broad sample of A_0 because we want to assess $u(y)$ over the whole set $g(A_0)$. Therefore if necessary the DSS should suggest to the decision maker that he add some alternatives to the set A_1 so that $g(A_1)$ "covers" $g(A_0)$. For a buying decision problem, A_1 might contain 5 to 15 alternatives, and A_0 might contain 10 to 100 or even more alternatives. When A_0 is small (say 5), then A_1 could be identical with A_0.

8.3 The Decision Support System

The DSS has been designed considering on one hand the main features of a semistructured decision problem as outlined in Section 8.1 and on the other hand the methodological framework presented in Section 8.2. The DSS uses a data base containing the set A of all available products, their objective characteristics (size, country of manufacture, etc.), and objective criteria including expert judgments considered "objective" by the users (price, gasoline consumption, seating comfort [an expert judgment]). Preferences on the objective criteria are either nondecreasing or nonincreasing. In addition, subjective criteria are used as explained below. The general decision process model (Figure 8.5) shows how to use the DSS.

The decision process is not linear; the DSS is designed to allow the DM to come back to earlier phases in the course of the learning process. Phase 1 is concerned with the definition of a set of criteria, constraint levels (a priori goal target information) defined on these criteria and the selection of the a priori admissible set of products A_0. Usually $A_0 \subseteq A$, where A is the initial set included in the data base. This is because each particular decision maker is obviously not interested in the whole set. Phase 2 enables those decision makers who wish to use their intuitive wholistic preference to do so, i.e., people who are intuitive can give an initial rank order preference to the alternatives in $A_1 \subseteq A_0$. They can then look for consistency of their wholistic preference with an analytical model of their preference as computed by UTA. The highly analytical DM may not want to use this wholistic preference and may wish to go directly to Phase 3 in order to make a direct assessment of his preferences. Therefore the DSS includes as a particular case of the usual

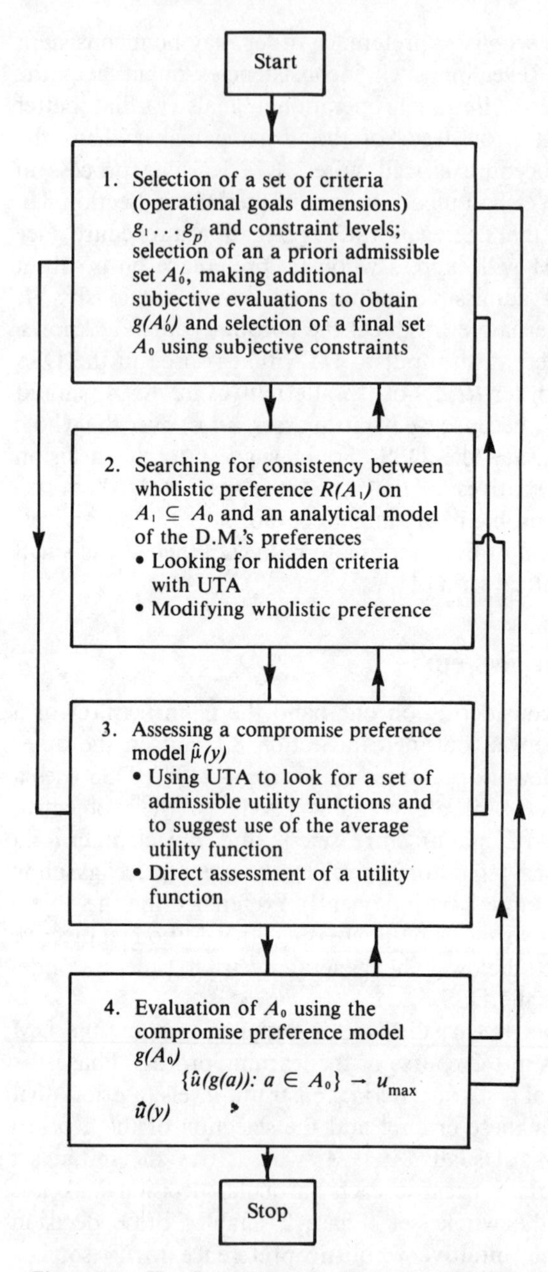

Figure 8.5. **The General Decision Process Model with Phases 1, 2, 3, and 4.**

methodological approach used in decision analysis to assess value functions in the certain case (Keeney and Raifa 1976). Most decision makers will probably use a combined intuitive-analytical, i.e., disaggregation-aggregation, approach and get into Phase 3 after a Phase 2. The utility function computed in Phase 3 by UTA and proposed as a compromise will considerably help direct assessment. The DM should enter Phase 4 only when the rank order of the alternatives in A_1 as computed from the compromise utility function $u(y)$ is consistent with his wholistic ranking $R(A_1)$. In Phase 4 the DM will use $u(y)$ to compute the utility of every alternative product in the admissible set A_0. We note that the ranking and/or utilities computed on A_0 might suggest feedback toward earlier phases.

We now present the four phases in detail.

PHASE 1. SELECTION OF AN ADMISSIBLE SET A_0 OF ALTERNATIVE PRODUCTS

1.1. DSS shows the list of objective criteria and other objective characteristics available in the data base.
DSS indicates the criteria, objective and subjective, usually considered as important.

1.2. DM selects the objective criteria to be used.
DM states a set of constraints:
defines if he wishes a constraint level b_i for each objective criterion:
$y_i \geq b_i$ when preference is a nondecreasing function of y_i (e.g., a constraint level on a performance criterion);
$y_i \leq b_i$ when preference is a nonincreasing function of y_i (e.g., a constraint level on price).
defines if he wishes constraints of objective characteristics (e.g., two-door cars are to be eliminated),
DSS applies the set of constraints to A and DM gets a list A'_0.

1.3. If A'_0 is not considered too small by DM, then go to 1.5.

1.4. DSS suggests some constraints to relax.
DM chooses constraints to be relaxed. Go to 1.2.

1.5. DM defines subjective criteria that use objective characteristics (e.g., classifying the country of manufacture on a scale such as: preferred, acceptable),
DM defines subjective criteria that will require of him a personal evaluation of all alternatives in A'_0. He might use a predefined scale, including a constraint level when the criterion is commonly used (e.g., the shape of a car could be evaluated on a scale as: unacceptable [constraint], ordinary, attractive, outstanding)
DSS applies these new constraints to A'_0 giving A_0, the set of a priori admissible product alternatives.

1.6. If the admissible set A_0 is considered too small by DM, then go to 1.4.

1.7. DSS computes $y_i^* = \max y_i(a)$ and $y_{i*} = \min y_i(a)$:
If $y_{i*} = y_i^*$ such a nondiscriminant criterion is deleted. (In practice more than 10 remaining criteria will be hard to work with.)
If only one criterion is remaining, go to 4.1. DSS uses a nonlinear additive utility function with marginal utilities defined by 1 to 3 linear pieces (piecewise linear) (See Section 8.6.)

1.8. DM may define, if he prefers, another number of linear pieces (1 to 5).

1.9. The assessment of $u(y)$ starts at this stage.
DSS will ask the DM to start by giving a wholistic preference (a ranking of some alternatives he knows). If DM prefers to give a direct assessment of $u(y)$, then go to 3.3.

PHASE 2. SEARCHING FOR CONSISTENCY

2.1. DM selects a subset $A_1 \subseteq A_0$ that he is willing to rank order according to his intuitive wholistic preference.
DSS suggests adding some alternatives to A_1 if necessary. (There must exist at least one alternative of A_1 in each of the piecewise-linear intervals computed in step 1.)

2.2. DM gives a rank-order $R(A_1)$ on A_1 based on his intuitive preference and noting the values $g(A_1)$.

2.3. DSS uses UTA (see Jacquet-Lagrèze and Siskos 1982) to check consistency between $R(A_1)$, and $R^*(A_1)$, the rank-order computed by UTA (optimality step in Jacquet-Lagrèze and Siskos 1982).

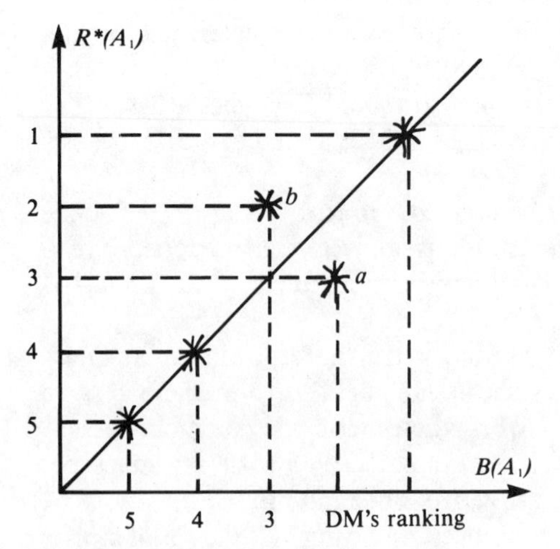

Figure 8.6. **Graphical Consistency Check.**

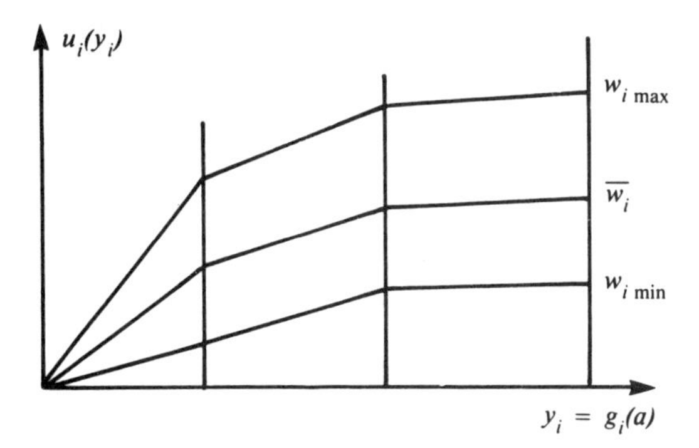

Figure 8.7. **Range of Admissible Utility Functions.**

DSS plots the alternatives A_1 in the ranking diagram where consistency is indicated by a straight line (see Figure 8.6).
If there is consistency, then go to step 3.1.
If there is inconsistency, then:
For an alternative such as "a" DSS searches for a missing criterion on which "a" has a high value and suggests the DM consider adding this criterion.
For an alternative such as "b" DSS searches for a missing criterion on which b has a low value and suggests the DM consider adding this criterion.
These suggestions are one way to move "a" up and "b" down so that $R^*(A_1)$ becomes more equal to $R(A_1)$.

2.4. DM has several options:
DM is willing to change $R(A_1)$ into $R^*(A_1)$: "A" moves left and "b" moves right in Figure 8.6. Go to step 3.1.
DM is willing to change $R(A_1)$ into a new one different from $R^*(A_1)$: Go to 2.2.
DM is not willing to change $R(A_1)$: He should look for missing criteria, adding perhaps one suggested in step 2.3 and then go to step 1.2, or adding a subjective one and go to 1.5. If DM is not willing to make any of the above decisions, he might increase the number of linear pieces on some of the marginal utility functions for better preference modeling and go to step 1.8. Otherwise it means that the numbers of linear pieces will remain insufficient or that the assumption of an additive utility function in DSS is not suitable (in this case see Section 8.6).

PHASE 3. ASSESSING A COMPROMISE PREFERENCE MODEL

3.1. DSS uses UTA with $R(A_1) = R^*(A_1)$ in order to show the DM in Figure 8.7 the range of admissible utility functions (shape of $u_i(y_i)$

with maximum weights $w_{i\,max}$, minimum weights $w_{i\,min}$, average weights $\bar{w}_i$; and the average utility functions $\bar{u}_i(y_i)$ (postoptimality analysis in Jacquet-Lagrèze and Siskos 1982).
DSS suggests an average utility function as a compromise utility function $\bar{u}(y) = \Sigma_i \bar{u}_i(y_i)$.

3.2. If DM thinks that the average utility function $\bar{u}(y)$ reflects his preferences, then go to step 3.6.

3.3. Direct assessment of $u'(y) = \Sigma_i u'_i(y_i)$
If DM comes from step 3.2, he reacts to $\bar{u}(y)$. He can modify the weights $\bar{w}_i \rightarrow w'_i$ (such that $\Sigma_i w'_i = 1$) and/or modify the shape of the marginal utilities $u_i(y_i)$.
If he wishes, DM can ask DSS to compute the values of the trade-off between any pair of criteria for $\bar{u}(y)$ and/or $u'(y)$.
If DM comes from 1.9. or 3.5., he must specify the weights w'_i and the shapes of the marginal utilities $u'_i(y_i)$ (answering questions on trade-offs, indifference points, . . .)

3.4. DSS computes the ranking $R'(A_1)$ using $u'(g)$.
If the DM agrees with $R'(A_1)$ and/or the utility values $u'(g(a))$: $a \in A_1$ then go to step 3.6.

3.5. DM might choose between
modify $u'(g)$ in a direct assessment way: go to step 3.3.
modify the set of criteria $g_1, \ldots, g_p$ and/or constraints: go to step 1.2 or step 1.5.
modify some personal evaluations: go to step 1.5.
modify the ranking $R(A_1)$: go to step 2.2.
modify the number of linear pieces: go to step 1.8.

3.6. Let $\hat{u}(y)$ be the compromise utility function, which is either $\bar{u}(y)$ or $u'(y)$.

PHASE 4. EVALUATIONS OF A_0 USING THE COMPROMISE PREFERENCE MODEL

4.1. Using $g(A_0)$ (from phase 1) and $\hat{u}(y)$ (from phase 3), DSS computes the utilities $\hat{u}(g(a))$: $a \in A_0$ and the rank-order $\hat{R}(A_0)$.

4.2. If DM agrees with $\hat{R}(A_0)$ and/or $\hat{u}(g(a))$: $a \in A_0$, then go to step 4.5.

4.3. If DM nonetheless has enough information to make a decision, then go to step 4.5.

4.4. DM might choose between:
modify $\hat{u}(g)$ in a direct way: go to step 3.3.
add some alternatives of A_0 into A_1 and use the indirect assessment procedure with UTA: go to step 2.1.
modify the set of criteria $g_1, \ldots, g_p$ and/or constraints: go to step 1.2 or step 1.5.
modify some personal evaluations: go to step 1.5.

4.5. DM makes his decision: buy or go and try the best or some of the best products.

STOP.

8.4 An Illustrative Example: The Decision to Purchase a Car

We shall present an illustrative example of the above methodology of the decision to purchase a car. We shall use a small data base involving a set A of 10 alternative cars. We shall follow step by step the decision process model given in Section 8.3.

STEPS 1.1 AND 1.2 The first two steps have been simplified for this example. We here assume that the DM has chosen the following three criteria:

1. g_1: consumption at 120 km/h expressed in liters/100 km
2. g_2: space measured in square meters (length $\times$ width)
3. g_3: price expressed in French francs.

The DM does not specify any constraints, so $A_0' = A$. Of course, in a real application A would contain a few hundred alternative cars and DM would no doubt want to specify some constraints.

STEP 1.3. No: A_0' is a reasonable size (10 cars).

STEP 1.5. DM does not use any subjective criteria for this simplified illustration, so $A_0 = A_0'$.

STEP 1.6. Yes: A_0 is a reasonable size (10 cars).

STEP 1.7. Maximum and minimum values are the following:

1. g_1: 12.95 and 6.75 [1/100 km]
2. g_2: 8.47 and 5.11 [m^2]
3. g_3: 75700 and 24800 [FF]

STEP 1.8. DSS will use three linear pieces for g_1 and g_2, but DM prefers to use four linear pieces for g_3, because price is important to him, and its range is quite large.

STEP 1.9. DM is willing to assess indirectly his utility function by giving a wholistic preference.

STEP 2.1. DM chooses the following subset A_1 = {DYANE, M230, P505, P104}, because he is familiar with these cars. DSS suggests adding BMW or VOLVO in order to have in A_1 a car with a high consumption. DM

Table 8.1 **Wholistic Ranking and Criteria Values for Cars in Set A_1.**

A_l	$R(A_l)$	C_{120} 1 / 100 km	*Space* m^2	*Price* *FF*	*Max. speed km / h (added in Step 2.4.)*
P505	1st	10.01	7.88	49500	173
P104	2nd	8.42	5.11	35200	161
DYANE	3rd	6.75	5.81	24,800	117
BMW	4th	12.26	7.81	68593	182
M230	5th	10.40	8.47	75700	180

chooses BMW. Therefore $A_1 = \{\text{DYANE, M230, P505, P104, BMW}\}$.

STEP 2.2. Considering the criteria values for the car in A_1, DM gives the rank order $R(A_1)$—see Table 8.1.

STEP 2.3. UTA gives the results shown in Figure 8.8 $R^*(A_1) \neq R(A_1)$: The rank orders are not consistent. Note: It is not possible to have a utility function with $u(g(\text{P104})) > u(g(\text{DYANE}))$ because, actually, DYANE dominates P104 on the three criteria (see Table 8.1). Therefore the best UTA could do is to calculate a utility function where these two cars are tied.

DSS suggests adding one of the following criteria, for which P104 is better than DYANE: maximum speed, horse power.

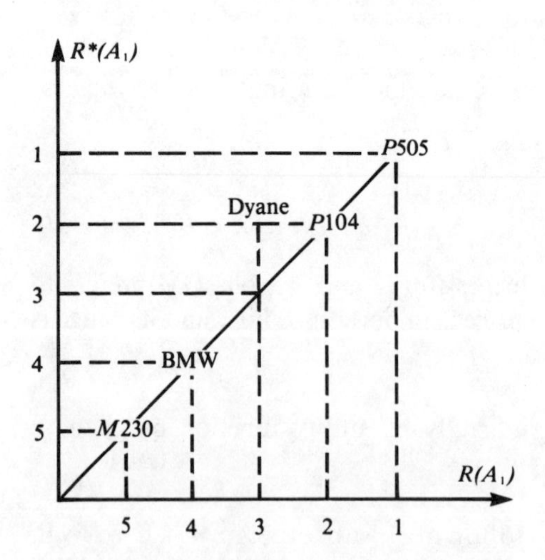

Figure 8.8. **Inconsistency Between $R(A_1)$ and $R^*(A_1)$.**

STEP 2.4. DM selects the third option. He strictly prefers P104 to DYANE and agrees to add a criterion g_4 (maximum speed expressed in km/h) and goes to 1.2.

STEP 1.2. DM is going to work with the four criteria and does not give a constraint level on the new criterion (see Table 8.1).

STEP 1.7. $g_4^* = 182$ km/h and $g_{4*} = 117$ km/h.

STEP 1.9. DM is willing to assess indirectly his utility function.

STEPS 2.1 AND 2.2. DM does not want to modify A_1 or $R(A_1)$.

STEPS 2.3. AND 2.4. UTA yields $R^*(A_1) = R(A_1)$, i.e., there is consistency.

STEP 3.1. The four marginal average utility functions $u_i(g_i)$ suggested as a compromise after the postoptimality analysis are shown by the solid lines on Figure 8.9 (ignore the other lines for the time being).

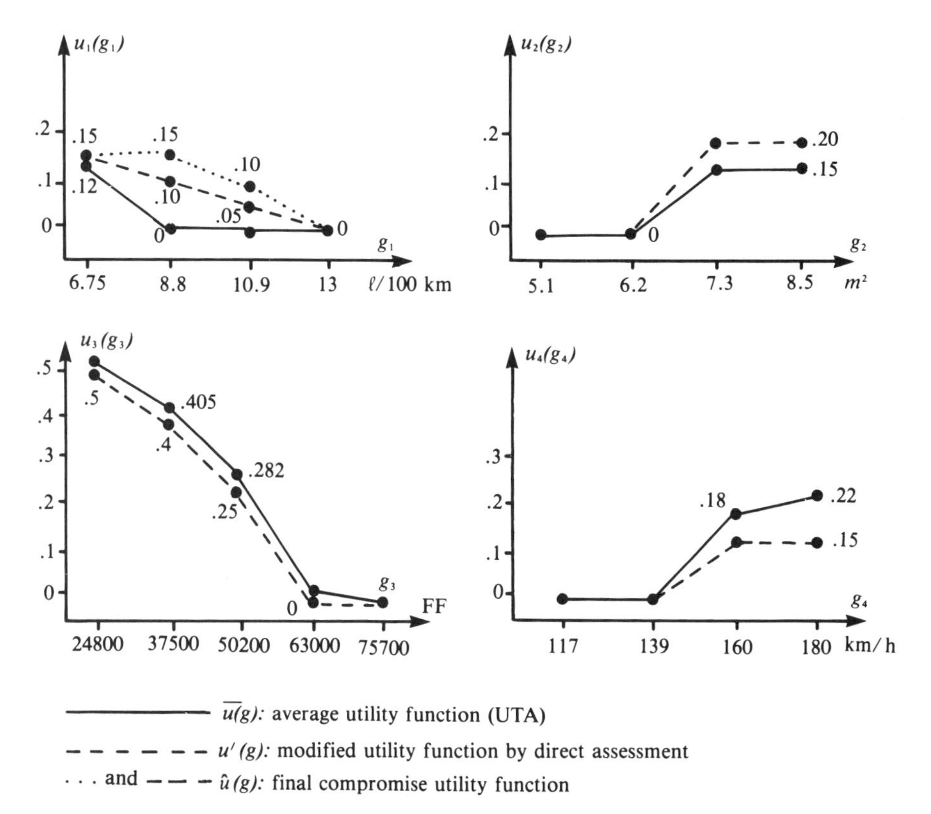

Figure 8.9. **Evolving Additive Utility Functions.**

Table 8.2 **Average Modified, and Final Rank-Orders and Utilities.**

	Average		*Modified*			*Final compromise*	
A_l	$R(A_l)$	$\bar{u}(A_l)$	$R'(A_l)$		$u'(A_l)$	$\hat{R}(A_l)$	$\hat{u}(A_l)$
P505	1	0.650	1	No	0.680	1	0.730
P104	2	0.637	2		0.678	2	0.718
DYANE	3	0.627	3		0.650	3	0.650
BMW	4	0.382	5	Yes	0.367	5	0.383
M230	5	0.371	4		0.412	4	0.462

STEP 3.2. DM does not like the shape of $u_1(y_1)$.

STEP 3.3. DM modifies the shape of $u_1(y_1)$ and changes slightly the weights ($w_1' = 0.15$, $w_2' = 0.2$, $w_3' = 0.5$, $w_4' = 0.15$), giving more importance to the space criterion g_2 to obtain the modified utility function $u'(y)$ shown by the dashed lines in Figure 8.9.

STEP 3.4. $R'(A)$ and $u'(A)$ are shown in Table 8.2. DM agrees with the modified ranking $R'(A_1)$ where the rankings of M230 and BMW are reversed but does not agree with the computed utilities for P505 and P104, which are almost numerically the same.

STEPS 3.5. AND 3.3. Looking at Table 8.1 and Figure 8.9, DM realizes that the impact of changing the shape of utility function $u_1(y_1)$ to $u_1'(g_1)$ has been to penalize relatively the overall utility u' of BMW, which has a high gas consumption, and to narrow the overall utility gap between the P505 and the P104. Because he does not like this narrow gap, DM decides to change the shape of $u_1'(y_1)$ even further, as shown by the dotted line in Figure 8.9 in order to avoid excessively penalizing the P505 on the consumption criterion.

STEP 3.4. This widens the overall computed utility gap between the two cars, as shown by the final compromise utilities $\hat{u}$ (see Table 8.2), with which DM agrees.

STEP 3.6. DM adopts the newly modified utility function as the final compromise $\hat{u}(g)$—the one with dashed lines for g_2, g_3, g_4, and the dotted line for g_1 in Figure 8.9.

STEP 4.1. DSS computes the ranking $\hat{R}(A_0)$ and utilities $\hat{u}(g(a))$: $a \in A_0$ (see Table 8.3).

STEP 4.2. Noting the overall utilities and values, DM agrees with $\hat{R}(A_0)$. He is especially interested in finding out that the OPEL, which he did not know before, ranks first ahead of P505.

Table 8.3 Computed Ranking, Overall Utility, and Criteria Values for Cars in Set A_0.

a	$\hat{R}(A_0)$	$\hat{u}(g(a))$	C_{120}	*Space*	*Price*	*Max. speed*
OPEL	1	0.752	10.48	7.96	46700	176
P505	2	0.730	10.01	7.88	49500	173
P104	3	0.718	8.42	5.11	35200	161
DYANE	4	0.650	6.75	5.81	24800	117
VISA	5	0.616	7.30	5.65	32100	142
GOLF	6	0.576	9.61	6.15	39150	148
M230	7	0.462	10.4	8.47	75700	180
CX	8	0.442	11.05	8.06	64700	178
VOLVO	9	0.401	12.95	8.38	55000	145
BMW	10	0.383	12.26	7.81	68593	182

STEP 4.5. DM would go and try the OPEL.

STOP

NOTE. Suppose in step 4.2 DM had not agreed with $\hat{R}(A_0)$ because there is no nearby service for OPEL. Then from step 4.3 he might have gone directly to step 4.5 and bought the second-ranked P505, a car he already knew. Alternatively, if there had been several unknown cars ranked ahead of the P505, DM might have gone to step 4.4, formally introducing the new "service-availability" criterion and going to step 1.5. Later in the procedure he would probably add OPEL and possibly some other cars to set A_1.

8.5 Concluding Remarks

Some remarks on implementation are in order. We plan to put the DSS on a microcomputer. So far phases 2, 3, and 4 have been programed in BASIC for microcomputers using a simplified version of UTA.* Further work is needed to implement phase 1, notably to establish a large-scale data base for particular products. Our initial product will be automobiles, perhaps followed by microcomputers themselves as a product-buying decision.

We plan at least initially to supplement the computerized support system with a consultant to aid the decision maker in using the system. Later, as we gain more knowledge of user behavior and as users become more familiar with microcomputers, the DSS can become even more conversational, and perhaps the consultant's role could be eliminated. In any case we would need ongoing expertise for designing and updating the data base.

In specializing the general ESD methodology for use in DSS for semi-structured buying decisions, we have treated in this chapter a situation

*A microcomputer program, PREFCALC, is available for IBM-PC, Zenith 100, AJILE (Hyperion). For more information, contact: EURO-DECISION, BP 57 78530 Buc, France, Tel (33-1) 39563705.

involving only one decision maker (buyer). In Chapters 9 and 10 the case of two or more decision makers within the ESD framework is discussed. In this case, for example, each decision maker can use the above DSS to assess his own preferences. Then, depending on the results, decision makers could decide to bargain and/or attempt to integrate their preference structures by exchanging information on their preferences and judgments. Alternatively, decision makers could use the DSS together from the start for joint learning and assessment of a coalition preference structure.

Finally, we note that the ESD framework provides a general methodology for designing specific DSS. The latter, in turn, as operational systems clarify the meaning of the general methodology and can provide suggestions for its evolution.

8.6 Relaxing the Additivity and/or Piecewise Linearity Assumptions

Whenever a marginal utility function $v_i(y_i)$ (normalized such that $v_i(y_{i*}) = 0$ and $v_i(y_i^*) = 1$) is known or assessed through any direct procedure (see Keeney and Raiffa (1976)) we could still use the UTA procedure and estimate indirectly the weights w_i by replacing $g_i(a)$ by $g_i'(a) = v_i(g_i(a))$ using one linear piece for $g_i \ldots$. Thus after step 1.9 the user could go to step 2.1. However, behaviorally speaking, most users would find it easier to proceed with piecewise-linear utilities, as suggested in step 1.7.

This procedure permits us also to use a nonadditive utility function. For example, if $u(y) = \Sigma_i w_i v_i(y_i) + \Sigma_{i,j} w_{ij} v_i(y_i) v_j(y_j)$, the weights w_i and w_{ij} could be indirectly estimated by using 1 linear piece for the new criteria $g_i', \ldots, g_{ij}'(a)$ defined by $g_i'(a) = v_i(g_i(a))$ and $g_{ij}'(a) = v_i(g_i(a)) v_j(g_j(a))$.

REFERENCES

Assael, H. 1981. *Consumer Behavior and Marketing Action*. Kent, Boston.

Bonczek, R. H., Holsapple, C. W., and Whinston, A. G. 1981. *Foundations of Decision Support Systems*. Academic Press, New York.

Chan, S. J., Park C. W., and Yu, P. L. 1980. "High-Stake Decision Making—An Empirical Study based on House Purchase Processes." Working Paper No. 141, School of Business, University of Kansas, Lawrence.

Engell, J. F., Kollat, D. T., and Blackwell, R. D. 1978. *Consumer Behavior*. Dryden Press.

Fishburn, P. 1967. "Methods for Estimating Additive Utilities." *Management Science*, 13.

Jacquet-Lagrèze, E. 1979. De la logique d'agrégation de critères à une logique d'agrégation—désagregation de préférences et de jugements. *Cahiers ISMEA 13* (4, 5, 6), pp. 839–859.

______. 1982. "A Behavioral Model of the Decision Process and an Application for Designing a Decision Support System." LAMSADE, University of Paris-Dauphine and New York University.

______, and Siskos, J. 1982. "Assessing a Set of Additive Utility Functions for Multicriteria Decision-Making—The UTA Method." *European Journal of Operational Research*, 10, No. 2.

Janis, J. L., and Mann, L. 1977. *Decision-Making—A Psychological Analysis of Conflict, Choice and Commitment*. Free Press, New York.

Keen, P. G. W., and Scott Morton, M. S. 1978. *Decision Support Systems: An Organizational Perspective*. Addison-Wesley, Reading, MA.

Keeney, R. L., and Raiffa, M. 1976. *Decisions with Multiple Objectives: Preferences and Value Trade-Offs*. Wiley, New York.

Shakun, M. F. 1981a. "Formalizing Conflict Resolution in Policy Making." *International Journal of General Systems*, 7, No. 3.

Shakun, M. F. 1981b. "Policy Making and Meaning as Design of Purposeful Systems." *International Journal of General Systems*, 7, No. 4.

Shakun, M. F., and Sudit, E. F. 1982. "Effectiveness, Productivity and Design of Purposeful Systems: The Profit-Making Case." New York University.

Sprague, R. M., Jr., and Carlson, E. D. 1982. *Building Effective Decision Support Systems*. Prentice-Hall, Englewood Cliffs, NJ.

CHAPTER 9

Decision Support Systems for Negotiations*

Evolutionary systems design (ESD) is a methodology for problem definition and solution in complex contexts involving multiparticipant, multicriteria, ill-structured, dynamic problems and provides a basis for decision support systems (DSS) (see Shakun 1981a, 1981b, and Chapters 1 and 2). Using the ESD framework Jacquet-Lagreze and Shakun (1984), and Chapter 8, develop a DSS for multicriteria decision making (MCDM) involving one decision maker, which they apply to semistructured buying decisions illustrated by car buying. This chapter treats the case of two or more decision makers (group) for which DSS for negotiations—negotiation support systems (NSS)—are the central focus. Here, too, car buying serves as a concrete application.

Under ESD the negotiation decision process for a group (coalition) involves searching for a single point intersection (solution) between two sets: (1) a group target (what the group wants) and (2) a group technologically feasible set (what the group can get). In negotiating, these sets may change, i.e., expand or contract until a single-point intersection is found. In general there is a mapping from a current set to a new set, by which the current set is redefined. These ideas are discussed in Section 9.2 following the one–decision maker case presented in Section 9.1.

*An earlier version of this chapter appeared in the *Proceedings of the 1985 IEEE International Conference on Systems, Man and Cybernetics*, Tucson, Arizona, November 12–15, 1985.

A basic idea in decision support systems is to show the decision problem —graphically or as relational data in matrix form—in three spaces, as a mapping from control space to goal space to utility space. (For example, in car buying controls are available cars; goals are gasoline consumption, space, price, etc.; utilities are measures of group members' satisfaction with goals delivered by various cars.) The decision (negotiation) process is represented correspondingly in each of these spaces as the search for a single-point intersection between the group target and the group technologically feasible set. Changes in these sets in any one space are shown by the DSS as corresponding changes in sets in the other two spaces. Section 9.2 also develops these ideas.

The DSS can also suggest various axioms and negotiation concession-making models as aids to changing sets. These are discussed in Section 9.3. Section 9.4 presents concluding remarks.

9.1 The Case of One Decision Maker

A DSS for MCDM involving one decision maker and applied to car buying is discussed in detail in Jacquet-Lagreze and Shakun (1984), and in Chapter 8. Consider a set A of strategies (controls, inputs, decisions, choices, actions). In the car-buying decision, A is the set of available cars representable by positive integers in R^1, car space. Let g be a function from A to R^p, the p-dimensional real vector space, which characterizes outcomes (goals, outputs, consequences, characteristics, criteria). In case of cars the criteria include price, gas consumption, space, etc. Then $y = g(a)$ for $a \in A$ is a vector of R^p representing the outputs of a particular input choice, a; $g(A)$ is the set of possible outputs representing technologically feasible performance. These outputs are generally constrained a priori by preliminary goal target Y_o information. For example, these constraints could be $Y_o = \{y \in R^p\colon y_i \geq b_i,\ i = 1, \ldots, p\}$. The intersection of Y_o and $g(A)$ is called $g(A_o)$, a set of a priori admissible outputs; $A_o = \{a \in A\colon g(a) \in Y_o\}$ is the corresponding set of a priori admissible inputs.

In addition to the admissible sets of cars, A_o, and goals, $g(A_o)$, we have a preference structure defined on $g(A_o)$. Here we assume a utility function $u(y)$ that is nonlinear and additive:

$$u(y) = \sum_{i=1}^{p} u_i(y_i) \tag{1}$$

With the UTA utility assessment procedure (Jacquet-Lagreze and Siskos 1982) implemented in the microcomputer program, PREFCALC (Jacquet-Lagreze 1985), the marginal functions $u_i(y_i)$ are taken as piecewise linear and nondecreasing or nonincreasing. Based on UTA, a disaggregation-aggregation learning process involving both wholistic and analytical judgments is implemented. Working with a small sample $A_1 \subseteq A_o$ as a reference set, a decision support system (Jacquet-Lagreze and Shakun 1984, and Chapter 8) can aid a decision

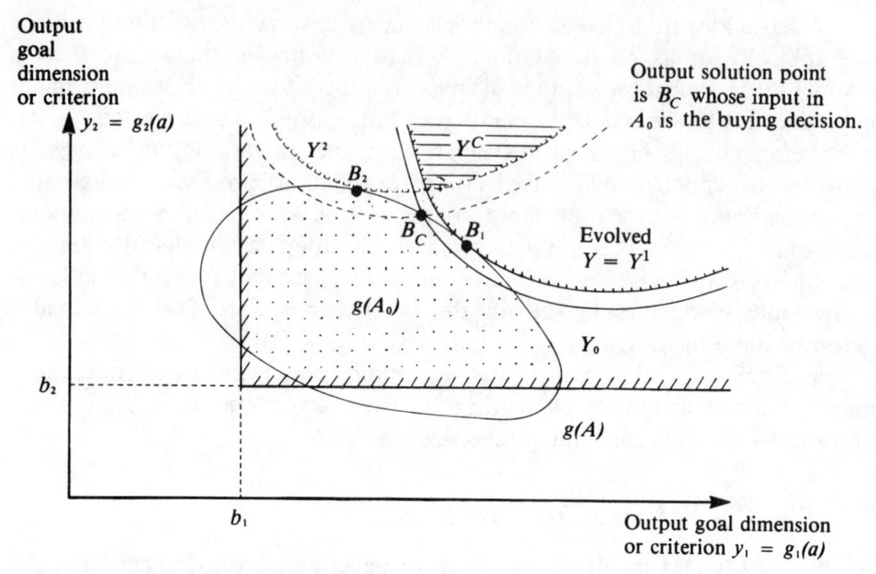

Figure 9.1. **Output, Goal or Criteria Space.**

maker in defining his utility function (1). Applying the utility function to the set of cars A_o results in a ranking of cars according to their numerical utilities. The car with the maximum utility is the buying decision. Figure 9.1 shows the criteria space.

The technologically feasible set $g(A)$ intersects the a priori goal target Y_o to give the a priori admissible set $g(A_o)$ (where $g(A_o) = g(A) \cap Y_o$), which typically in car buying has many points. In order to find a single-point solution in criteria space the intersection set $g(A_o)$ must be reduced in size. This can be done by contracting either Y_o or $g(A)$. Because for cars the latter set is fixed at a particular time, we contract Y_o by using the user's utility function. By maximizing utility, we contract the target Y_o to evolved goal target Y, which has a single-point intersection (solution) with $g(A_o)$ at point B_1 whose preimage in A_o is the car-buying decision. (Ignore dotted curves, B_2, B_o, Y^2, and Y^C in Figure 9.1 for the moment.)

9.2 Group Decision Making: Negotiations

Assume each decision maker (player) in a group called coalition C has worked individually with the single-user DSS procedure outlined in Section 9.1. If the same car does not have the highest utility for all players, then there is a conflict. Referring to Figure 9.1, with two players (e.g., husband and wife), if B_1 is player 1's output (highest utility) solution and B_2 is player 2's, there is a conflict. Note geometrically that Y^C, the coalition (group) goal target—the intersection of the goal targets Y^1 and Y^2 for players 1 and 2, respectively, i.e., $Y^C = Y^1 \cap Y^2$—has an empty intersection with $g(A_o^C)$, the group admissible output set. In Figure 9.1, for simplicity $g(A_o^C) = \cap g(A_o^j)$ for players $j = 1, 2$ is simply shown as $g(A_o)$. If group goal target Y^C expands, e.g., by expansion through negotiations of goal targets Y^1 and Y^2, there could be a solution at output point B_C, the intersection between expanded Y^C and $g(A_o^C)$.

The discussion of Figure 9.1 shows that the search process for a solution involves contracting or expanding sets. "Expansion" (contraction) means that some new (old) points are added (dropped) to (from) a set; this expansion (contraction) does not preclude dropping (adding) some other points from (to) the set. Thus expansion/contraction involves a mapping from an original (current) set to a new set. For a group C two sets are subject to expansion/contraction mapping. They are (1) $g(A^C) = g(A)$, the group technologically feasible output set or, more precisely, the admissible set $g(A_o^C) = g(A) \cap Y_o^C$ where $Y_o^C = \cap Y_o^j$, and (2) the group goal target $Y^C = \cap Y^j$. In searching for a solution, i.e., searching for a single-point intersection between $g(A_o^C)$ and Y^C, note the following:

1. For the group goal target Y^C, higher utility aspirations (or goal demands) by players contract the target; lower utility aspirations (expressed in concession making) expand the target. Goal target expansion/contraction involves negotiations.
2. For the group admissible technologically feasible set $g(A_o^C)$, axioms can contract the feasible set and new technology can expand it. For example, with nondecreasing (or nonincreasing) marginal utility functions, the Pareto optimality axiom for utilities (Harsanyi 1977; Luce and Raiffa 1957; Owen 1982) constrains (contracts) the feasible goal set to the upper right boundary in Figure 9.1 when searching for solutions. New technological cars on the market can expand the feasible set. In other words feasible set contraction can employ solution concepts involving specification of axioms imposing agreed-upon properties on the solution; expansion can involve withdrawal of axioms previously specified or creation of new technological inputs.

The previous search focusing on goal space is paralleled in car space and utility space because of the mapping from car space to goal space to utility space (via the marginal utility functions). Figure 9.2 shows utility space for two players corresponding to the goal space of Figure 9.1.

Consider the group utility target $U^C = \cap U^j$, where U^j is player j's utility target. In arriving at a solution at point $P_C = (u_1(B_C), u_2(B_C))$, the group utility target—initially U^C (initial) based on individual player use of the single-user DSS—has expanded to U^C (final) intersecting the feasible set at P_C. (Ignore other items on Figure 9.2 for the moment.) The progress of negotiations, here concession making in goal and utility spaces and corresponding concessions in car space, can be shown by a DSS either graphically, as in Figures 9.1 and 9.2, or as relational data in matrix form, as in Table 9.1.

In Table 9.1, car $a \in A_o^C = \cap A_o^j$, the group joint set of a priori admissible cars, is specified by name. The goals are: $y_1 = C120$ is the gasoline

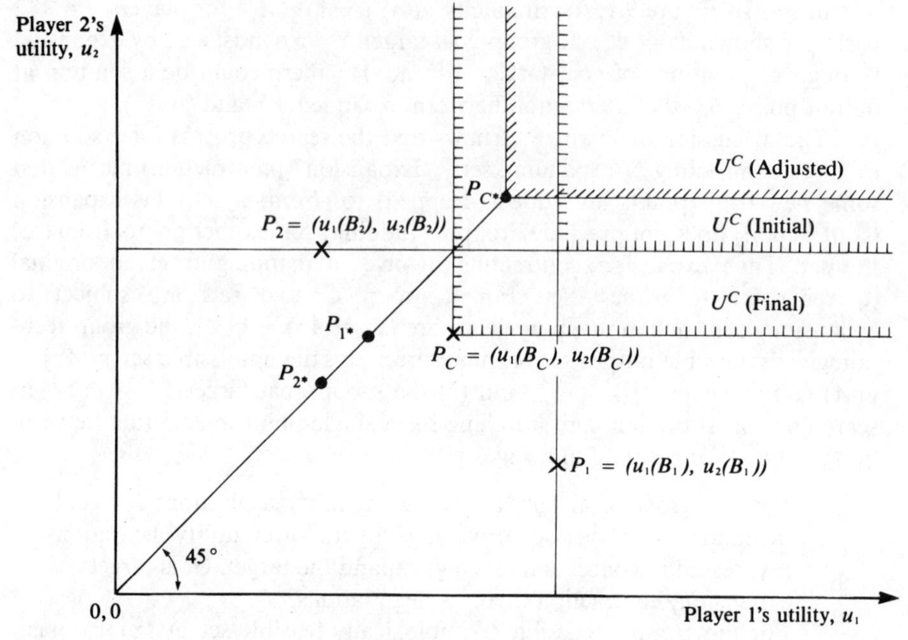

Figure 9.2. **Utility Space for Two Players.**

consumption, liters/100 km, at 120 km/hr; y_2 = space is in square meters; y_3 = price is in French francs; y_4 = maximum speed is in km/hr. Utilities u_1 and u_2 are the utilities of players 1 and 2, respectively. For exchanging information the DSS could display the larger set $a \in \cup A_o^j$, which includes cars a priori admissible to at least one player. In this case a car inadmissible for player j would be listed as "inadmissible" in the utility column u_j, but it conceivably could become admissible in the course of negotiations.

Table 9.1 **Relational Data Corresponding to Figures 9.1 and 9.2**

Car *a*	*Goal* $y_1 = C_{120}$	*Goal* y_2 = *space*	*Goal* y_3 = *price*	*Goal* y_4 = *max. speed*	*Utility* u_1	*Utility* u_2
Opel	10.48	7.96	46700	176	.752	.383
Peugeot 505	10.01	7.88	49500	173	.730	.401
Peugeot 104	8.42	5.11	35200	161	.718	.442
Dyane	6.75	5.81	24800	117	.650	.462
Visa	7.30	5.65	32100	142	.616	.576
Golf	9.61	6.15	39150	148	.576	.616
Mercedes 230	10.40	8.47	75700	180	.462	.650
CX	11.05	8.06	64700	178	.442	.718
Volvo	12.95	8.38	55000	145	.401	.730
BMW	12.26	7.81	68593	182	.383	.752

Thus Table 9.1 shows a set of ten cars and their corresponding goal and utility values for two players. The utility values u_1 for player 1 are taken from Jacquet-Lagreze and Shakun (1984) based on use of the single-user DSS. For illustration the utility values u_2 for player 2 are listed in reverse order of those for player 1. In row 1 of Table 9.1 we see the goal point $B_1 = (10.48, 7.96, 46700, 176)$ of Figure 9.1, and utility point $P_1 = (.752, .383)$ of Figure 9.2, corresponding to player 1's first car choice, Opel. Similarly in row 10 of Table 9.1 we see $B_2 = (12.26, 7.81, 68593, 182)$, $P_2 = (.383, .752)$, corresponding to player 2's first choice, BMW. Thus to begin with, player 1's feasible target is defined by row 1, similarly row 10 for player 2. Concession making involves players adding more rows to their respective targets, thereby expanding them. Given the symmetry of the situation, the solution is likely to be Visa with $P_C = (.616, .576)$, Golf with $P_C = (.576, .616)$, or a random choice between them.

In addition to the above displays, the DSS can show graphically the marginal utility functions. For output goal y_i, $u_{ij}(y_i)$ gives the marginal utility function of player j. If for a particular i the DSS shows both $u_{i1}(y_i)$ and $u_{i2}(y_i)$ on the same graphical axes, then the two players can compare, exchange information (perhaps leading toward consensus), and negotiate on their marginal utility functions. The marginal utilities u_{ij} can also be included in the relational data of Table 9.1 by inserting columns u_{i1} and u_{i2} for $i = 1, 2, 3, 4$, i.e., eight columns of the u_{ij} inserted, say, between the y_4 and u_1 columns. The DSS could display the projection of the relational data of Table 9.1 onto goal y_i, u_{i1}, and u_{i2} to enable the players to compare their marginal utility values for a particular goal y_i.

If players change their marginal utility functions so that they approach one another, the feasible set in utility space approaches a positive-sloping 45° line whose highest utility point is the solution, P_{C*} (Figure 9.2), thus achieving consensus. Of course, U^C (initial) is readily adjusted to U^C (adjusted) to give a single-point intersection at P_{C*}. In other words, in utility space, Figure 9.2, there is a function F: $P_C \to P_{C*}$, $P_1 \to P_{1*}$, $P_2 \to P_{2*}$ mapping the original feasible set to points along the dotted straight line with solution at point P_{C*}. U^C (adjusted) is also shown following the mapping: U^C (initial) $\to$ U^C (adjusted). The arrival at a common coalition utility function (through exchange of information and negotiation until players' marginal utility functions are identical) means in goal space, Figure 9.1, that individual players' goal targets Y^1 and Y^2 have become the same. In other words, although not drawn on Figure 9.1, now $Y^1 = Y^2 = Y^C$, the coalition goal target that intersects $g(A_o^C)$ at a solution point B_{C*} whose preimage in car space is the car-buying decision.

In addition to exchanging information and negotiating to expand targets, players can consider the use of axioms to contract the feasible region, e.g., (1) to a single-solution point in utility space—in Figure 9.2, Nash axioms (Harsanyi 1977; Luce and Raiffa 1957; Owen 1982) might give solution point P_C, which is accommodated by the mapping: U^C (initial) $\to$ U^C (final), or (2) to a constrained set of points (e.g., the Pareto optimal set might be $\{P_1, P_C, P_2\}$ in Figure 9.2). The latter could be followed by compromise (concessions) to select a single point from this set, e.g., P_C, or perhaps consensus leading to P_{C*} might be realized.

The preceding discussion has been in terms of two players who together form a group, coalition C. More generally, we let $\eta = \{j\}$, where $j = 1, 2, \ldots, N$ be the set of all players and $C \subseteq \eta$ be a subset of players who can form a coalition C. In general a group (coalition) DSS is intended to aid a coalition C in decision making with the possibility that players not in coalition C can form a coalition $\bar{C}$ (see Shakun 1981a). Here for car buying we assume $C = \eta$, the grand coalition, i.e., the group consists of all N players.

For group negotiations a group p-dimensional goal space is defined to include all the goal dimensions (criteria) of interest to any decision maker in the group. Of course any particular decision maker can place a weight of zero on any criterion if it is of no importance to him. Under this group procedure a criterion i could be introduced that a player j has previously deleted from his utility function because it was nondiscriminatory, i.e., for player j, $y_{i*} = y_i^*$ where

$$y_{i*} = \min_{a \in A_o} y_i(a) \text{ and } y_i^* = \max_{a \in A_o} y_i(a)$$

If the equality $y_{i*} = y_i^*$ is a result of player j's a priori goal fixing y_i at a certain value (e.g., admissible cars must have exactly four doors), then unless other players include this y_i value in their admissible outputs, no agreement is possible. Of course player j can also change his mind. In other words, for agreement, players' admissible outputs with respect to y_i must have a non-empty intersection. As long as player j has $y_{i*} = y_i^*$, then criterion i is not included in computing his utility function, although it is a criterion dimension in the group p-dimensional goal space and is included in the utility function of other players.

9.3 Negotiations: Axioms and Concession Making

As discussed in Section 9.2, negotiations in group decision making involve expansion/contraction mapping of targets and feasible sets to obtain corresponding single-point solutions (intersections) in car, goal, and utility spaces. As illustrated, the expansion/contraction process can be aided by graphical or relational–data matrix displays of these spaces and players' marginal utility functions to facilitate consensus seeking and compromise.

For a given goal space we noted in Section 9.2 that the expansion/contraction process involves negotiations and axioms. Use of axioms may in fact be considered part of negotiations in that players may negotiate the choice of axioms or in any case must accept them if they are to be used. From this viewpoint expansion/contraction is a negotiation process in which axioms can play a part. Finally we note that this discussion pertains to a given goal space.

However in the mapping from a current set to a new set the goal space can be redefined while using the DSS (Jacquet-Lagreze and Shakun 1984). For use of a goals/values referral process in negotiations to redefine the goal space, see Shakun (1981a) and Chapter 2. Sometimes a better agreement resulting in higher utilities to all players may result when the goal space is redefined. The Pareto optimal frontier may be pushed further out when the goal space is redefined so as to take advantage of asymmetries of player interest (Barclay and Peterson 1976).

NEGOTIATION AXIOMS

Focusing on the group DSS for cars, players may choose from some well-known negotiation axioms to contract this feasible set of cars. As discussed in the game theory literature (e.g., see Harsanyi 1977; Heckathorn 1980; Luce and Raiffa 1975; Owen 1982; Sen 1970), these axioms include (1) individual rationality, (2) Pareto optimality (joint rationality or efficiency), (3) independence of irrelevant alternatives, (4) independence of linear transformations, (5) symmetry, and (6) monotonicity. Such axioms contract the feasible set and in some cases, if sufficient compatible ones are used, give a single joint solution.

Use of axiom 2, Pareto optimality, contracts the feasible region to the upper right boundary in utility space, the Pareto optional frontier. Axioms 1 and 2 together define a set (called the negotiation set) of utility payoff vectors $u = (u_1, \ldots, u_j, \ldots, u_N)$ where u_j is the payoff to player j; u is called an imputation. Additional axioms can further restrict the set of imputations to yield the core and stable sets. The Shapley axioms (Owen 1982) yield the Shapley value, a single-point solution, which gives a payoff to player j equal to the average payoff contribution he makes to all the coalitions he can join. The Shapley solution is always an imputation.

Axioms 1 through 5 give the single-point Nash solution if the utility feasible set is convex; if mixed strategies are allowed, convexity is assured (see Owen 1982). For two players the Nash solution is the utility point (u_1, u_2), which maximizes the product $(u_1 - u_{1CON})(u_2 - u_{2CON})$ over the utility admissible feasible set $S = u[g(A_o^C)]$, where $(u_{CON} = u_{1CON}, u_{2CON})$ is the conflict payoff that players receive if they cannot agree on a solution. Felsenthal and Diskin (1982) argue that for $j = 1, 2$ the minimum utility $u_{MIN}^s = (u_{jMIN}^s)$ —where $u_{jMIN}^s = Min(u_j)$ for $u = (u_j) \in s$ and $u \geq u_{CON}$ where s is the Pareto optional boundary of S—should be used instead of u_{CON}. They argue that because $u_{MIN}^s \geq u_{CON}$, at worst it is rational for a player to take u_{MIN}^s rather than u_{CON} so that u_{MIN}^s should be used as a reference point in the Nash product, i.e., $max(u_1 - u_{1MIN}^s)(u_2 - u_{2MIN}^s)$. In general for N players the Nash product contains N terms rather than only two.

Axioms 1, 2, 4, 5, 6 give the Kalai and Smorodinsky (1975) solution. For a two-person game we define an ideal utility vector $u_{IDEAL} = (u_{jMAX})$ for $j = 1, 2$ where $u_{jMAX} = max(u_j)$ and $u = (u_j) \in S$. We also specify the conflict payoff $u_{CON} = (u_{jCON})$. Then the Kalai-Smorodinsky solution $\bar{u} = (\bar{u}_j)$ is

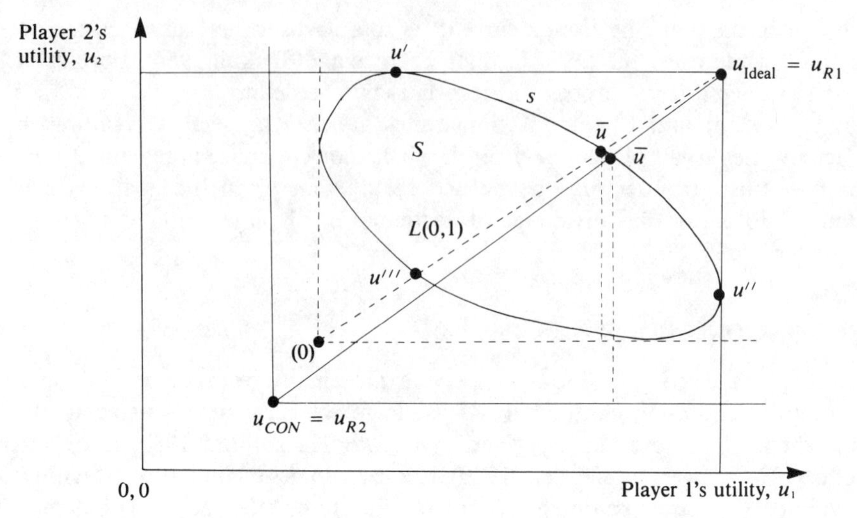

Figure 9.3. **Kalai-Smorodinsky Solution for Two-Person Game and General Solution for *N*-Person Game.**

the intersection of the line joining U_{CON} to U_{IDEAL}, and the Pareto optimal boundary, s of the feasible set, S. Thus the solution is given by:

$$\frac{\bar{u}_2 - u_{2CON}}{\bar{u}_1 - u_{1CON}} = \frac{u_{2IDEAL} - u_{2CON}}{u_{1IDEAL} - u_{1CON}} \tag{2}$$

with the geometry shown in Figure 9.3. In other words, measured relative to the conflict point, the Kalai-Smorodinsky solution (2) gives utility gains to the two players proportional to the maximum gains they could ideally get, i.e., equal percent of maximum gain. Felsenthal and Diskin (1982) replace u_{CON} by u_{MIN}.

Kalai (1977) assumes axioms 1, 2, 3, 5, and 6 in an N-person game and derives a proportional solution involving interpersonal utility comparisons (note that axiom 4 is not included). Figure 9.3 is generalized to N players with solution $\bar{u}$ again the intersection of the line joining u_{CON} to some reference point, u_{R1}, and the Pareto optimal boundary of the feasible set. The proportion of utility gains among the N players is not determined by the Kalai model, i.e., u_{R1} is not fixed by the model. It could be chosen equal to u_{IDEAL}. However, because interpersonal utility comparisons are involved, it may be natural to normalize the players' utilities over a 0 to 1 interval. Then $u_{IDEAL} = (1)$, the unit vector, and $u_{CON} = (0)$, the zero vector.

Interpreting Figure 9.3 more generally, we suggest a single-point general solution $\bar{u}$ to the N-person game as the intersection of the line joining two reference points, u_{R1} and u_{R2}, with the Pareto optional boundary of the feasible set. Figure 9.3 shows $u_{R1} = u_{IDEAL}$, $u_{R2} = u_{CON}$. Alternatively, one reference point can be specified, plus a condition that serves to specify the second point; in the Nash solution $u_{R2} = u_{CON}$ is specified plus the condition of maximizing the Nash product.

Suppose we normalize utilities, choosing $u_{R1} = u_{IDEAL} = (u_{jMAX}) = (1)$ and $u_{R2} = u_{MIN} = (u_{jMIN}) = (0)$ where $u_{jMIN} = min(u_j)$ for $u = (u_j) \in S$. Then, thinking of cars as an example, we define a coalition C utility function, u_C as follows:

$$\begin{aligned} u_C &= u_C(u_1(y), \ldots, u_N(y)) = u_C(u_1[g(a)], \ldots, u_N[g(a)]) \\ &= \underset{u_j|a}{min}\left(u_j[g(a)]\right) \quad \text{for } j = 1, 2, \ldots, N;\ a \in A_o \end{aligned} \tag{3}$$

The coalition maximizes u_C by choosing car a:

$$\underset{a \in A_o}{max}\, u_C = \underset{a \in A_o}{max}\ \underset{u_j|a}{min}\left(u_j[g(a)]\right) \tag{4}$$

We note that (3) and (4) are defined over discrete sets, which may be convexified through the use of mixed strategies. Thus a maximin axiom involving interpersonal utility comparisons based on 0–1 normalization is used to choose a solution. Rawls (1971) uses such a maximin approach in his theory of justice (see also the discussion in Sen 1970). Kalai (1977) notes that this maximin solution is a proportional solution. This may be shown by reference to Figure 9.3 using the dotted coordinate axes whose origin is (0), the zero vector.

With respect to the dotted coordinate axes, because we have now chosen $u_{R1} = (1)$ and $u_{R2} = (0)$, points $u = (u_j)$ on dotted line $L(0, 1)$ joining them have equal values of u_j for $j = 1, 2, \ldots, N$. Thus the proportional solution $\bar{u} = (\bar{u}_j)$—the intersection of $L(0, 1)$ and s—has equal $\bar{u}_j$, i.e., gives equal utility to all N players. That the maximin solution is also at $\bar{u}$ is outlined as follows: Consider the subset $S_j^* \subset S$ over which a particular player j's utility $u_j = u_j^*$ is the minimum in (3). By symmetry, the line $L(0, 1)$ having points with equal u_j is on the boundary of S_j^* for all j. $L(0, 1)$ is the intersection of hyperplanes, which together with the Pareto optimal surface s are involved in defining the boundaries of the subsets S_j^*. Let $s_j^* \subset s$ be the Pareto optimal frontier of S_j^*. Because s_j^* dominates all interior points in S_j^*, the maximum of the minimum utilities u_j^* over S_j^* must lie on s_j^* and in fact must be $\bar{u}$ because $\bar{u}_j \geq u_j^*$ for all $u_j^* \in s_j^*$. Because the same argument applies to all S_j^*, $\bar{u}$ is the maximin solution over S. Note that when $N = 2$ in Figure 9.3, s_1^* is the Pareto boundary from u' to $\bar{u}$ and s_2^* from $\bar{u}$ to u''. The line $L(0, 1)$ is the boundary partitioning S into S_1^* (defined by boundary $u'\bar{u}u'''$) and S_2^* (defined by $\bar{u}u''u'''$). $Max\, u_1^* = \bar{u}_1$ and $max\, u_2^* = \bar{u}_2$; so $\bar{u}$ is the maximin solution over S.

We note that for the data of Table 9.1 the maximin solution is Visa or Golf (or a random choice between them).

Harsanyi (1977) discusses the question of interpersonal utility comparisons. Distinguishing between arbitration models and bargaining models, he argues that arbitration models, which define "fair" or "moral" solutions based on ethics and welfare economics, necessarily must use interpersonal comparisons of utility. He argues that bargaining models and game theory in general must not use interpersonal utility comparisons—solutions must be independent of linear transformations of any player's utility function (axiom 4). Harsanyi feels the principal mechanism of conflict resolution in society is bargaining, for which game theory is the appropriate modeling approach. Game theory requires that moral preference always be incorporated into each player's utility (payoff) function rather than being used in the formal definition of the solution. Each player then maximizes his own payoff. Under the game theory approach interpersonal comparison of utility is not allowed. However, Harsanyi acknowledges that in the real world solutions themselves are influenced by moral considerations that do use interpersonal comparisons of utility. Hence we feel that in a practical DSS players should be free to draw on both types of axioms—those permitting and those not permitting interpersonal utility comparisons.

If interpersonal utility comparisons are allowed, Harsanyi advocates a social welfare function that is the arithmetic mean of individual utilities. (A social welfare function is a function from individual player preference structures, R_j for $j = 1, 2, \ldots, N$—which, for example, can be an ordering or a utility function—to a group preference structure, R_C, i.e., $R_C = f(R_1, \ldots, R_j, \ldots, R_N)$. (See Luce and Raiffa 1957.) Thus an average utility axiom would define coalition C's utility function u_C as:

$$u_C = (1/N)\sum_j u_j[g(a)] \quad j = 1, 2, \ldots, N;\ a \in A_o \tag{5}$$

and a would be chosen to maximize this u_C. For related discussions of social welfare functions see Keeney and Kirkwood (1975) and Keeney (1976). The average utility axiom expressed by equation (5) does not deal directly with distributional issues among players, whereas the maximin axiom expressed by equation (4) does. However all social choice axioms appear to raise difficulties in some social situations (Sen 1970).

As noted, after discussing interpersonal utility comparisons, Harsanyi (1977) comes down on the side of game theory models that are independent of these. For classical games he obtains single-point solutions related to the Nash-Zeuthen solution, to the modified Shapley value, and to their generalizations. This is based on the principle of mutually expected rationality, i.e., if a player follows certain rationality postulates, and is a rational player, he must expect and act on the expectation that other rational players will do likewise. Harsanyi shows that this restricts the subjective probabilities a rational player can entertain about another's behavior and leads to the Zeuthen principle, which says that at any given stage of bargaining between two rational players the next concession must always come from the player less willing to risk a conflict as measured by the highest probability of conflict a player would be willing to face rather than accept the terms his opponent offers. Zeuthen's principle in turn leads to the Nash solution.

However in real negotiations (see discussion in Fogelman-Soulie, Munier, and Shakun 1983, and Chapter 7) the key principle of mutually expected rationality may apply only imperfectly because real players have cognitive limits or bounds to their rationality. Consequently, subjective probability may be far less restricted than mutual expected rationality indicates, resulting in Zeuthen's principle not in fact being realized. Further, Harsanyi's general discussion is restricted to games with complete information where players know fully their own and other players' utility functions. Fogelman-Soulie, Munier, and Shakun (1983) and Chapter 7 propose an alternative modeling approach to bypass these limitations. It uses subjective probabilities to guide concession making and does not require use of rationality postulates (although in principle some axioms that partially restrict subjective probabilities can be used).

Other axiomatic solution concepts that could be available in a DSS are discussed in the literature (e.g., see Owen 1982; Shenoy 1980). Finally, the question of axioms remains open. Axioms are part of negotiations, and where agreed upon by players they assist negotiations by contracting the feasible set. As such they should be available to players in a group DSS.

NEGOTIATIONS: CONCESSION MAKING

In addition to axioms, negotiations (Figure 9.1) can involve concession making by players to expand their goal targets Y^j and consequently to expand coalition C's goal target Y^C until it has a single-point intersection with the admissible set $g(A_o^C)$. Concession making may be viewed as a problem of stochastic terminal control (Fogelman-Soulie, Munier, and Shakun [1983] in which players make concessions sequentially. Using the model, a player who must now decide on a concession (which includes no concession) estimates the negotiation trajectory in goal space associated with his making a given current goal concession followed by assumed probabilistic concessions by other players. For the latter the estimating player uses his subjective probabilities. Each goal concession considered when projected by the trajectory gives a terminal payoff probability distribution in goal space. The player chooses a preferred terminal payoff distribution and thus goal concession (hence corresponding car concession). Visual (gestalt) and multicriteria methods (including UTA and others, e.g., ELECTRE—see Goicoechea, Hansen, and Duckstein 1982; Keeney and Raiffa 1976; Zeleny 1982)—can be used to choose. When UTA or other multicriteria utility methods are used to choose a preferred terminal payoff distribution, utility is assessed on risky terminal goal outputs using such criteria as means and standard deviations. The riskiness of terminal goal outputs associated with a current goal concession is due to the riskiness of negotiations as expressed by the subjective probabilities.

Fogelman-Soulie, Munier, and Shakun (1983) develop the two-player case in detail for bivariate negotiations. Their approach is not limited to preference structures using a utility function; nor does it use the expected utility hypothesis based on decision theory axioms (Luce and Raiffa 1957) under which a concession is chosen by maximizing expected utility. The approach of Fogelman-Soulie, Munier, and Shakun (1983), and Chapter 7, may be used where maximizing expected utility is inapplicable. For example, if an outcome's utility depends on the probabilities of other outcomes, we cannot normally assess a multicriteria utility function on the goal valuables (say by UTA) for maximizing expected utility. Where maximizing expected utility applies and a player can assess his multicriteria utility function (say by UTA), maximizing expected utility can be used to choose a concession (see Rao and Shakun 1974, and Chapter 6).

Thus at any stage of negotiations a player chooses a goal concession (here in effect a control variable) to direct the negotiation trajectory toward a preferred terminal payoff. A DSS can help the player do this. The Rao and Shakun (1974) and Fogelman-Soulie, Munier, and Shakun (1983) approaches provide dynamic programing models for this purpose. Finally, this discussion pertains to a given goal space that can also be redefined (Fogelman-Soulie, Munier, and Shakun 1983; Jacquet-Lagreze and Shakun 1984; Shakun 1981a).

The preceding discussion focused on concessions in goal space that are reflected in control (car) and utility spaces. Let us now focus on concessions (expansions of targets) in control (car) space that are reflected in expansions of goal and utility targets. We consider control (car) target expansion by stages. In the first stage a_{1j} is the car target of player j, a member of coalition C. Naturally, a_{1j} would be the car preimage of the highest utility, u_j for player j from the single-user DSS. If the a_{1j} are the same car for all j (i.e., the intersection $A_1^C = \bigcap_j a_{1j}$ is nonempty), then this car is the group choice. If not, each player j adds a second car a_{2j} so that his car target is the set (a_{1j}, a_{2j}). If the intersection over j, i.e., $A_2^C = \bigcap_j (a_{1j}, a_{2j})$ is nonempty, then the group has at least one and at most two cars as solutions. If there is more than one solution, another criterion, e.g., maximin over these solutions, can be used to choose one car. If the intersection A_2^C is empty, then consider $A_3^C = \bigcap_j (a_{1j}, a_{2j}, a_{3j})$, then, if necessary, $A_4^C = \bigcap_j (a_{1j}, a_{2j}, a_{3j}, a_{4j})$, etc., until a nonempty intersection is found. If this nonempty intersection first occurs for some set A_Δ^C $(\Delta = 1, 2, 3, \ldots)$, then the group has at least one and at most *min* (Δ, N) cars as solutions. A_Δ^C is the coalition car target at stage Δ where $A_\Delta^C = \bigcap_j A_\Delta^j$ and player j's car target is $A_\Delta^j = (a_{ij}, a_{2j}, \ldots, a_{\Delta j})$. The A_Δ^C coalition car target expansion process in car (control) space maps into an expansion of the coalition goal target Y^C in goal space and utility target U^C in utility space. These expansions continue until A_Δ^C, Y^C, and U^C have nonempty intersections with the feasible sets in the respective spaces. In control space, if the coalition car target A_Δ^C is nonempty, it is by definition feasible because it consists of feasible cars.

The following question arises: In what order should player j add cars (he need not communicate utilities) to his expanding car target, $A_\Delta^j = (a_{1j}, a_{2j}, \ldots a_{\Delta j})$? Because the group solution is found when for the first time $A_\Delta^C = \bigcap_j A_\Delta^j$ is nonempty, in order to get his highest utility car consistent with the A_Δ^C solution process, it is in player j's interest to add cars in order of his true descending utilities (not to manipulate) if he thinks he does not know anything about the other players' rank orders or if players know each other's rank orders. Otherwise player j may be tempted—for his own benefit in getting a group car decision that is higher up on his own true list—to manipulate the procedure by using false rank orders. (Of course, player j may be nonmanipulative in principle, especially if he thinks other players are also nonmanipulative.) Thus the procedure could be troubled, as are other solution approaches by players reporting false or imcomplete preference information. (Nurmi [1984] discusses strategic manipulation of five group decision methods; Barclay and Peterson [1976] and Yager [1980] discuss the effect on the Nash solution of players reporting false utilities.) Player j can refuse to make a concession at any stage by not adding a car, i.e., by choosing $a_{\Delta j} = 0$. He can also make a concession on his part conditional on other players making concessions. If all players adopt this conditional principle, then at any stage Δ either they all make concessions or they all do not. If all players persist in no concession, then negotiations under this process will break off.

For the data of Table 9.1, this process of conditional car target expansion with no manipulation results in solutions Visa and Golf at the $\Delta = 6$th expansion.

As another example suppose players' utilities for a set of nine cars give the following true rank orders (descending utility):

Player 1	Player 2
Opel	Visa
Peugeot 505	Dyane
Golf	Peugeot 104
Peugeot 304	Peugeot 304
Dyane	Peugeot 505
Peugeot 104	Golf
Visa	Opel
Car 1	Car 1
Car 2	Car 2

Under conditional car target expansion with no manipulation and strong ordering, the solution is Peugeot 304 at the $\Delta = 4$th expansion. The Peugeot 304 is Pareto optimal. In general, with strong ordering, solutions under car target expansion with no manipulation are Pareto optimal, which we now show.

For a car a' to be an interior point—not on the Pareto optimal boundary in utility space—there must be another car a'' that is Pareto optimal whose utility is higher for at least one player and not lower for the others. With strong ordering the Pareto optimal point a'' has utility higher than a' for all players. Golf, Peugeot 104, car 1, and car 2 are interior points, and all other cars are on the Pareto optimal boundary. An interior point a' cannot be a solution because by definition there is another car a'' whose utility is higher for both (all) players, and under car target expansion with no manipulation this car a'' will show up before a' in the expanding targets of both (all) players and therefore in their intersection. Hence, an interior point a' cannot be a solution, i.e., with strong ordering solutions under car target expansion with no manipulation are Pareto optimal.

With weak ordering of cars and no manipulation, in expanding his car target a player adds a *group* of cars at any stage when he is indifferent among them, and other players must add an equal number of cars to match the largest number added because of indifference by any player j. In this case the first nonempty intersection A_Δ^C will contain at least one Pareto optimal car.

If player 1 knows player 2's rank order but not vice versa, then player 1 may be tempted to manipulate, e.g., as follows:

Player 1's manipulative rank order	Player 2's true rank order
Car 1	Visa
Car 2	Dyane
Opel	Peugeot 104
Peugeot 505	Peugeot 304
Golf	Peugeot 505
Peugeot 304	Golf
Dyane	Opel
Peugeot 104	Car 1
Visa	Car 2

Under conditional car target expansion with manipulation by player 1, the solution (at the $\Delta = 5$th expansion) is Peugeot 505, which, according to player 1's true rank order above, he prefers to Peugeot 304, the solution with no manipulation. This is at the expense of player 2, who prefers Peugeot 304 to Peugeot 505.

If the set of car alternatives has only seven cars—assume cars 1 and 2 are not in the set—then even if player 1 wants to manipulate, there is no manipulative rank order for player 1 that gives a solution he prefers to Peugeot 304, the nonmanipulative solution.

Returning to the nine-car case, under conditional car target expansion with manipulation by player 1, the latter could declare ("concede") his true first choice Opel at stage 1 followed by a manipulative order, i.e., car 1, car 2, Peugeot 505, Golf; and the manipulated solution would still be the same, Peugeot 505. This manipulation, the addition by player 1 to his target of car 1 and car 2 in false rank order—cars he knows are at the bottom, below Opel, of player 2's true rank order—in effect amounts to making no concession to player 2. It is equivalent under unconditional car target expansion—where a player j can refuse to make a concession at any stage Δ by not adding a car to his target (i.e., $a_{\Delta j} = 0$)—to player 1 choosing $a_{\Delta 1} = 0$ for $\Delta = 2, 3$ after a first-stage input of $a_{11} =$ Opel. What under conditional car target expansion is manipulation on the part of player 1 is simply a bargaining choice not to concede at stage $\Delta = 2, 3$ under unconditional expansion. Thus the question of manipulation does not come up there.

9.4 Concluding Remarks

A basic idea in decision support systems is to show the decision problem—graphically or as relational data in matrix form—in three spaces as a mapping from control space to goal space (and through marginal utility functions) to utility space. Within each of these spaces the solution process is characterized by adaptive change, i.e., expansion/contraction of sets. In general there is a mapping from a current set to a new set by which the current set is redefined. In each space two sets are subject to expansion/contraction: (1) a coalition (group) target and (2) a coalition (group) feasible set. For a solution (decision) a single-point (or single-set—see Section 1.3) intersection between these two sets is required.

With one decision maker being a group of one, this description applies to a DSS for one decision maker (Chapter 8) or for a group (this chapter). For the latter, decision support for negotiations—negotiation support systems (NSS)—is the central focus. Negotiations are identified with the expansion/contraction (mapping) process previously noted involving consensus seeking and compromise. The use of axioms to contract the feasible set has been discussed in detail. For expanding targets several negotiation concession-making models have been suggested.

Whether the current negotiation (expansion/contraction or mapping) is focused in control (car) space, goal space, or utility space, the DSS enables showing the process—including players' marginal utility functions—graphically or as relational data in all three spaces at user command, thereby facilitating user understanding. The DSS can also suggest various axioms and negotiation concession-making models as requested by users. A prototype of the single-user DSS for car buying (Chapter 8) has been implemented on the IBM microcomputer. Computer implementation of the group DSS for negotiations (NSS) is discussed by Jarke, Jelassi, and Shakun in Chapter 10. For an application of group DSS to design and negotiation of new products see Chapters 11 and 12.

This chapter uses a utility function u_j (assessed by UTA) for each individual player j's preference structure; the u_j then are used in making a group decision. Other individual preference structures could be considered, which then could be used in group choice—e.g., see Heidel and Duckstein (1983) for ELECTRE, Hall and Haimes (1976) and Haimes (1979) for the surrogate worth trade-off (SWT) method, Fraser and Hipel (1979, 1984) for the conflict analysis (improved metagame analysis) technique. For a discussion of some problems of strategic manipulation in group decision making, see Nurmi (1984).

Finally, we note that Faure, Le Dong, and Shakun (1988) include relational (socioemotional) aspects of negotiations as well as substance (task) in the control/goal/preference problem representation.

REFERENCES

Barclay, S., and Peterson, C. R. 1976. "Multiattribute Utility Models for Negotiations." Technical Report 76-1. Decisions and Designs, Inc., McLean, VA.

Fandel, G., and Gal, T. (eds.). 1979. *Multiple Criteria Decision Making Theory and Application*. Springer-Verlag, Berlin.

Faure, G. O., Le Dong, V., and Shakun, M. F. 1988. "Social-Emotional Aspects of Negotiations." Graduate School of Business Administration, New York University, New York.

Felsenthal, D. S., and Diskin, A. 1982. "The Bargaining Problem Revisited." *Journal of Conflict Resolution*, 26, No. 4, pp. 664–691.

Fogelman-Soulie, F., Munier, B., and Shakun, M. F. 1983. "Bivariate Negotiations as a Problem of Stochastic Terminal Control." *Management Science*, 29, No. 7, pp. 840–855.

Fraser, N. M., and Hipel, K. W. 1979. "Solving Complex Conflicts." *IEEE Transactions on Systems, Man and Cybernetics*, SMC-9, No. 12, pp. 805–816.

______. 1984. *Conflict Analysis: Models and Resolutions*. North Holland, New York.

Goicoechea, A., Hansen, D. R., and Duckstein, L. 1982. *Multiobjective Decision Analysis with Engineering and Business Applications*. Wiley, New York.

Haimes, Y. Y. 1979. "The Surrogate Worth Trade-Off (SWT) Method and Its Extensions." In Fandel, G., and Gal, T. (eds.). *Multiple Criteria Decision Making Theory and Application*. Springer-Verlag, Berlin.

Hall, W. G., and Haimes, Y. Y. 1976. "The Surrogate Worth Trade-Off Method with Multiple Decision-Makers." In Zeleny, M. (ed.). *Multiple Criteria Decision Making, Kyoto 1975*. Springer-Verlag, New York.

Harsanyi, J. C. 1977. *Rational Behavior and Bargaining Equilibrium in Games and Social Situations*. Cambridge University Press, Cambridge.

Heckathorn, D. 1980. "A Unified Model for Bargaining and Conflict." *Behavioral Science*, 25, pp. 261–285.

Heidel, K. J., and Duckstein, L. 1983. "Extension of ELECTRE Technique to Group Decision Making: An Application to Fuel Emergency Control." Department of Systems and Industrial Engineering, University of Arizona, Tucson.

Jacquet-Lagreze, E. 1985. *PREFCALC, Version 2.0*. EURO-DECISION, BP 57, 78530 Buc, France.

______, and Shakun, M. F. 1984. "Decision Support Systems for Semi-Structured Buying Decisions." *European Journal of Operational Research*, 16, No. 1, pp. 48–58.

______, and Siskos, J. 1982. "Assessing a Set of Additive Utility Functions for Multiple Criteria Decision Making—The UTA Method." *European Journal of Operational Research*, 10, No. 2, pp. 151–164.

Kalai, E. 1977. "Proportional Solutions to Bargaining Situations: Interpersonal Utility Comparisons." *Econometrica*, 45, No. 7, pp. 1623–1630.

______, and Smorodinsky, M. 1975. "Other Solutions to Nash's Bargaining Problem." *Econometrica*, 43, No. 3, pp. 513–518.

Keeney, R. L. 1976. "A Group Preference Axiomatization with Cardinal Utility." *Management Science*, 23, No. 2, pp. 140–145.

______, and Kirkwood, C. W. 1975. "Group Decision Making Using Cardinal Social Welfare Functions." *Management Science*, 22, No. 4, pp. 430–437.

______, and Raiffa, H. 1976. *Decisions with Multiple Objectives: Preferences and Value Tradeoffs*. Wiley, New York.

Luce, R. D., and Raiffa, H. 1957. *Games and Decisions*. Wiley, New York.

Nurmi, H. 1984. "On the Strategic Properties of Some Modern Methods of Group Decision Making." *Behavioral Science*, 29, pp. 248–257.

Owen, G. 1982. *Game Theory*, 2nd ed. Academic Press, New York.

Rao, A. G., and Shakun, M. F. 1974. "A Normative Model for Negotiations." *Management Science*, 20, No. 10, pp. 1364–1375.

Rawls, J. 1971. *A Theory of Justice*. Harvard University Press, Cambridge, MA.

Sen, A. K. 1970. *Collective Choice and Social Welfare*. Holden-Day, San Francisco.

Shakun, M. F. 1981a. "Formalizing Conflict Resolution in Policy Making." *International Journal of General Systems*, 7, No. 3, pp. 207–215.

______. 1981b. "Policy Making and Meaning as Design of Purposeful Systems." *International Journal of General Systems*, 7, No. 4, pp. 235–251.

Shenoy, P. 1980. "On Committee Decision Making: A Game Theoretical Approach." *Management Science*, 26, No. 4, pp. 387–399.

Yager, R. R. 1980. "On the Cheating Problem in Nash's Bargaining Model." *Journal of Information and Optimization Sciences*, 1, No. 3, pp. 281–290.

Zeleny, M. (ed.). 1976. *Multiple Criteria Decision Making, Kyoto 1975*. Springer-Verlag, New York.

______. 1982. *Multiple Criteria Decision Making*. McGraw-Hill, New York.

CHAPTER 10

MEDIATOR: Toward a Negotiation Support System

Matthias Jarke, M. Tawfik Jelassi, and Melvin F. Shakun*

10.1 Introduction

Negotiation support systems (NSS)—computer-assisted negotiations—provide decision support in problems involving multiple decision makers, thus extending decision support systems (DSS)—e.g., see Keen and Scott-Morton (1978), Sprague and Carlson (1982), Bonczek, Holsapple, and Whinston (1981)—where the initial emphasis has been on single decision maker situations.

Shakun (1981a, 1981b, 1988) develops evolutionary systems design (ESD) as a methodology for problem definition and solution (design) in complex contexts involving multiplayer, multicriteria, ill-structured, dynamic problems. In general, in NSS we are interested in similar problems. In particular, in this chapter ESD is used as a basis for MEDIATOR, a system designed to support negotiations in a setting that we now describe in overview form and develop in detail in the following sections.

NEGOTIATION SETTING OVERVIEW

A group of *N* players is involved in negotiations. A human mediator supports these negotiations, and he in turn is supported by the negotiation support system, MEDIATOR. The (human) mediator supports negotiations by assist-

*All three authors are affiliated with the Graduate School of Business Administration, New York University.

ing the players in a process of consensus seeking within which compromise is possible. Using MEDIATOR, the mediator aids in consensus seeking by helping the players build a common (group) joint problem representation of the negotiations. The negotiation problem representation is shown by MEDIATOR—graphically or as relational data in matrix form—in three spaces as mappings from control space to goal space (and through marginal utility functions) to preference (here utility) space. (In some cases involving risk a fourth space, criteria space, can be used between goal space and preference space—see Giordano, Jacquet-Lagreze, and Shakun [1985] and Shakun [1988]). These spaces can be redefined while using MEDIATOR. For use of a goals/values referral process to redefine goal space see Shakun (1981a, 1988).

At each stage of the negotiations the common joint problem representation shows the acknowledged degree of consensus (or conflict) among the players, i.e., at each stage players may show different individual problem representations. The evolution of problem representation can be described as a process of consensus seeking—through sharing of views, which constitutes exchange of information—within which compromise is possible. The mediator can support compromise through use of axiomatic solution concepts and/or concession-making procedures in the MEDIATOR model base. Computer display of the evolving problem representation can be used to support continued consensus seeking. In each space (control, goal, and preference) the negotiation process represents adaptive change, i.e., mappings of group target and feasible sets in seeking a solution—a single-point intersection between them. For further methodological discussion see Shakun (1985, 1988) and Giordano, Jacquet-Lagreze, and Shakun (1985).

In the basic scenario as described we think of the mediator as supporting the negotiations and in turn being supported by MEDIATOR, but not himself deciding on them. However, MEDIATOR should also be useful in compulsory arbitration where the mediator decides (chooses) the solution. In some contexts the mediator can be a group leader, e.g., the president of a company, who finally makes a decision supported by MEDIATOR. In other contexts MEDIATOR could support the players directly without the use of a human mediator. Here we work with the basic scenario as noted.

DATA BASE–CENTERED DSS DESIGN OVERVIEW

A number of DSS design strategies have been proposed, including those that start from the decision models used, from the user interfaces required, or from a task analysis. In organizations in which decisions are based on large amounts of existing data, it seems more natural to follow a data base–centered approach. This method embeds the decision models and user interfaces of a DSS in a data base management environment that provides them with data, stores their execution sequences, and retains their results. A data base approach to DSS was first proposed by Donovan (1976) for single-user DSS and later extended by Jarke (1981) and others to cover not only the data management but also the model management and multiuser aspects of DSS.

In the negotiation support setting discussed in this chapter the data base is also used as a communication center among the mediator and the players. Besides providing the initial data underlying the problem to be solved, the data base management system (DBMS) also manages the evolving group joint problem representation. Furthermore it provides a large number of tools for generating this joint problem representation and protecting it against unauthorized or erroneous access.

In the following sections we develop this negotiation support system concept in detail. In Section 10.2 we summarize the single decision maker case as background for the group negotiation problem discussed in Section 10.3, based on Shakun (1985). In Section 10.4 we illustrate the use of MEDIATOR by an application to group car buying. The data base–centered system architecture for MEDIATOR is developed in Section 10.5. Section 10.6 presents concluding remarks.

10.2 The Case of One Decision Maker

A DSS for multicriteria decision making (MCDM) involving one decision maker and applied to car buying is discussed in detail in Jacquet-Lagreze and Shakun (1984). Consider a set A of strategies (controls, inputs, decisions, choices, actions). In the car-buying decision, A is the set of available cars representable by positive integers in R^1, car space. Let g be a function from A to R^p, the p-dimensional real vector space that characterizes outcomes (goals, outputs, consequences, characteristics, criteria). In the case of cars the criteria include price, gas consumption, space, etc. Then $y = g(a)$ for $a \in A$ is a vector of R^p representing the output of a particular input choice, a; $g(A)$ is the set of possible outputs representing technologically feasible performance. These outputs are generally constrained a priori by preliminary goal target Y_o information. For example, these constraints could be $Y_o = \{y \in R^p \colon y_i \geq b_i,$ $i = 1, \ldots, p\}$. The intersection of Y_o and $g(A)$ is called $g(A_o)$, a set of a priori admissible outputs. $A_o = \{a \in A \colon g(a) \in Y_o\}$ is the corresponding set of a priori admissible inputs.

In addition to the admissible sets of cars, A_o, and goals, $g(A_o)$, we have a preference structure defined on $g(A_o)$. Here we assume a utility function $u(y)$ that is nonlinear and additive:

$$u(y) = \sum_{i=1}^{p} u_i(y_i) \tag{1}$$

With the UTA utility assessment procedure (Jacquet-Lagreze and Siskos 1982) implemented in the microcomputer program, PREFCALC (Jacquet-Lagreze 1985), the marginal functions $u_i(y_i)$ are taken as piecewise linear and nondecreasing or nonincreasing. Based on UTA, a disaggregation-aggregation learning process involving both wholistic and analytical judgments is imple-

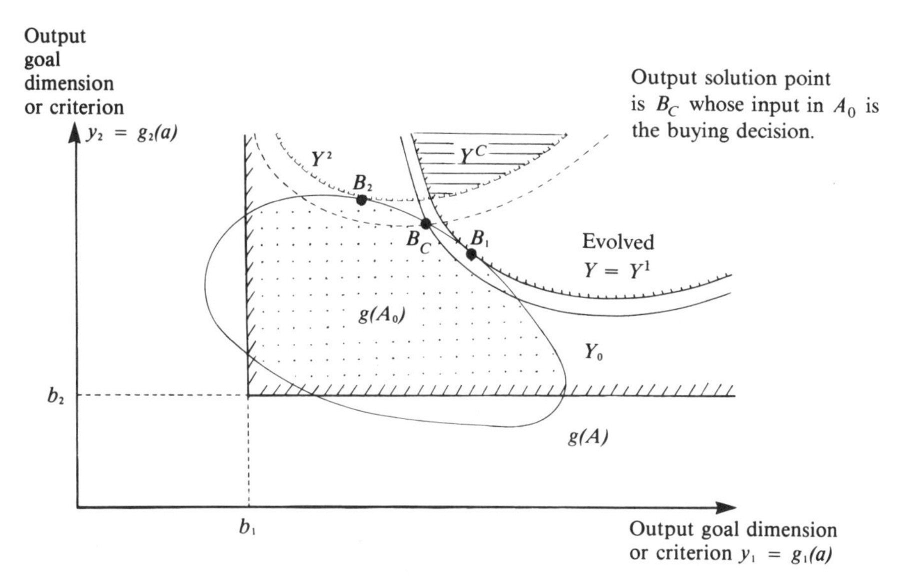

Figure 10.1. **Output, Goal or Criteria Space**

mented. Working with a small sample $A_1 \subseteq A_o$, a decision support system (Jacquet-Lagreze and Shakun 1984) can help a decision maker define his utility function (1). Applying the utility function to the set of cars A_o results in a ranking of cars according to their numerical utilities. The car with the maximum utility is the buying decision. Figure 10.1 shows the criteria space.

The technologically feasible set $g(A)$ intersects the a priori goal target Y_o to give the a priori admissible set $g(A_o) = g(A) \cap Y_o$, which in car buying typically has many points. In order to find a single-point solution in criteria space, the intersection set $g(A_o)$ must be reduced in size. This can be done by contracting either Y_o or $g(A)$. Because for cars the latter set is fixed at a particular time, we contract Y_o by using the user's utility function. By maximizing utility the target Y_o contracts to evolved goal target Y, which has a single-point intersection (solution) with $g(A_o)$ at point B_1, whose preimage in A_o is the car-buying decision. (Ignore dotted curves, B_2, B_C, Y^2, and Y^C in Figure 10.1 for the moment.)

10.3 Group Decision Making: Negotiations

Assume that each decision maker (player) in a group called coalition C has worked individually with the single-user DSS procedure outlined in Section 10.2. If the same car does not have the highest utility for all players, there is a conflict. Referring to Figure 10.1, with two players (e.g., husband and wife), if B_1 is player 1's output (highest utility) solution and B_2 is player 2's, there is a conflict. Note geometrically that Y^C, the coalition (group) goal target—the intersection of the goal targets Y^1 and Y^2 for players 1 and 2, respectively, i.e.,

$Y^C = Y^1 \cap Y^2$—has an empty intersection with $g(A_o^C)$, the group admissible output set. In Figure 10.1, for simplicity $g(A_o^C) = \cap g(A_o^j)$ for players $j = 1, 2$ is simply shown as $g(A_o)$. If group goal target Y^C expands, e.g., by expansion through negotiations of goal targets Y^1 and Y^2, there could be a solution at output point B_C, the intersection between expanded Y^C and $g(A_o^C)$.

Our discussion of Figure 10.1 clarifies that the search process for a solution involves contracting or expanding sets. By expansion (contraction) we mean that some new (old) points are added (dropped) to (from) a set; this expansion (contraction) does not preclude dropping (adding) some other points from (to) the set. Thus expansion/contraction involves a mapping from an original (current) set to a new set. For a group C two sets are subject to expansion/contraction mapping: (1) $g(A^C) = g(A)$, the group technologically feasible output set or more precisely the admissible set $g(A_o^C) = g(A) \cap Y_o^C$ where $Y_o^C = \cap Y_o^j$, and (2) the group goal target $Y^C = \cap Y^j$. In searching for a solution, i.e., searching for a single-point intersection between $g(A_o^C)$ and Y^C, we note the following:

1. For the group goal target Y^C, higher utility aspirations (or goal demands) by players contract the target; lower utility aspirations (expressed in concession making) expand the target. Goal target expansion/contraction involves negotiations.
2. For the group admissible technologically feasible set $g(A_o^C)$, axioms can contract the feasible set, and new technology can expand it. For example, with nondecreasing (or nonincreasing) marginal utility functions, the Pareto optimality axiom for utilities (Harsanyi 1977; Luce and Raiffa 1957; Owen 1982) constrains (contracts) the feasible goal set to the upper right boundary in Figure 10.1 when searching for solutions. New technology cars on the market can expand the feasible set. In other words feasible set contraction can employ solution concepts involving specification of axioms imposing agreed-upon properties on the solution; expansion can involve withdrawal of axioms previously specified or creation of new technological inputs.

The previous search focusing on goal space is paralleled in car space and utility space because of the mapping from car space to goal space to utility space (via the marginal utility functions). Figure 10.2 shows utility space for two players corresponding to the goal space of Figure 10.1.

Consider the group utility target $U^C = \cap U^j$ where U^j is player j's utility target. In arriving at a solution at point $P_C = (u_1(B_C), u_2(B_C))$, the group utility target—initially U^C (initial) based on individual player use of the single-user DSS—has expanded to U^C (final), intersecting the feasible set at P_C. (Ignore other items on Figure 10.2 for the moment.) The progress of negotiations, here concession making in goal and utility spaces and corresponding concessions in car space, can be shown by a DSS either graphically, as in Figures 10.1 and 10.2, or as relational data in matrix form, as in Table 10.1.

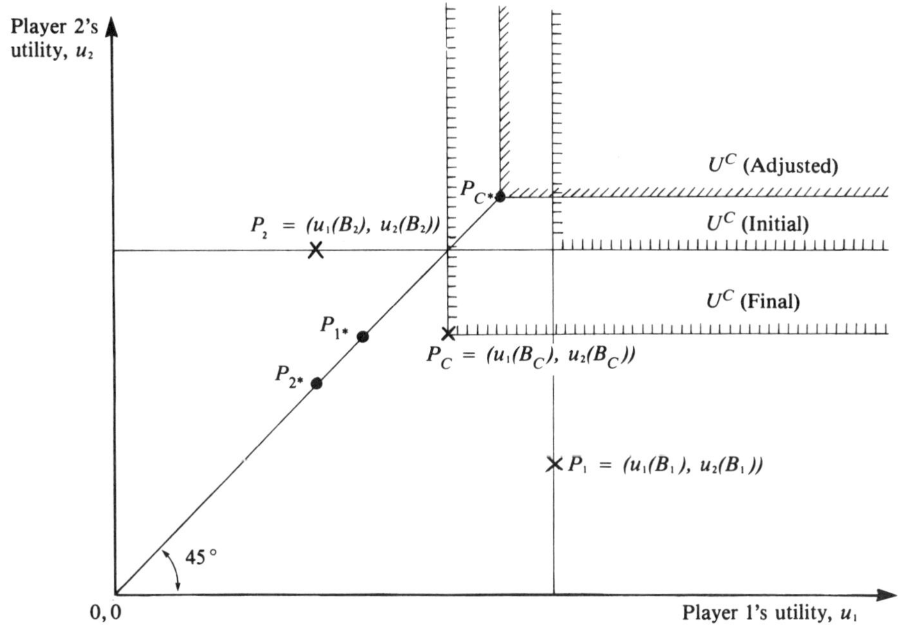

Figure 10.2. **Utility Space for Two Players**

Table 10.1 **Group Relational Data (Reverse Utilities) Corresponding to Figures 10.1 and 10.2**

Control		*Goal*	*Goal*	*Goal*	*Goal*	*Initial*	
Make	*Model*	y_1	y_2	y_3	y_4	*utilities*	
		C120	*Space*	*Price*	*Speed*	u_1	u_2
Opel	Record	10.48	7.96	46700	176	.75	.38
Peugeot	505	10.01	7.88	49500	173	.73	.40
Peugeot	104	8.42	5.11	35200	161	.72	.44
Citroen	Dyane	6.75	5.81	24800	117	.65	.46
Citroen	Visa	7.30	5.65	32100	142	.62	.58
VW	Golf	9.61	6.15	39150	148	.58	.62
Mercedes	230	10.40	8.47	75700	180	.46	.65
Citroen	CX	11.05	8.06	64700	178	.44	.72
Volvo	244	12.95	8.38	55000	145	.40	.73
BMW	520	12.26	7.81	68593	182	.38	.75

In Table 10.1, car $a \in A_o^C = \cap A_o^j$, the group joint set of a priori admissible cars, is specified by name. The goals are as follows: y_1 = C120 is the gasoline consumption, liters/100 km, at 120 km/hr; y_2 = space is in square meters; y_3 = price is in French francs; y_4 = maximum speed is in km/hr. Utilities u_1 and u_2 are the utilities of players 1 and 2, respectively. For exchanging information the DSS could display the larger set $a \in \cup A_o^j$, which includes cars a priori admissible to at least one player. In this case a car inadmissible for player j would be listed as "inadmissible" in the utility column u_j, but it conceivably could become admissible in the course of negotiations.

Thus Table 10.1 shows a set of ten cars and their corresponding goal and utility values for two players. The utility values u_1 for player 1 are taken from Jacquet-Lagreze and Shakun (1984) based on use of the single-user DSS. For illustration the utility values u_2 for player 2 are listed in reverse order of those for player 1.* In row 1 of Table 10.1 we see the goal point $B_1 = (10.48, 7.96, 46700, 176)$ of Figure 10.1 and utility point $P_1 = (.752, .383)$ of Figure 10.2 corresponding to player 1's first car choice, Opel. Similarly in row 10 of Table 10.1 we see $B_2 = (12.26, 7.81, 68593, 182)$, $P_2 = (.383, .752)$ corresponding to player 2's first choice, BMW. Thus to begin with, player 1's feasible target is defined by row 1, similarly row 10 for player 2. Concession making involves players adding additional rows to their respective targets, thereby expanding them. Given the symmetry of the situation, the solution is likely to be Visa with $P_C = (.616, .576)$, Golf with $P_C = (.576, .616)$, or a random choice between them.

In addition to these displays, the NSS can show graphically the marginal utility functions. For output goal y_i, $u_{ij}(y_i)$ gives the marginal utility function of player j. If for a particular i the DSS shows both $u_{i1}(y_i)$ and $u_{i2}(y_i)$ on the same graphical axes, then the two players can compare, exchange information (perhaps leading toward consensus), and negotiate on their marginal utility functions. The marginal utilities u_{ij} can also be included in the relational data of Table 10.1 by inserting columns u_{i1} and u_{i2} for $i = 1, 2, 3, 4$, i.e., eight columns of the u_{ij} inserted, say, between the y_4 and u_1 columns. The DSS could display the projection of the relational data of Table 10.1 onto goal y_i, u_{i1}, and u_{i2} to enable the players to compare their marginal utility values for a particular goal y_i.

If players change their marginal utility functions so that they approach one another, the feasible set in utility space approaches a positive-sloping 45° line whose highest utility point is the solution, P_{C*} (Figure 10.2), thus achieving consensus. Of course, U^C (initial) is readily adjusted to U^C (ad-

*Technically, for the y_1 through y_4 values given in Table 10.1 PREFCALC shows these reverse utilities u_2 to be inconsistent. However if player 2 is not at all interested in goals y_1 through y_4 but only in a fifth goal y_5, style, then utilities u_2, measuring the utility of style, can be consistent for player 2. (If player 1 is not at all interested in style, then his utilities u_1 would not change if style is introduced.)

justed) to give a single-point intersection at P_{C*}. In other words, in utility space, Figure 10.2, there is a function F: $P_C \rightarrow P_{C*}$, $P_1 \rightarrow P_{1*}$, $P_2 \rightarrow P_{2*}$ mapping the original feasible set to points along the dotted straight line with solution at point P_{C*}. U^C (adjusted) is also shown following the mapping: U^C (initial) $\rightarrow$ U^C (adjusted).

The arrival at a common coalition utility function (through exchange of information and negotiation until players' marginal utility functions are identical) means in goal space, Figure 10.1, that individual players' goal targets Y^1 and Y^2 have become the same. In other words, although not drawn on Figure 10.1, now $Y^1 = Y^2 = Y^C$, the coalition goal target that intersects $g(A_o^C)$ at a solution point B_{C*} whose preimage in car space is the car-buying decision.

In addition to exchanging information and negotiating to expand targets, players can consider the use of axioms to contract the feasible region, e.g., (1) to a single solution point in utility space—in Figure 10.2, Nash axioms (Harsanyi 1977; Luce and Raiffa 1957; Owen 1982) might give solution point P_C, which is accommodated by the mapping: U^C (initial) $\rightarrow$ U^C (final), or (2) to a constrained set of points (e.g., the Pareto optimal set might be $\{P_1, P_C, P_2\}$ in Figure 10.2). The latter could be followed by compromise (concessions) to select a single point from this set, e.g., P_C, or perhaps consensus leading to P_{C*} might be realized.

10.4 Using MEDIATOR: Application to Group Car-Buying Decisions

As noted in the negotiation setting overview, a human mediator supports group negotiations, and he in turn is supported by the negotiation support system, MEDIATOR. The (human) mediator supports negotiations by assisting the players in a process of consensus seeking within which compromise is possible. Using MEDIATOR, the mediator helps in consensus seeking by aiding the players to build a group joint problem representation of the negotiations—in effect joint mappings from control space to goal space (and through marginal utility functions) to utility space.

Assume each decision maker (player) in a group has worked individually with the single-user multicriteria DSS as discussed in Section 10.2. Using PREFCALC, he has established his initial individual mappings from control space to goal space (and through the marginal utility functions) to utility space. For this illustration of car buying we assume a negotiation setting between two players (e.g., husband and wife) wherein the players respond positively to the mediator's suggestion that players build a joint problem representation with the help of MEDIATOR.

For the group representation MEDIATOR uses a common set of dimensions—the union of the individual player dimensions—to define group (joint) control, goal and utility spaces. The evolving problem representation is shown—graphically or as relational data in matrix form—in the three group spaces, as discussed in Section 10.3.

***Table 10.2* Group Problem Representation**

Control		Goal	Goal	Goal	Goal	Initial utilities		First evolved utilities		Second evolved utilities	
Make	*Model*	y_1	y_2	y_3	y_4						
		C120	*Space*	*Price*	*Speed*	u_1	u_2	u_1	u_2	u_1	u_2
Opel	Record	10.48	7.96	46700	176	.75	.62	.74	.66	.74	.70
Peugeot	505	10.01	7.88	49500	173	.73	.58	.72	.63	.72	.57
Peugeot	104	8.42	5.11	35200	161	.72	.15	.67	.29	.67	.27
Citroen	Dyane	6.75	5.81	24800	117	.65	.31	.64	.45	.64	.40
Citroen	Visa	7.30	5.65	32100	142	.62	.24	.60	.38	.60	.34
VW	Golf	9.61	6.15	39150	148	.58	.21	.55	.33	.55	.42
Mercedes	230	10.40	8.47	75700	180	.46	.76	.45	.65	.45	.70
Citroen	CX	11.05	8.06	64700	178	.44	.59	.43	.54	.43	.49
Volvo	244	12.95	8.38	55000	145	.40	.75	.44	.65	.44	.60
BMW	520	12.26	7.81	68593	182	.38	.49	.39	.47	.39	.52

Table 10.2 shows the initial group mappings from control (car) to goal to utility spaces (ignore first and second evolved utilities for the moment). Suppose that player 2's initial individual problem representation had only three goal dimensions, say y_1, y_2, and y_3, whereas player 1's had all four goals. The common set of goal dimensions—the union—has all four goals with player 2 placing zero weight on y_4. Note that in this example there is no conflict in group control and goal space, i.e., players have the same individual problem representation in these spaces. They differ only in their representations in group utility space as shown under "initial utilities" in Table 10.2. (For a more general development see Giordano, Jacquet-Lagreze, and Shakun 1985). A look at the initial individual marginal utility functions, Figure 10.3, reveals the underlying preference conflict. We consider several scenarios based on play by student/faculty players.

SCENARIO 1

Players look at the initial utilities in Table 10.2 and perhaps, at the mediator's suggestion, at the utility functions in Figure 10.3. From Table 10.2 MEDIATOR displays the car rank orders and utilities shown in Table 10.3. The mediator asks whether players would now like to consider compromise or to seek consensus further by exchanging information. In scenario 1 we assume that either immediately or after viewing and discussing the marginal utility functions (Figure 10.3) but not changing them players are interested in compromise. The mediator can support compromise through use of axiomatic solution concepts (Nash, Kalai-Smorodinsky, etc.) and/or concession-making procedures (Rao-Shakun, etc.) in the MEDIATOR model base—see Shakun (1985) for a detailed discussion.

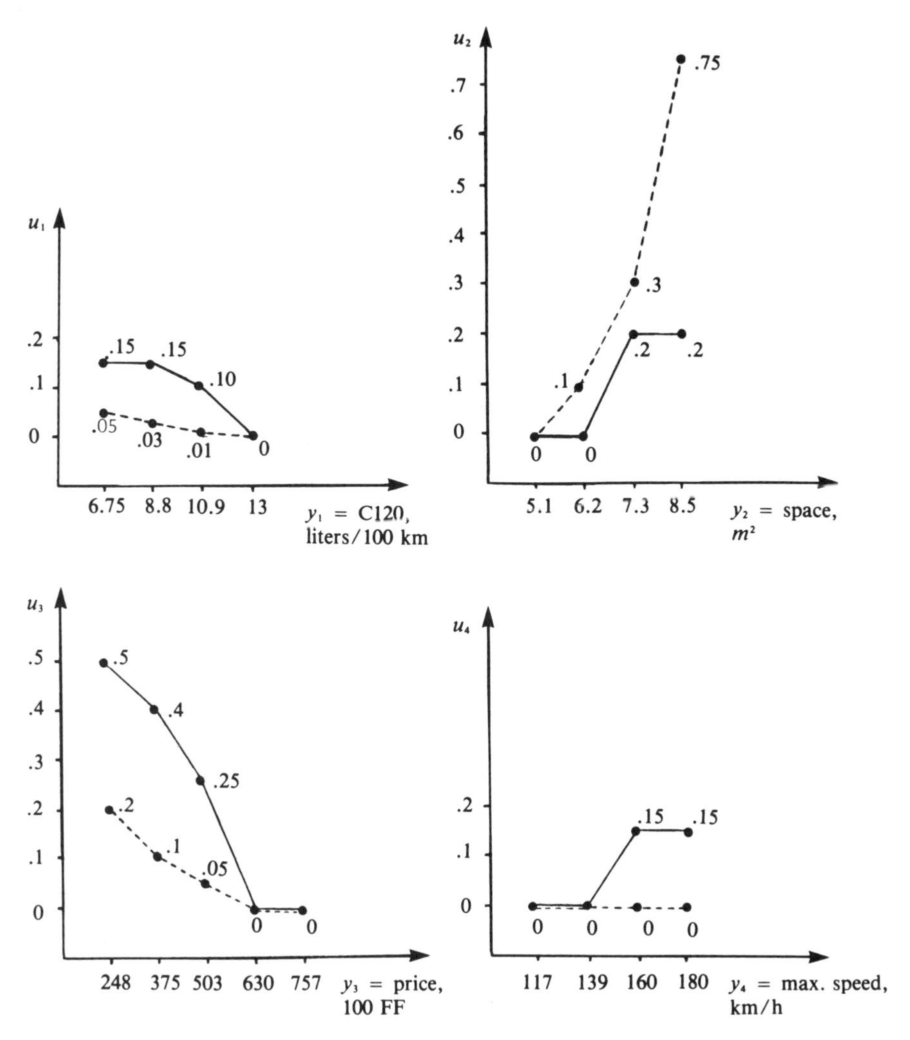

Figure 10.3. **Initial Marginal Utility Functions for Player 1 (Solid) and for Player 2 (Dotted)**

As an example the mediator can suggest concession making following conditional car target expansion. Under this procedure each player successively expands the list (target) of cars that he would be willing to accept. At stage 1, player 1's car target would be his first choice, Opel, and player 2's, his first choice, M230. The intersection of these two car targets is empty. Using Table 10.3, if players continue to expand their individual car targets by stages until a nonempty intersection is achieved, then concession making will continue to the third stage, with Opel as the intersection and compromise solution.

Table 10.3 **Initial Car Rank Orders and Utilities**

	Player 1			*Player 2*	
Opel	Record	.75	Mercedes	230	.76
Peugeot	505	.73	Volvo	244	.75
Peugeot	104	.72	Opel	Record	.62
Citroen	Dyane	.65	Citroen	CX	.59
Citroen	Visa	.62	Peugeot	505	.58
VW	Golf	.58	BMW	520	.49
Mercedes	230	.46	Citroen	Dyane	.31
Citroen	CX	.44	Citroen	Visa	.24
Volvo	244	.40	VW	Golf	.21
BMW	520	.38	Peugeot	104	.15

As another attempt at compromise the mediator can ask MEDIATOR to compute the maximin solution concept. First MEDIATOR normalizes each player's utilities between 0 (for his last car choice) and 1 (for his first car choice). Using Table 10.2, the normalized utility for each player and minimum utility comparing normalized utilities between players for each car are computed and shown in Table 10.4.

The car that maximizes the minimum utility giving a maximin utility of .77 is Opel, which is the maximin solution. Thus both concession making following conditional car target expansion and the maximin solution concept give Opel.

SCENARIO 2

After looking at the initial utilities in Table 10.2 and the marginal utility functions, Figure 10.3, players decide to discuss their marginal utility functions and modify them to those shown in Figure 10.4. This leads to an evolved group problem representation. Thus in Table 10.2 the overall utilities evolve from the initial utilities to the first evolved utilities. Here consensus on a car

Table 10.4 **Initial Normalized Utilities**

		Player 1	*Player 2*	*Minimum utility*
Opel	Record	1	.77	.77
Peugeot	505	.95	.71	.71
Peugeot	104	.92	0	0
Citroen	Dyane	.73	.26	.26
Citroen	Visa	.65	.15	.15
VW	Golf	.54	.10	.10
Mercedes	230	.22	1	.22
Citroen	CX	.16	.72	.16
Volvo	244	.05	.98	.05
BMW	520	0	.56	0

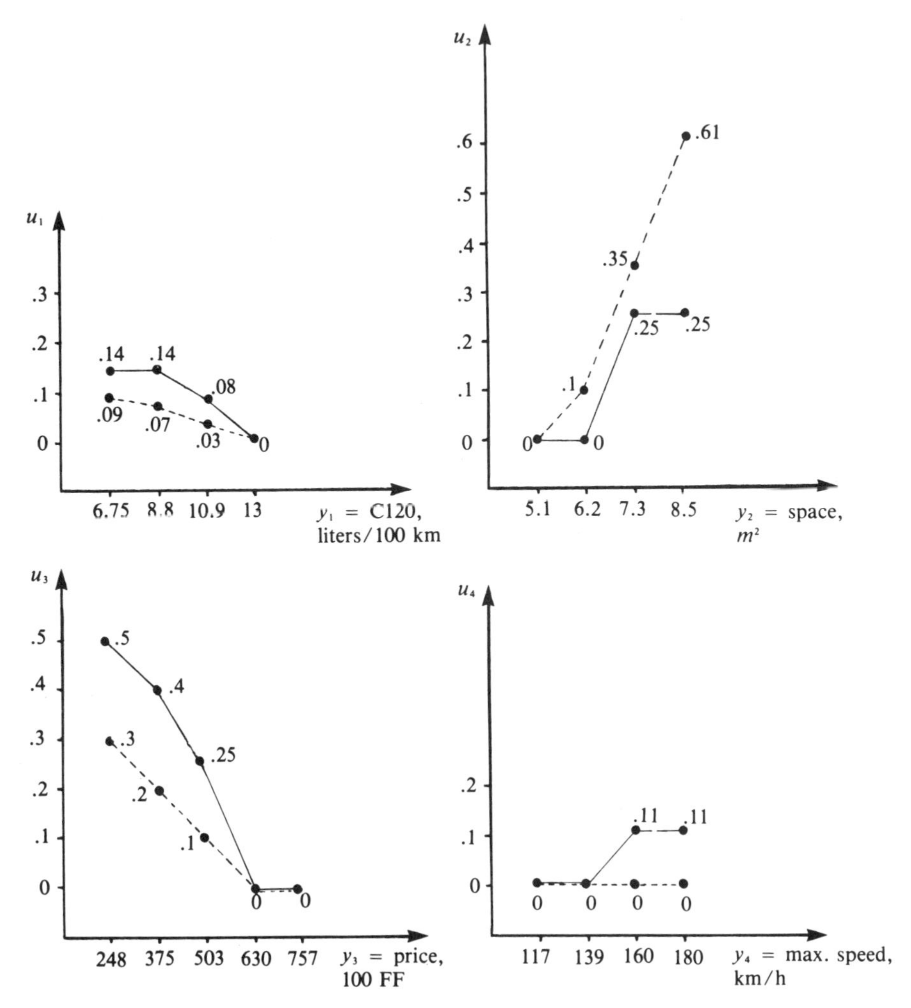

Figure 10.4. **First Evolved Marginal Utility Functions for Player 1 (Solid) and for Player 2 (Dotted)**

decision—Opel—has been achieved because Opel gives each player his highest utility.

SCENARIO 3

In evaluating the results of scenario 2, player 2 realizes that, although Opel has the highest computed utility, .66, he is not at all familiar with this car and so doesn't want to buy it. He prefers M230 or Volvo, which have utility of only .65. Player 1 says that M230 and Volvo are low down on his preference list. He suggests P505 as a compromise being his second choice (with a utility of .72 and close to Opel at .74) and giving player 2 a utility of .63, which is almost as high as .65 associated by player 2 with M230 and Volvo. The scenario now divides into two subscenarios.

In subscenario A player 2 feels that player 1's proposed compromise of P505 is reasonable and accepts it. In subscenario B player 2 says that, although player 1's compromise suggestion of P505 is reasonable in the light of the first evolved utilities, he now realizes that something about the problem representation is bothering him. It does not include the nationality (country of

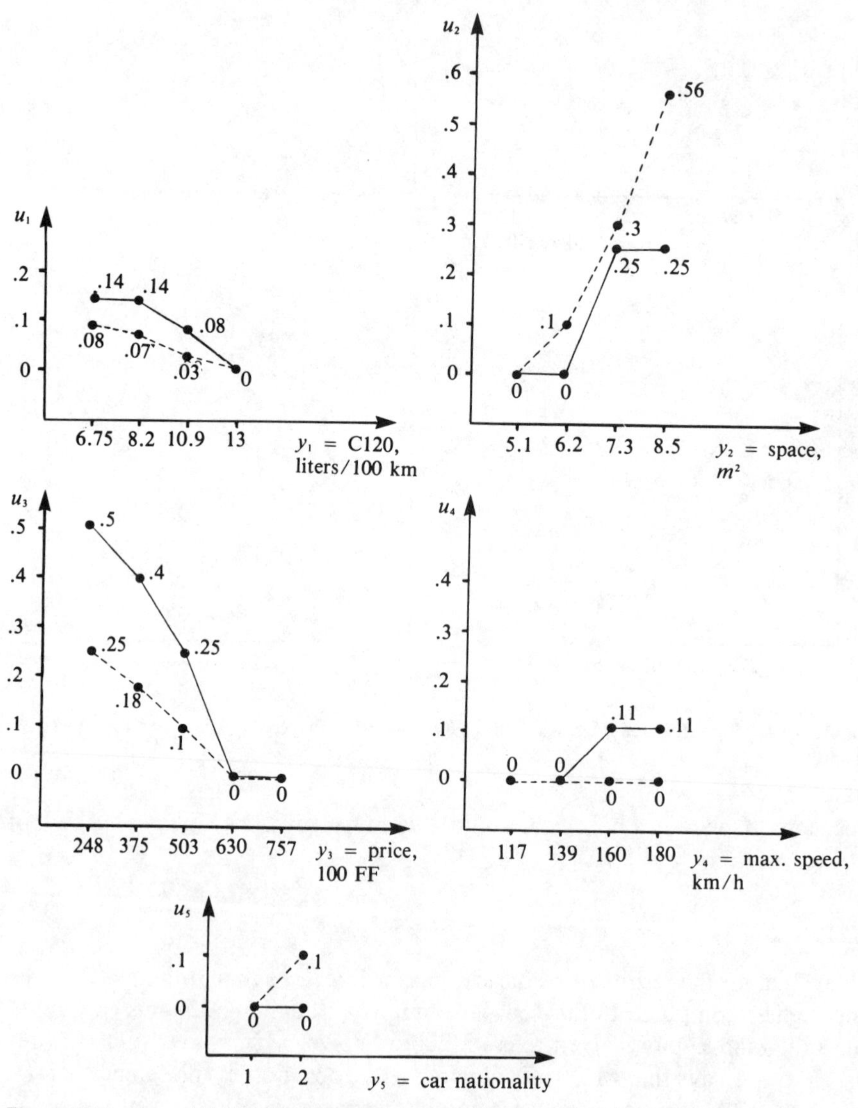

Figure 10.5. **Second Evolved Marginal Utility Functions for Player 1 (Solid) and for Player 2 (Dotted)**

manufacture) of the cars. He would like to assign a preference weight to German cars. Thus player 2 has introduced a fifth goal dimension, y_5, car nationality, to the group problem representation—imagine a column for goal y_5 after goal y_4 in Table 10.2 where German cars (Opel, Golf, M230, BMW) are assigned a nominal value of, say, 2 and all other cars, a value of 1.

Using PREFCALC, player 2 modifies his marginal utility functions to those shown by the dotted lines in Figure 10.5. He places a weight of .1 on car nationality, y_5, and modifies the relative weights on the other criteria so that the sum of the weights equals 1. Note that with PREFCALC's utility normalization the criterion weight equals the marginal utility at the most preferred goal value considered. Player 1 places zero weight on car nationality, y_5, so that his marginal utilities in Figure 10.5 are in effect the same as in Figure 10.4.

The marginal utility functions in Figure 10.5 give overall utilities shown undcr "second evolved utilities" in Table 10.2. For player 1 there is no change —his second evolved utilities are the same as his first evolved utilities. For player 2 the second evolved utilities show the German Opel with utility .70, the German M230 with .70, the Swedish Volvo with .60, and the French P505 with .57. Player 2 still rules out the Opel as an unfamiliar car. He now prefers the M230 (utility .70) over the Volvo (.60); before introducing car nationality, they were tied at .65 each. Although before the P505 utility .63 was close to player 2's first choices of M230 and Volvo, each with utility .65, now the utility gap between P505 (utility .57) and the first choice M230 (utility .70) is large; so player 1's compromise suggestion of P505 represents a large utility drop for player 2 from his first choice. The mediator asks MEDIATOR to compute the maximin solution. Using Table 10.2, MEDIATOR computes for each car the second evolved normalized utility for each player and, comparing these, the minimim utility—see Table 10.5.

The maximin solution is Opel. With player 2 ruling it out, player 1 argues that P505, as the maximin solution over the remaining cars, is fair. Besides, he would have to drop to low utility levels of .45 and .44 (Table 10.2) if he were

Table 10.5 **Second Evolved Normalized Utilities**

		Player 1	*Player 2*	*Minimum utilities*
Opel	Record	1	1	1
Peugeot	505	.94	.70	.70
Peugeot	104	.80	0	0
Citroen	Dyane	.71	.30	.30
Citroen	Visa	.60	.16	.16
VW	Golf	.46	.35	.35
Mercedes	230	.17	1	.17
Citroen	CX	.11	.51	.11
Volvo	244	.14	.77	.14
BMW	520	0	.58	0

to consider M230 or Volvo, respectively. Player 2 is convinced and accepts P505 as the solution.

10.5 System Architecture for MEDIATOR

In this section we describe software requirements and a system architecture for MEDIATOR. Basically, MEDIATOR integrates a collection of software components used by the players and the human mediator through the use of a shared data base. The need for analyzing such components in a DSS, in addition to the description of operational research *models*, arises from two sources. Any DSS must offer a *user-friendly interface* and *efficient data access*. Otherwise it will not be used by computer-naive decision makers. More specifically, however, a multiperson DSS like MEDIATOR must also facilitate and structure the *communication* among the players and with the human mediator.

One approach to implementing such a communication facility is a direct message exchange subsystem based on electronic mail (Bui and Jarke 1984). Another approach—the so-called decision room—leaves the responsibility for player communication outside the system: The players are assembled in a single room and can communicate without computer aids (Huber 1982).

In contrast to these two methods our approach is *data base–centered*. The data base–centered approach was introduced for single-user DSS by Donovan (1976) and extended to hierarchically organized distributed DSS by Jarke (1981, 1982). This chapter extends the approach further to negotiation support systems. MEDIATOR achieves communication mostly through sharing data stored in a common data base. This data base would typically be located on a mainframe or on a separate file server accessible to all players and the mediator. It contains base data underlying the decision-making process as well as intermediate results of the negotiation process—the sequence of group joint problem representations.

The rules of communication (also called *communication protocols*—Tanenbaum 1981) are implemented through granting different access rights to players and mediator. In the following subsections we first motivate this approach by a requirements analysis and then provide a more detailed technical description. As a running example we shall once more use the two-player car-buying application.

SYSTEMS REQUIREMENTS FOR MEDIATOR

MEDIATOR is designed to provide user-friendly interfaces, efficient data and model access, and structured communication facilities to both the players and the mediator. Systems requirements for MEDIATOR can be grouped into two categories.

The first class of requirements is derived from the method itself. Because negotiation is viewed as an evolutionary process of information exchange leading to consensus or compromise, the system has to provide efficient support for interactive use by decision makers and mediators with limited computer skills. Moreover the system has to offer at least two kinds of representations for information display. A relational data base system must support the matrix representation needed in the detailed display of criteria and utilities versus alternatives (see Table 10.2). Additionally, the systems can present data graphically (e.g., piecewise linear marginal utility functions, see Figure 10.3). Finally, the system must be able to analyze and present the consequences of changes in control space, goal space, and utility space.

The second class of systems requirements results from some implicit assumptions in the proposed method. MEDIATOR has to satisfy these assumptions prior to the actual negotiation procedure. The main assumption is that players accept the idea of building a *group joint problem representation*. The remainder of this subsection investigates in detail the consequences of this assumption for MEDIATOR's design. The assumption has three facets: a jointly acceptable data base of underlying facts, jointly acceptable definitions of alternatives, and mutually understood definitions of criteria and preferences.

JOINTLY ACCEPTABLE DATA BASE The method assumes that the players agree on a common underlying set of facts about the domain of decision. The example of arms control negotiations shows that such an agreement may be very difficult to reach. The players may not even agree on a common scope of alternatives for a particular negotiation (e.g., strategic versus Euro-strategic versus space weapons). Moreover disagreement on the underlying facts is almost certain. Therefore MEDIATOR allows the players to agree that each will use his own data separately, i.e., an agreement exists that the players cannot agree on a common set of data. This version of the assumption may be the only way to get negotiations started if there is deep distrust among the players—witness again arms control negotiations.

JOINTLY ACCEPTABLE ALTERNATIVE DEFINITIONS This assumption requires that the players agree on a common definition of the dimensions (not necessarily feasible regions) of the control space. We distinguish syntactic and semantic disagreements that have to be resolved in prenegotiations.

Syntactic disagreements involve a different understanding of terms. For example, two players may give the same alternative a different name (synonyms) or two different alternatives the same name (homonyms). It is very important and usually not too difficult to resolve such misunderstandings.

Semantic disagreements involve a different partitioning of the control space. Different partitioning may mean varying degree of detail, or it may mean completely different dimensions. Disagreement often results from differences in knowledge or from basically inconsistent views of the problem. The former is easier to resolve than the latter.

For example, in a car-buying decision one player may distinguish cars by their engine type, another one by their make, yet another one by their make, model, and version. The obvious solution is to define alternatives by combining all suggested partitionings. However this may lead to an intolerably large number of alternatives. It is the mediator's task to help the players define a mutually understandable, acceptable, yet manageable set of alternatives. MEDIATOR can help in this task using certain concepts of data base theory. (See "DSS of the Mediator" later in this chapter.)

MUTUALLY UNDERSTOOD CRITERIA AND PREFERENCE DEFINITIONS Although the previous assumptions concerned the control space, this one involves possible misunderstandings in the goal space definition. Of course the method does not require players to use the same criteria. However it is important that the mediator and his support system, MEDIATOR, understand the meaning of criteria in order to make useful suggestions. Again, there may be syntactic or semantic problems. The former involve the naming problems of synonyms and homonyms; the latter could mean different units of measure or different ways to compute criteria from the available data.

As an example, one player may compute a criterion "space" in square meters; another may use "space" or its synonym "roominess" but measure it in cubic feet. On the semantic side both players could define "space" as the size of the inner sitting room; then it would be desirable to merge the two criteria. Alternatively, one of the players may use "space" for the outer size of the car. This could lead to the apparent paradox that one player tries to minimize "space" while the other is maximizing it. Clearly, it is appropriate here to rename and separate the criteria.

Each player defines preferences on criteria, e.g., a utility preference measure. However preference measures used by players need not be the same.

As a consequence of these assumptions, MEDIATOR supports a *two-phase negotiation process*. In the first phase, called *view integration*, the human mediator is supported in achieving a joint problem representation in the three steps of data base selection, alternative definition, and criteria and preference definition. Upon successful completion of this phase, the second phase, called *negotiation*, proceeds as described in Sections 10.3 and 10.4.

SOFTWARE CAPABILITIES AND COMPONENTS

An architecture for the MEDIATOR DSS should offer some software capabilities to support the systems requirements described in "Systems Requirements for MEDIATOR." We shall first review the major components of single-user DSS and then propose a specific architecture for MEDIATOR.

Each player and the mediator employ a single-user personal DSS that has the traditional three components of model management, data management, and dialog management (Sprague and Carlson 1982). For MEDIATOR this single-user DSS is a data-based version of PREFCALC (Jacquet-Lagreze 1985) for the players and an enhanced version for the mediator. For example, in an enhanced version the marginal utility functions for two or more players can be shown on the same graphical axes, as in Figures 10.3, 10.4, 10.5, to facilitate comparison.

The *dialog manager* is responsible for effective interaction between the DSS and its users, namely, each player and the human mediator. It provides menu management, screen composition, and graphics as well as relational representation facilities (Jarke, Jelassi, and Stohr 1984).

The *model manager* consists of executable modules together with modeling language facilities and execution management. In particular the negotiation models in the mediator DSS allow mappings of user changes (or adaptations) in all three spaces (control, goal, and utility space).

The *data manager* accesses and maintains the user's private as well as the jointly acceptable mainframe data bases. It contains a standard DBMS with enhanced data dictionary and view management facilities (Jarke, Jelassi, and Stohr 1984; Jelassi 1985; Jelassi, Jarke, and Stohr 1985). The "data dictionary" stores metadata such as alternative definitions, criteria definitions, function definitions, and units of measure. A "generalized view processor" helps the user define his personal customized view of the underlying data base. In particular, alternatives and criterion values can be derived automatically from the stored data base records and their attributes.

For n players there are $n + 1$ DSS of this nature. In addition group decision (Bui and Jarke 1984) or negotiation support systems require a *communications manager* to integrate the single-user DSS (Figure 10.6). In MEDIATOR this is accomplished in the following manner (Figure 10.7).

Each player and the mediator retain their private data bases, typically stored on a personal computer. The jointly acceptable data are stored in a common data base located on a mainframe or larger minicomputer and accessible to all the personal computers. The model/method base may contain different tools for each player, but they share the PREFCALC method. Conceptually, this method could be stored in a common model base associated with the common data base. From an implementation viewpoint it is more efficient to have copies on each microcomputer in order to avoid communication delays.

After establishing their individual preferences using single-user PREFCALC, players transfer their definitions of alternatives and criteria and their matrix and utility function representation to the common data base. Each player occupies a private section of that data base, which can be accessed only by himself and by the mediator. The mediator will then start the process of integrating these personal problem representations into the group joint problem representation.

Once this is accomplished the joint problem representation is stored in the publicly accessible area of the common data base. From then on the "official" negotiation will work only with the joint representation. The players

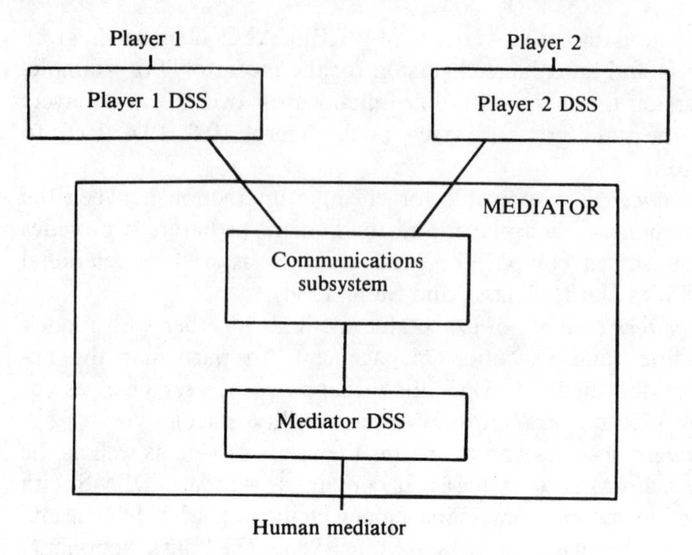

Figure 10.6. **General NSS Structure—Players' and Mediator's DSS**

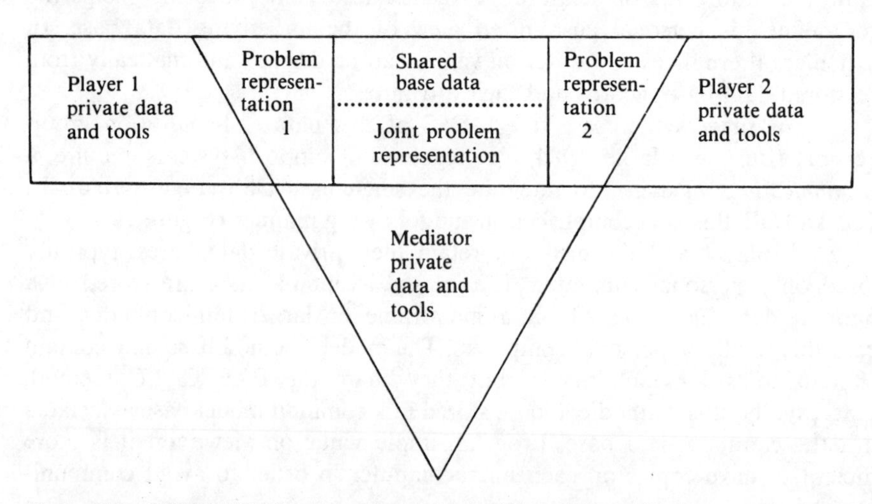

Figure 10.7. **MEDIATOR Design—Communication Through Data Sharing**

are free to continue using their local representation and other decision support tools for personal deliberations.

From a computer science point of view MEDIATOR's design poses the following research questions. How do we provide:

1. efficient data base access for each player
2. data base and model base facilities to support the mediator
3. communication between players and mediator
4. user interfaces for players/mediator

The following two subsections address our solutions to these problems, first for the individual player DSS, then for the mediator DSS and its communication with the player DSS.

DSS OF THE INDIVIDUAL PLAYERS

The DSS of the individual players are based on a stand-alone version of PREFCALC (Jacquet-Lagreze 1985). PREFCALC is a single-user DSS implemented on a personal computer; its underlying algorithms have already been described in Section 10.2. The method assumes its input to be stored in a relational format where the records (rows of the matrix) correspond to alternatives and the columns, to criteria. The table entries are criterion values.

In earlier work (Jarke, Jelassi, and Stohr 1984; Jelassi 1985; Jelassi, Jarke, and Stohr 1985) we enhanced the system with user-friendly capabilities to (1) access external mainframe data bases, (2) define alternatives from sets of records rather than from single records, and (3) use composite criteria computed from stored attributes by user-defined or selected functions. The following description is based on Jarke, Jelassi, and Stohr (1984).

Figure 10.8 shows how the PREFCALC input is generated from the data base. The method starts from a set of ALTERNATIVES, each characterized by a number of properties or attributes. For example, in a car-buying example the alternatives are types of cars, and the base relation would contain relevant attributes such as "maximum speed," "fuel consumption at speed 120 km/h," etc. However the stored relation may differentiate the type of car in many more classes than needed for the decision. Thus a decision alternative may correspond to a set of several data base records.

Some (but not necessarily all) of the attributes of the data base records may be important to a particular user's decision problem. The attributes in this subset are used as CRITERIA. However the user may also wish to derive more complex criteria from the stored attributes or ask for a presentation in different units of measure (e.g., horsepower instead of kilowatts). Therefore some computation may be necessary to derive criteria from the stored data.

Both tasks (alternative and criteria definitions) are accomplished by the aforementioned "generalized view processor" in conjunction with a menu interface generator. The resulting user view is called the DECISION MATRIX. In order to create a decision matrix the user has to define—through a sequence of menus—how decision alternatives and decision criteria are derived from the underlying data base.

Alternatives are defined in two steps. In the *data-staging* step the user selects a subset or CATEGORY of alternatives to be considered. For example, in a car-buying application the user may be interested only in trucks but not in other types of cars. Data staging extracts data from one or more mainframe data bases and constructs from them a single selected subrelation (the CATEGORY) on which all further processing will be performed, using the microcomputer DSS data base. Users may either choose from a menu of category

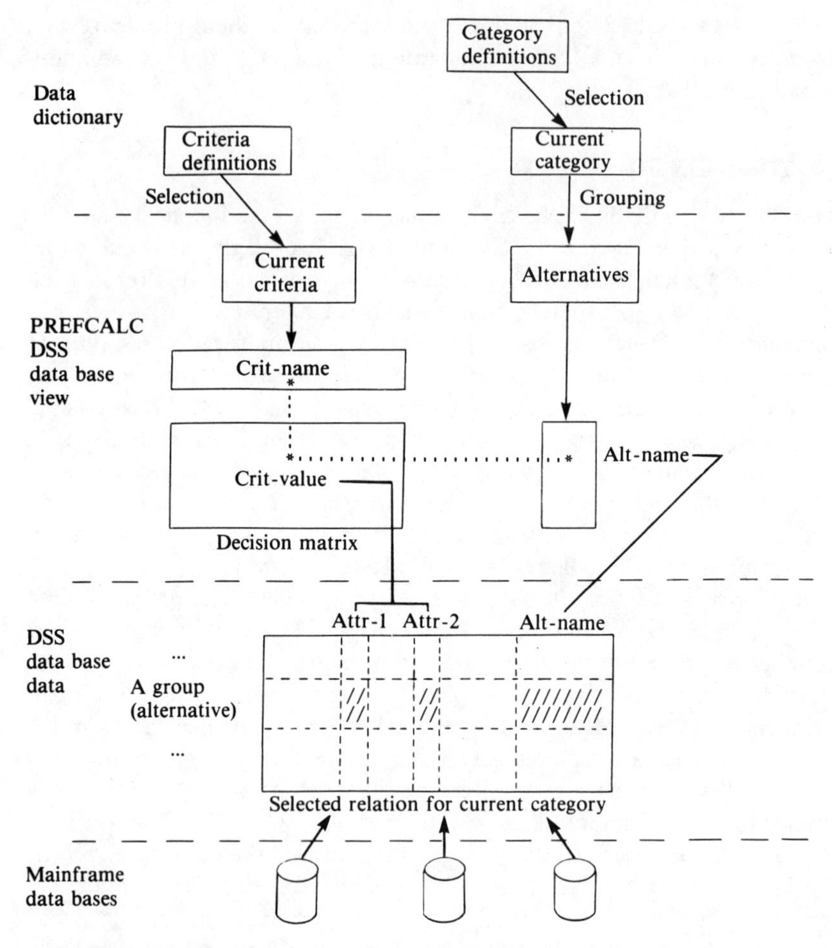

Figure 10.8. **Data Extraction in MEDIATOR (Simplified from Jarke, Jelassi, and Stohr, 1984)**

names defined in the CATEGORY DEFINITIONS section of the data dictionary or define their own category via a distributed data base query.

In the *grouping* step the user chooses a grouping of data base records within the CURRENT CATEGORY relation such that each group constitutes an ALTERNATIVE. Groups are defined by common values of certain attributes (the ALT-NAME). For example, some player may be interested in distinguishing cars only by their make and model, but not by details such as number of doors, engine power, etc.

Criteria are derived from attributes of the data base records. In the simplest case an attribute value can directly serve as a criterion value (e.g., maximum speed). Frequently, however, the criterion value may be a function of one or several attribute values. For example, a criterion "consumption" may be defined as the average of the stored data base values of fuel consumption in the city and on highways.

Moreover, whenever alternatives correspond to groups of records rather than to single records, criterion values must be based on aggregate functions over these records (e.g., average, minimum, maximum, forecast for next year). The data dictionary contains a library of such CRITERIA DEFINITIONS from which the user can choose the CURRENT CRITERIA using a menu of CRIT-NAMES. (Of course a more sophisticated user can also add functions to the library.)

Finally, the combination of alternative definitions (grouping) and criteria definitions (computations) allows the derivation of criterion values for alternatives (CRIT-VALUE) from the data base. All these operations can be performed within an extended relational data base framework discussed in detail in Jarke, Jelassi, and Stohr (1984) and Jelassi (1985).

Figure 10.9 gives an example of decision matrix construction for player 2 in the example of Section 10.4. This player wants to use the car mostly in the city and is therefore interested in a spacious but not too expensive car with little consumption in the city. This player ignores other known car characteris-

1) Player 2 decides to look at subcompacts only. A category "subcompact" has been previously defined for the "cars" mainframe database as follows:

```
DEFINE CATEGORY subcompact FOR cars
       WHERE length < 480;
```

The player can invoke this category simply by menu selection:

CURRENT CATEGORY: subcompact

Make	Model	Version	DIN-Cons	Length	Width	Price	Speed ...
Opel	Record	2000	11.4	462	173	46700	176
Peugeot	505	GR	9.4	458	172	49500	173
Peugeot	104	ZS	8.7	337	152	35200	161
Citroen	Dyane		6.8	387	150	24800	117
Citroen	Visa	Super E	9.2	369	154	32100	142
VW	Golf	GLS	9.7	398	158	39150	148
VW	Golf	GTI	11.2	398	158	45000	181*
Mercedes	230	E	13.6	472	179	75700	180
Citroen	CX	PALLAS	14.1	465	177	64700	178
Citroen	CX	GTI	15.3	465	177	69700	178*
Volvo	244	GLE	16.9	479	171	51000	140^{+}
Volvo	244	TURBO	12.5	479	171	59000	150^{+}
BMW	520		12.5	462	170	68593	182
...							

Figure 10.9. **Generating a Problem Representation for Player 2**

2) Player 2 does not like "GTI" sports versions and therefore adds the following category definition to the CATEGORIES model base:

```
DEFINE CATEGORY no-sport FOR subcompact
       WHERE version <> "GTI";
```

the application of which eliminates the rows marked "*"

3) Player 2 is not interested in the "length" and "width" of the car but only in the "space" it provides. The CRITERIA DEFINITIONS model base contains the definition of a virtual attribute "space":

```
DEFINE CRITERION space FOR cars
       AS avg(length*width)
```

which is inherited by all subrelations of cars and can again be applied by simple menu selection. avg(x) is a "vertical" function that computes the average value of x over all rows in a particular alternative (as defined in step 4, below).

4) Player 2 is not interested in the differences between the different versions but wants to define alternatives only via make and model (as shown in the GROUP-BY below). Therefore, the two rows marked '+' must be combined into one. In order to do this, player 2 defines the DECISION-MATRIX as follows (again, MEDIATOR uses menu selection to make this process more user-friendly):

```
DEFINE VIEW  decision-matrix (make, model, consumption, space, price)
AS SELECT    make, model, avg(DIN-cons), space, avg(price)
   FROM      no-sports
   GROUP-BY  make, model
```

which results in the following decision matrix for Player 2:

Make	Model	Consumption	Space	Price
Opel	Record	11.4	7.96	46700
Peugeot	505	9.4	7.88	49500
Peugeot	104	8.7	5.11	35200
Citroen	Dyane	6.8	5.81	24800
Citroen	Visa	9.2	5.65	32100
VW	Golf	9.7	6.15	39150
Mercedes	230	13.6	8.47	75700
Citroen	CX	14.1	8.06	64700
Volvo	244	14.7	8.38	55000
BMW	520	12.5	7.81	68593

Figure 10.9. **(Continued).**

tics (e.g., speed, number of doors, etc.); moreover the data base consulted by player 2 contains only information about the DIN consumption. Player 2 is not very interested in the differences among various versions of a car but rules out sports versions (e.g., the Golf GTI in the example table). Note that the space criterion must be computed from the stored values of car length and width.

After the construction of the input matrix, the CATEGORY relation will be needed only if the multicriteria decision method proposes the inclusion of a new criterion in order to resolve apparent inconsistencies in user preferences. Otherwise the method proceeds to construct utility functions through aggregation and disaggregation of preferences, as described in Section 10.2 (see also Jacquet-Lagreze and Shakun 1984).

DSS OF THE MEDIATOR

In the previous subsection it was demonstrated that a minor extension of the relational model of data bases (Codd 1970; Ullman 1982) is sufficient to support the single-player data preparation process for the multiple criteria DSS, PREFCALC. In this subsection this result will be extended. Relational operations, enhanced by redefinitions of terms, can also efficiently support the *view integration* phase of mediation. The discussion will follow the same sequence (data base selection, alternative definition, criteria, and preference definition) as before. Subsequently, we review the mediation support tools used in the negotiation phase.

The first task in establishing a group joint problem representation is the choice of underlying data bases upon which the definition and evaluation of alternatives can be based. In general the mediator will have to start with the union of all such data bases as far as they are made available to him. In this case it is crucial to establish the logical relationships between different sources of information about the same entity. For example, one catalog may refer to cars by make, model, version, etc., another one by product number. A translation table must be used to permit a join among the different entity identifiers. This task may be complicated if no one-to-one mapping exists; in this case the least common superset must be constructed to permit mappings. In most intraorganizational negotiations, however, the data base selection step will be simple because players access the same organizational (mainframe) data bases to begin with.

The data base selection step establishes a logical view of the mainframe data bases as a large "universal relation" (Ullman 1982). From this the group joint CATEGORY relation can be easily defined. As mentioned in Section 10.3, either the union (if there are few feasible alternatives) or the intersection (if there are many group-feasible alternatives) of the individual CATEGORY relations can be chosen as the joint representation. This means in a relational query language that the individual CURRENT CATEGORY definitions are simply conjunctions and disjunctions of restriction predicates.

Next the individual ALTERNATIVES definitions must be integrated. Because the group CURRENT CATEGORY relation contains all the attributes of the individual CURRENT CATEGORY relations, grouping will simply use all grouping attributes of the individual ALTERNATIVES definitions simultaneously. Unfortunately, this solution may result in very long alternative names and a large number of alternatives to be considered. For example, if one player distinguishes cars by their maximal speed, another one by their make, and a third one by make, model, and version, a particular group alternative could be named: "180–190 km/h, Mercedes (user2), Mercedes (user3), M190, E."

The concept of a "functional dependency" as developed by data base theory (see, e.g., Ullman 1982) can be exploited to simplify this naming problem. A data base attribute is called functionally dependent on a set of other data base attributes if for each combination of values of these attributes the dependent attribute can assume at most one value. Obviously, each attribute is functionally dependent on itself. Therefore the first simplification is to unify the two occurrences of "Mercedes" (provided both players mean the same thing—the mediator DSS can test this by looking at the data bases and groupings used by both players).

Assume that the data base schema in the data dictionary also states that maximal speed is functionally dependent on make, model, and version; i.e., for each version of a car there is only one maximal speed. In this case MEDIATOR can automatically simplify the group joint alternative grouping to make, model, and version. The simplified example alternative name then becomes just "Mercedes M190 E." For the sake of player 1 maximal speed will be retained as a criterion (but not as an alternative name) in the decision matrix.

If such automatic simplification proves insufficient, the human mediator will make other suggestions. One option was already mentioned: reducing the set of alternatives by presenting only the intersection-feasible ones. (In a many-player situation the requirement of mutual feasibility can be relaxed to, e.g., "acceptable to at least 50 percent," etc.) As another option consider the case of different degrees of specialization among the players. If the large number of alternatives is created by varying degrees of detail (i.e., one alternative definition is a subset of the other), the mediator may suggest postponing the decision about detailed alternatives until after a preselection of "good" higher level alternatives.

Next the group joint *criteria* definitions (columns of the group decision matrix) must be established. This step starts formally by executing a relational join operation over all the players' decision matrices understood as derived relations. The join columns are the alternative names as established in the previous step. This will result in a preliminary version of the group decision matrix in which the alternative names are common to all players but all criteria are disjunct. This means that players are assumed to assign weights of

0 to all columns but those stemming from their own decision matrix. If there are *n* players each with *m* criteria, there will be *nm* criteria in the preliminary group decision matrix.

The mediator will now try to collapse criteria that appear more than once and to unify similar criteria. In "Systems Requirements for MEDIATOR" we have already illustrated the pitfalls. The function definitions stored in the CRITERIA DEFINITION section of the players' data dictionaries are the major MEDIATOR tool to assist in criteria integration. If the function definitions of two criteria are equal, proportional (possibly different units of measure), or reciprocal, there is a good chance that two criteria mean the same thing (respectively, one is the negation of the other), even if they have different names.

By contrast, if criteria have the same name but differ significantly in their function definition, they may mean different things. An indicator of such semantic disagreements may be the players' marginal utility functions. Therefore MEDIATOR offers overlay of marginal utility curves for any pair (or small group) of players. Usually, one would expect that players' utility curves for the same criterion differ in weight and steepness but will rarely cross (one monotonically increasing, the other one decreasing). If they do cross, this may mean severe value disagreements or simply misunderstanding of terms.

Human mediator intervention to resolve such questions remains necessary even in the presence of an NSS. If one looks at the function definitions alone, it may be very difficult or even impossible to prove or disprove equivalence of functions (even though the form of the function definitions in our SQL extension are very standardized). Therefore the integration step will only be supported but not completely automated.

Once the alternatives and criteria have been integrated as far as appropriate, MEDIATOR constructs the group joint problem representation using the player's preference information.

As an example consider the view integration process preceding the negotiations described in Section 10.4. In contrast to player 2, who wanted a city car, player 1 wants to use the car mostly for business trips. He is therefore initially interested in highway consumption, high speed, a limited price, and much space. His data base has more detailed information on consumption at various speeds but is otherwise identical to that of player 2. In the first step of view integration both players agree to use the intersection of the two sets of acceptable cars, ruling out sports versions, which were not acceptable to player 2. Both players name one of their criteria "consumption," but a review of the criteria definitions by MEDIATOR reveals that one means the DIN consumption, the other one, the highway consumption. Because both measures are highly correlated and player 1's business trips will account for most of the kilometers anyway, the players agree in the criteria integration phase to work on the basis of highway consumption and to call this criterion C120. The criterion, space, has identical definitions in both decision matrices and will simply be merged. Based on this information, both players reconsider their utility evaluations and come up with the initial group joint problem representation shown in Table 10.2.

An interesting design question for negotiation support systems in general arises after the view integration. In which form should the result be fed back to the players? MEDIATOR's answer was chosen for reasons of simplicity: The same view of the joint problem representation is offered to all players.

Alternatively, one could try to adapt the joint representation to each player's language, i.e., trying to translate the global view back to the individual view as far as possible. Data base theory has shown the impossibility to do this automatically, but practical ways around this problem have also been devised (Furtado and Casanova 1985). However these methods are very difficult to implement, and it is not clear from an application standpoint whether this solution is even desirable—consider, e.g., a second channel of communication among the players that might result in considerable confusion unless a common language is enforced.

The remaining MEDIATOR tools support the human mediator in the actual *negotiation phase* as described in Section 10.3. The implementation of the comprehensive example presented in Section 10.4 is based on the group joint problem representation as just developed. Here we summarize the major software tools grouped by the problem space in which they apply. We do not consider behavioral tools such as Delphi and NGT. For an overview see Bui and Jarke (1984), DeSanctis and Gallupe (1984), or Huber (1984).

MEDIATOR allows the human mediator to perform *what-if* analyses of possible suggestions he might make. For example, before suggesting that players lower their utility threshold, the mediator must make certain that this will generate additional alternatives for discussion. Otherwise the players will feel that they made a concession for nothing, and the negotiation climate may deteriorate.

In the *control space* the relational query language offers the option of including or excluding sets of alternatives from consideration by certain attribute or criterion values. The mediator can apply such queries either directly at the level of the DECISION MATRIX (usually to make it smaller) or at the level of the group CURRENT CATEGORY relation (usually to increase the set of feasible alternatives that appear in the DECISION MATRIX). The same technique may be applied either to the reference set of alternatives presented to the users or to the CATEGORY as a whole.

Changes in the *goal space* involve a redefinition of the set of CURRENT CRITERIA, either by adding or by deleting CRIT-NAMEs from the DECISION MATRIX. Because the mediator DSS has full access to the group CURRENT CATEGORY relation, such changes can usually be effected without recomputing the whole DECISION MATRIX or reaccessing the mainframe data bases. The dialog manager may even suppress a criterion without changing the internal representation at all.

Before a change in the goal space is made, the mediator will frequently be interested in the importance of criteria. Would dropping a criterion change the ratings? Is the ranking by a particular criterion inconsistent with the overall utility ranking of alternatives? To answer such questions, certain display techniques for relational data are employed, most prominently, alternative ranking by some criterion.

The idea of ranking alternatives is also used to answer what-if questions in the *utility space*. Because player utilities are simply additional attributes of the DECISION MATRIX relation, they can be easily used as sorting criteria. Alternatives can be presented ordered by a particular player's utility either down to a certain rank ("present the five best alternatives for player 1") or down to a certain minimal utility ("present all alternatives above normalized player 1 utility .50"). PREFCALC already offers this facility for a single player. See Section 10.4 for an example of the display of two players' utility rank orders in concession making. For more than two players it may also make sense to display the aggregated utilities of certain coalitions.

The second representational tool at the utility level involves the overlayed marginal utility curves. What-if questions allow the mediator to vary the weights and numbers of linear pieces of each player tentatively. As stated before, all these what-if studies are intended to prevent having the players agree to useless concessions and redefinitions that would unnecessarily delay the decision process and destroy the players' trust in the mediator's (and in the DSS's) abilities. An underlying assumption made in this context is that players are actually interested in a fast decision while preserving their interest—an assumption that is usually justified in intraorganizational negotiations but may not hold in other cases.

In summary, the tools described here are mostly data base and display tools related to the algorithms presented in Section 10.3. Other mathematical or behavioral tools may also be needed, but their discussion would go beyond the scope of this chapter. See Shakun (1985) for a detailed presentation of axiomatic and concession-making procedures.

Once the human mediator has formed an opinion from the what-if analyses, he broadcasts messages to the players either directly or by notifying them of proposed changes tentatively made on the group problem representation. Broadcasting messages keeps the mediation process as unbiased and open as possible. (Our underlying hypothesis is that players may discontinue the use of MEDIATOR if it means loss of information to them.) Examples of such changes are changes in the joint set of alternatives, introduction of new criteria or changes in weights, areas where concessions of players may reduce differences in opinion, and changes in utility values.

10.6 Concluding Remarks

The system design for MEDIATOR supports building a group joint problem representation (view integration). Negotiation involves the evolution of this problem representation—consensus seeking—within which compromise is possible. At any stage of problem representation the mediator can support compromise through use of axiomatic solution concepts and/or concession-making procedures in the MEDIATOR model base.

With systems like MEDIATOR we are moving toward decision support systems for multiplayer, multicriteria, ill-structured, dynamic problems, thus implementing in a decision support context the methodology of evolutionary systems design (ESD)—policy-making under complexity.

REFERENCES

Bonczek, R. H., Holsapple C. W., and Whinston, A. B. 1981. *Foundations of Decision Support Systems*. Academic Press, New York.

Bui, X. T., and Jarke, M. 1984. "A DSS for Cooperative Multiple Criteria Group Decision Making." *Proceedings 5th International Conference on Information Systems*, Tucson, AZ, pp. 101–113.

Codd, E. F. 1970. "A Relational Model of Data for Large Shared Data Banks." *Communications of the ACM*, 13, No. 6, pp. 377–387.

DeSanctis, G., and Gallupe, B. 1985. "Group Decision Support Systems: A New Frontier." *Data Base*, 16, No. 1, pp. 3–10.

Donovan, J. J. 1976. "Database System Approach to Management Decision Support." *ACM Transactions on Database Systems*, 1, No. 4, pp. 344–369.

Furtado, A., and Casanova, M. 1985. "View Updates." In Kim, W., Reiner, D., and Batory, D. (eds.). *Query Processing in Database Systems*. Springer-Verlag, New York.

Giordano, J. L., Jacquet-Lagreze, E., and Shakun, M. F. 1985. "A Decision Support System for Design and Negotiation of New Products." Graduate School of Business Administration, New York University, New York. Also in Shakun, M. F., 1988. *Evolutionary Systems Design: Policy Making Under Complexity and Group Decision Support Systems*. Holden-Day, Oakland, CA.

Harsanyi, J. C. 1977. *Rational Behavior and Bargaining Equilibrium in Games and Social Situations*. Cambridge University Press, Cambridge.

Huber, G. P. 1982. "Group Decision Support Systems as Aids in the Use of Structured Group Management Techniques." *DSS-82 Transactions*, pp. 96–108.

______. 1984. "Issues in the Design of Group Decision Support Systems." *MIS Quarterly*, 7, No. 3, pp. 195–204.

Jacquet-Lagreze, E. 1985. PREFCALC, Version 2.0. EURO-DECISION, BP 57, 78530 Buc, France.

______, and Shakun, M. F. 1984. "Decision Support Systems for Semi-Structured Buying Decisions. *European Journal of Operational Research*, 16, No. 1, pp. 48–58.

______, and Siskos, J. 1982. "Assessing a Set of Additive Utility Functions for Multiple Criteria Decision Making: The UTA Method." *European Journal of Operational Research*, 10, No. 2, pp. 151–164.

Jarke, M. 1981. *Ueberwachung und Steuerung von Container-Transportsystemen*. Gabler, Wiesbaden, West Germany.

______. 1982. "Designing Decision Support Systems: A Container Management Example." *Policy Analysis and Information Systems*, 6, No. 4, pp. 351–372.

______, Jelassi, M. T., and Stohr, E. A. 1984. "A Data-Driven User Interface Generator for a Generalized Multiple Criteria Decision Support System." *Proceedings IEEE Workshop on Languages for Automation*, New Orleans, pp. 127–133.

Jelassi, M. T. 1985. "An Extended Relational Database for Generalized Multiple Criteria Decision Support Systems." Ph.D. dissertation, Department of Computer Applications and Information Systems, New York University.

______, Jarke, M., and Stohr, E. A. 1985. "Designing a Generalized Multiple Criteria Decision Support System. *Journal of MIS*, 1, No. 4, pp. 24–43.

Keen, P. G. W., and Scott-Morton, M. S. 1978. *Decision Support Systems: An Organizational Perspective*. Addison-Wesley, Reading, MA.

Luce, R. D., and Raiffa, H. 1957. *Games and Decisions*. Wiley, New York.

Owen, G. 1982. *Game Theory*, 2nd ed. Academic Press, New York.

Shakun, M. F. 1981a. "Formalizing Conflict Resolution in Policy Making." *International Journal of General Systems*, 7, No. 3, pp. 207–215.

______. 1981b. "Policy Making and Meaning as Design of Purposeful Systems." *International Journal of General Systems*, 7, No. 4, pp. 235–251.

______. 1985. "Decision Support Systems for Negotiations." *Proceedings of the IEEE International Conference on Systems, Man and Cybernetics*, Tucson, AZ, November. Also in Shakun, M. F. 1988. *Evolutionary Systems Design: Policy Making Under Complexity and Group Decision Support Systems*. Holden-Day, Oakland, CA.

______. 1988. *Evolutionary Systems Design: Policy Making Under Complexity and Group Decision Support Systems*. Holden-Day, Oakland, CA.

Sprague, R. M., and Carlson, E. D. 1982. *Building Effective Decision Support Systems*. Prentice Hall, Englewood Cliffs, NJ.

Tanenbaum, A. 1981. "Network Protocols." *ACM Computing Surveys*, 13, No. 4, pp. 453–489.

Ullman, J. D. 1982. *Principles of Database Systems*, 2nd ed., Computer Science Press, Rockville, MD.

CHAPTER 11

A Decision Support System for Design and Negotiation of New Products

Jean-Louis Giordano*, Eric Jacquet-Lagreze, and Melvin F. Shakun*****

11.1 Introduction

New product designs frequently involve major strategic commitments having important impacts for a company. For example, decisions on new cars are generally made three to five years in advance of their introduction into the market. The impact of such decisions on market share, profits, employment, and other goals is considerable.

New product decisions in major corporations are made within a group decision-making process. For example, with automobiles the group can consist of the company president, the directors of marketing, design engineering, manufacturing, finance, etc.

Although group participants can have different, even conflicting points of view, the atmosphere is generally one of problem solving. The decision is frequently a complex one involving risks and uncertainty in a multiparticipant, multicriteria, partially structured, dynamic context (see Giordano 1982).

Jacquet-Lagreze and Shakun (1984) develop a decision support system (DSS) for multicriteria decision making (MCDM) involving one decision maker. The system is implemented on microcomputers using the PREFCALC computer package (Jacquet-Lagreze 1985). Shakun (1985) discusses the case of multiple decision makers and how to use PREFCALC to support negotiations.

*Regie Nationale des Usines Renault, France
**LAMSADE, University of Paris-Dauphine, France
***New York University, New York, USA

In this chapter we develop a negotiation support system (NSS) for a group making new product decisions. The approach will be illustrated by the design of new cars, although it is, of course, useful for other similar types of decision problems.

In a DSS one usually considers data management, model management, and dialog management (Sprague and Carlson 1982). Here the data base contains information such as characteristics of the products in the market (objective features such as gas consumption, maximum speed for cars, etc.). It also contains ratings of the products on various consumer choice criteria that are generally more qualitative, such as style, comfort, and performance. Commercial information such as sales volumes and prices are also available. Access to such data assumes the connection of mainframes and microcomputers and the use of data base management systems available on the mainframes and micros. We shall not discuss the data manager or dialog manager here, but see Jarke, Jelassi, and Stohr (1984), Jelassi, Jarke, and Stohr (1985), and Jarke, Jelassi, and Shakun (1985). Neither will we develop the model manager as such—see the just mentioned references also—but will discuss the model base components.

The model base should contain, whenever possible, such things as simulation models of market share for new products, econometric models of the global market demand, and technological forecasting models for new products. The use of these kinds of models enables the user to determine the consequences of launching a new product. We shall not present such models here. We assume they are available, that some participants would like to use them, and that some others would not like to use them because they are not confident of their results.

We will focus our attention on the evaluation and the negotiation support models. From a practical viewpoint we shall use an MCDM model such as PREFCALC and a spreadsheet program such as LOTUS 123 to implement some of the graphic representation of the data, beliefs, values, and preferences.

The NSS is designed as a tool to synthetize information, simulate alternative new product decisions, and represent participant agreements and disagreements. It should be used in an interactive way by any one of the participants in order to increase his understanding of the process or by several participants in order to support them in designing a compromise solution.

Otherwise put, in this chapter we operationalize the evolutionary systems design (ESD) group problem representation discussed by Shakun (1988).

11.2 Controls, Goals, Criteria, Preferences: The Four Reference Spaces

THE CONTROL OR DECISION SPACE

Formally, we consider a control or decision space X with $x = (x1, \ldots xn)$ being a point in this space. The number and type of variables that define the decision space depend on the stage of the decision process. At an early stage we can consider variables at a rather highly aggregated level. For the design of a car we might use, for example:

$x1$: performance index (such as maximum speed or another index)
$x2$: gas consumption
$x3$: roominess
$x4$: noise level
$x5$: comfort
$x6$: ruggedness
$x7$: style (shape of the car)
$x8$: price

To make a decision at this level (i.e., to choose a point x in the space X) consists of defining design goals for the engineering department and a selling price.

To use such variables supposes that we already have employed submodels in order to define indexes related to more technical characteristics such as engine power and car weight. In case we do not have submodels we use the consumer ratings on these variables. The level of such an attribute has a meaning when comparing the value $x1$ of a hypothetical alternative (a new car) to existing products on the market. We can for instance fix the noise level $x4$ of the new car to be similar to that of a particular existing car.

THE GOAL SPACE

The goal dimensions introduced in this space are those the company uses in order to evaluate, compare, and eliminate different alternatives x. Such goal dimensions can be:

$y1$: sales volume or market share
$y2$: manufacturing cost per unit
$y3$: return on investment

As we shall see in "The Criteria Space," mapping between different spaces can be difficult. For instance it would be useful to have a marketing simulation model to map an alternative x into $y1$. However, when no explicit model is available to the participants or some participants do not trust a marketing or financial model others have used for mapping, these participants could directly use some of the decision variables as goal variables.

For this reason we shall allow the participants to use some of the xi as additional goal dimensions. It will be the NSS user's responsibility to employ a consistent goal space Y and not to count the same objective twice.

THE CRITERIA SPACE

In some situations the goal dimensions could be used directly as evaluative criteria. But, mainly because of the uncertainty, the lack of precision of the input data, and/or the output data of the submodels, the mapping of an alternative x into the goal space will generally not be a single point (deterministic).

For example, alternative x' may yield a better market share than alternative x'', but as a mean estimate only; x' could be a more risky choice than x''. In this example, if the market share simulation model gives as an output a probability distribution of the market share, we could use the mean and the standard deviation as the two corresponding criteria.

Another situation is when the experts (models and/or human experts) do not agree on the mapping of alternative x into a given yi. The evaluations we get could be in the form of a distribution or ratings, as in the DELPHI approach. To try to understand the reasons for such a conflicting position is an important step in the decision-making process. In the case where the experts themselves are various participants, j (e.g., marketing, engineering, etc.), we suggest splitting the initial goal dimension yi into different expert criteria, gij.

THE PREFERENCE (UTILITY) SPACE

The participants' overall preference for various alternative vectors x will usually differ; this is why we suggest using an NSS to reduce the initial conflicting positions.

One of the reasons for the difference can be a *conflict of expertise*, as suggested previously. But even with the same criteria values the overall preference can differ because the "weights" participants give to the criteria are not the same. The participants could even use different criteria to assess preferences explicitly (with the use of an MCDM model such as PREFCALC) or implicitly, using some implicit aggregation model. Let's call a *conflict of policy-making* a difference due to varied preference structures. In general the disparity of preference is due to a combination of both effects (conflict of expertise and conflict of policy-making).

In order to support participants in their negotiation process we suggest they use formally an MCDM model such as PREFCALC, which is particularly well adapted to the learning process of preferences. Using PREFCALC, the preference structure of a participant j at a certain stage of the process is given by:

- a set of criteria $gj = (g1j, \ldots, gnj)$
- evaluation of different existing products and new alternative decisions x using the set of criteria
- constraint levels gi^*j on some criteria eliminating inadmissible alternatives for participant j. (If $gij(x) < gi^*j$ for one criterion gij, then alternative x is not feasible or admissible for him.)
- a utility function aggregating the criteria $uj(gj)$, defined for admissible values of the criteria ($gij \geq gi^*j$)

11.3 Mapping Between the Four Spaces: Support for Negotiations

An important goal to achieve in order to support negotiation consists in being able to establish a correspondence or mapping between the different spaces as illustrated in Figure 11.1.

MAPPING THE CONTROL OR DECISION SPACE INTO THE GOAL SPACE

For mapping decision space into goal space we need technological forecasting, sales volume, market share, etc.

In case we do not have such models we can use experts' subjective ratings. One example is the feasibility of an alternative. If we consider a point (alternative) x in the decision space X, the limit of what will be technologically feasible in a few years (we have to decide now about a car to be marketed in a few years) is not clear. The technology makes progress every year, and we have to extrapolate, as in Figure 11.2, what will be the new frontier (see Giordano 1982).

If we lack a suitable technological forecasting model, we can ask experts in the engineering department to evaluate the new alternatives (products) on a criterion "feasibility" using a scale of 0–1 or 0–100.

For other criteria the mapping could have the support of a model, such as a market share simulation model, a financial model computing the return on investment, etc.

OTHER MAPPINGS: REPRESENTATION TOOLS FOR NEGOTIATION

Other mappings will be used extensively as tools to represent, at any stage of the process, the different participant views of the problem.

We assume here that a participant's beliefs and preferences are not given in the sense of being preexisting. They are progressively constructed, based on available data, confrontation with models, and communication and mutual influences among the participants.

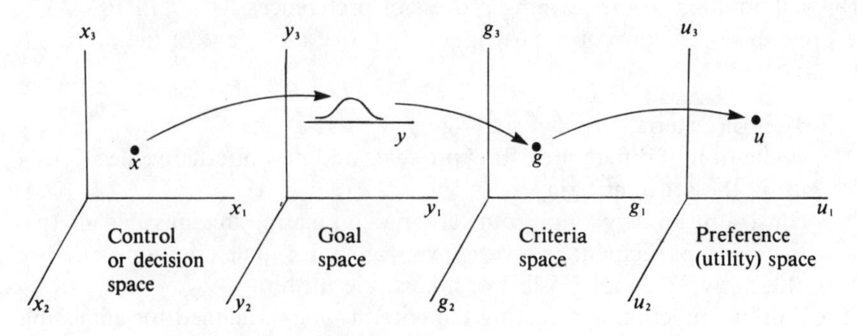

Figure 11.1. **Mapping Between Different Spaces**

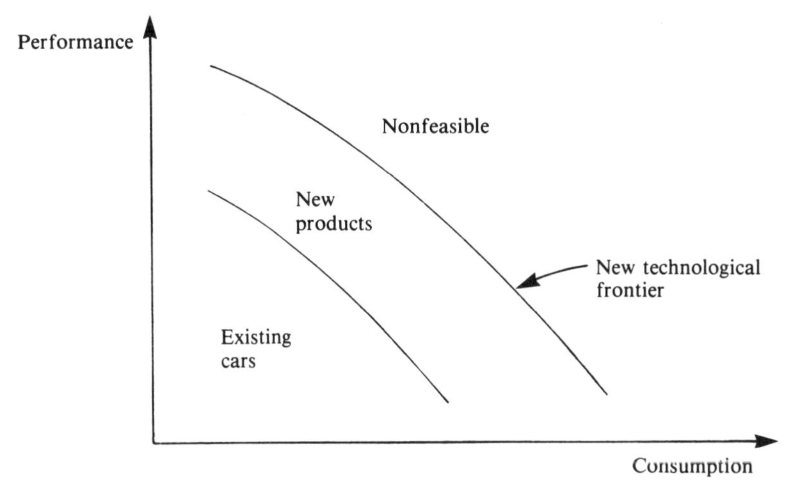

Figure 11.2. **Technological Frontier**

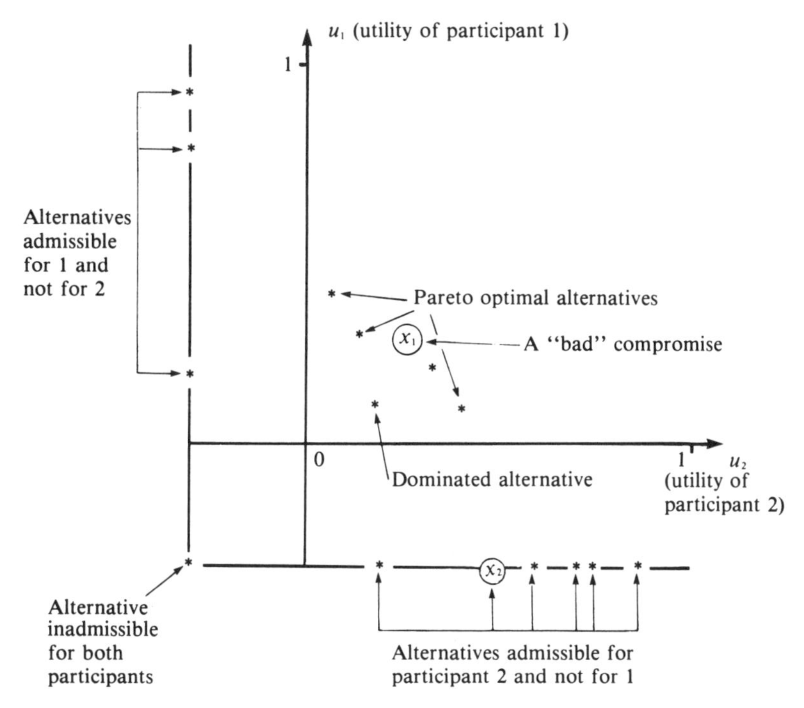

Figure 11.3. **Utility Space—Considerable Conflict**

If we accept this assumption, to support negotiation means to give a common language, more precise than the usual one, more quantitative where figures are available, or well established through the use of models. It is important to make available tools that show to what extent participants agree or disagree and the reasons they are in conflict in order to help them modify their beliefs (i.e., evaluate products on the criteria) and preferences (i.e., utilities).

To represent and analyze conflict of expertise we can use the mapping of the alternatives into the criteria space. Global mapping using techniques like component factor analysis gives a first joint representation of the evaluations. A more detailed representation can be given using a two-dimensional space representing the same goal dimension with the evaluations of two participants or experts.

To represent the global preferences, we suggest using a classical representation of the alternatives in utility space (Figure 11.3). This is particularly adapted to the situation of two participants, which is often the case (e.g., marketing and engineering departments). With three (or more) participants it is possible to use three-dimensional graphics, to use several two-dimensional graphs (e.g., utilities for participants 1 and 2, 2 and 3, 1 and 3), to use a joint representation obtained with a component analysis, or to display several one-dimensional spaces showing profiles of different alternatives.

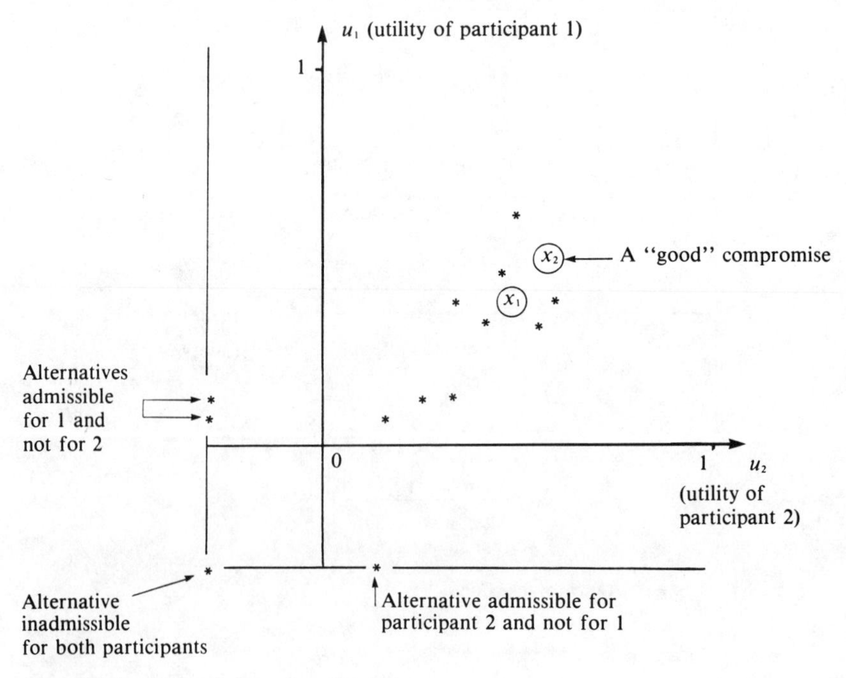

Figure 11.4. **Utility Space—Less Conflict**

At any stage of the process we could suggest some compromise solution using a normative concept such as the Nash solution, which would give solution $x1$ in Figure 11.3. But considering negotiation as a process of mutual influence, we can suggest that participants use the graphics representations of their conflict as stimuli for discussion and mutual understanding. They would enter a phase of "concensus seeking" (see Jacquet-Lagreze 1981; Shakun 1988), which could yield a less conflicting view such as the representation of Figure 11.4. A compromise solution such as $x2$ could appear attractive to both participants.

11.4 Implementation on Microcomputers

THE NSS MODEL BASE

It is planned to have an integrated package allowing the users to display any graphic representation discussed in the previous section. Meanwhile it is possible to implement the NSS in a rather convenient way using, besides specific marketing and financial models, which we suppose available on a microcomputer, the following programs:

- LOTUS 123 (or SYMPHONY) to manipulate the criteria evaluation tables (existing cars and new hypothetical alternatives) and their utility evaluations on different objective and subjective criteria (It is also used to prepare many of the graphic representations previously introduced.)
- PREFCALC, a single-user MCDM model that can easily read a range of data extracted from LOTUS (the user view of the problem) (The user can assess interactively his own preferences using an additive nonlinear utility function. He can also introduce his perception of another participant's preferences.)
- PC123, an interface program that enables users to recover results of PREFCALC such as utilities, create a file readable by LOTUS 123, and, using LOTUS, display graphics like those of Figures 11.3 and 11.4
- A multivariate analysis (component analysis) program easily interfaced with LOTUS 123

ILLUSTRATION IN A SIMPLIFIED AND HYPOTHETICAL SITUATION

Let us consider two participants, the marketing department and the engineering department. They want to compare six new models of cars: *XX*1, *XX*2, *XX*3, *XX*4, *XX*5, and *XX*6. They use as references some of the existing cars of the same category. We assume that by other means it was possible to determine the values of the criteria, and especially the feasibility of the new

Table 11.1 **Criteria Evaluation of Six Alternatives for a New Product**

Cars	*Power*	*Cons.*	*Price*	*MSpeed*	*Comfort*	*Style*	*Feasib.*
NB.L	1	−2	−2	2	2	2	1
MAX	10.71	8.8	52,000	170	8.5	8.7	100
MIN	5.315	5.3	32,900	126	7.1	7.3	90
RI85	10.71	8.3	50,475	163	8.4	8.1	100
131C	9.28	8.8	43,600	150	8.4	8.1	100
PA5S	6.05	6.2	34,300	140	7.1	7.6	100
RI65	8.71	8.6	43,541	150	7.7	8.1	100
127U	6.05	7.1	34,700	135	7.9	7.9	100
FI13	8.78	7.9	45,900	158	8.2	8.0	100
205R	8.58	6.0	48,900	154	8.2	8.5	100
104L	7.12	5.7	37,600	138	7.8	7.3	100
104G	7.12	5.7	41,400	138	8.1	7.5	100
104Z	7.12	5.6	34,100	138	7.5	7.8	100
205B	6.22	6.4	38,300	134	7.1	8.4	100
2051	7.12	5.3	44,800	142	8.2	8.4	100
104S	9.93	7.7	46,900	164	7.7	7.8	100
104R	7.91	7.2	43,600	146	7.7	7.6	100
2053	8.59	6.0	46,800	154	8.0	8.5	100
205T	9.94	7.1	51,900	170	7.9	8.7	100
205L	6.22	6.4	40,500	134	8.3	8.7	100
5_TS	8.91	6.6	45,000	154	8.4	7.8	100
5_BA	5.32	6.4	32,900	126	7.4	7.8	100
5_TL	6.73	6.0	40,400	137	7.4	8.0	100
5_TX	8.91	6.6	51,800	154	8.3	7.7	100
5BVA	8.60	7.4	47,800	142	8.4	7.7	100
XX1	7.00	7.0	38,000	145	7.5	8.0	100
XX2	7.00	7.9	40,000	145	7.8	8.3	100
XX3	7.00	5.5	40,000	145	7.5	8.0	100
XX4	8.00	5.5	43,000	150	8.0	8.0	95
XX5	9.00	6.5	47,000	160	8.5	8.7	98
XX6	10.00	6.5	52,000	170	8.5	8.7	90

models, on a scale of 0–100. The information is displayed in a LOTUS 123 spreadsheet (see Table 11.1).

Some of the numerous graphic representations will be presented. Figures 11.5 and 11.6 show the marginal utility functions (mostly hypothetical here) that we assume are those assessed by the participants.

We assume that participants do not want to share information on their marginal utility functions, at least not at the beginning of the discussions, and that they wish only to compare the overall results of their utility evaluations. In other words, assuming they agreed to use the same criteria evaluations given in Table 11.1, they do not wish to communicate at this stage the criteria they used or the "weights" given to them, but only the overall utility results.

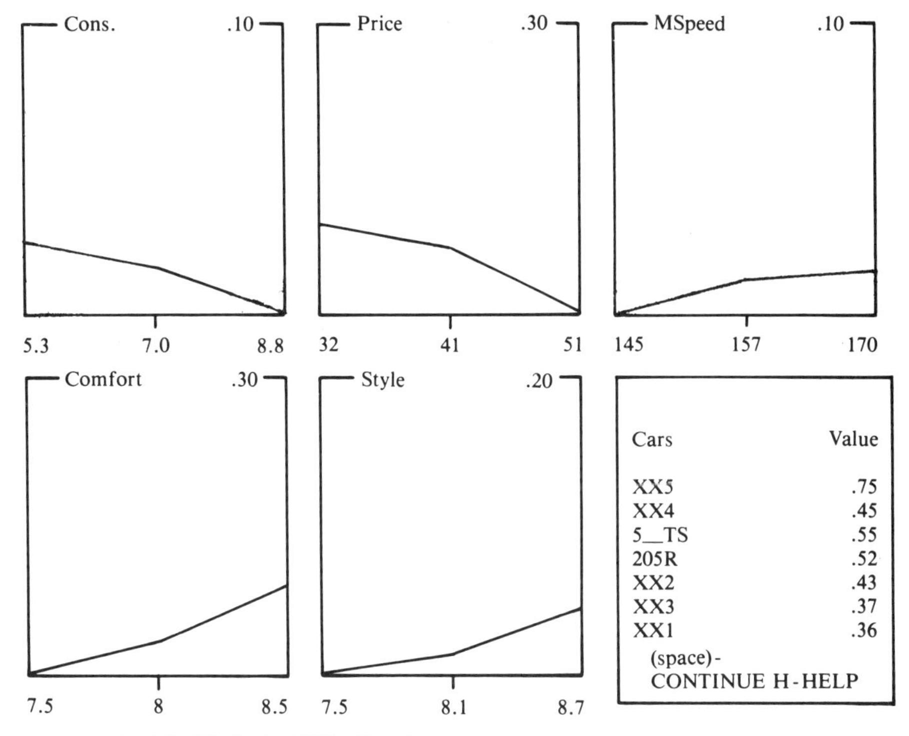

Figure 11.5. **The Marketing Utility Function**

Using PC123, we build a new spreadsheet. After eliminating all the cars not acceptable by both participants, except the alternative $XX6$, we get Table 11.2. Note that we now have two new columns that contain the overall utility evaluations of marketing (MKG) and engineering (ENG). The value -0.2 means that the corresponding alternative is not acceptable for the participant (as a result of using PREFCALC). This value is arbitrary and is chosen to have (using 123) the graphic of Figure 11.7 (see also Figures 11.3 and 11.4). Figure 11.7 shows that $XX6$ is not acceptable for both participants. $XX2$ is acceptable for marketing but not for engineering.

In order to better represent the acceptable alternatives, Figure 11.8 shows a zoom effect on Figure 11.7. It is a very easy task to develop Figure 11.8 from Figure 11.7 with 123. One needs only to define manual scales instead of the automatic scales used as the default option in building graphics. Of course, these graphics can be saved using the "name" option in the graph menu of 123.

In order to get some insight as to the reasons why $XX3$ and $XX5$ are so much opposed in the preference space (Figures 11.7 and 11.8), the participants can use a graphic of the profiles of these alternatives as shown in Figure 11.9. Because the criteria are scaled differently, we need some, but easy, manipula-

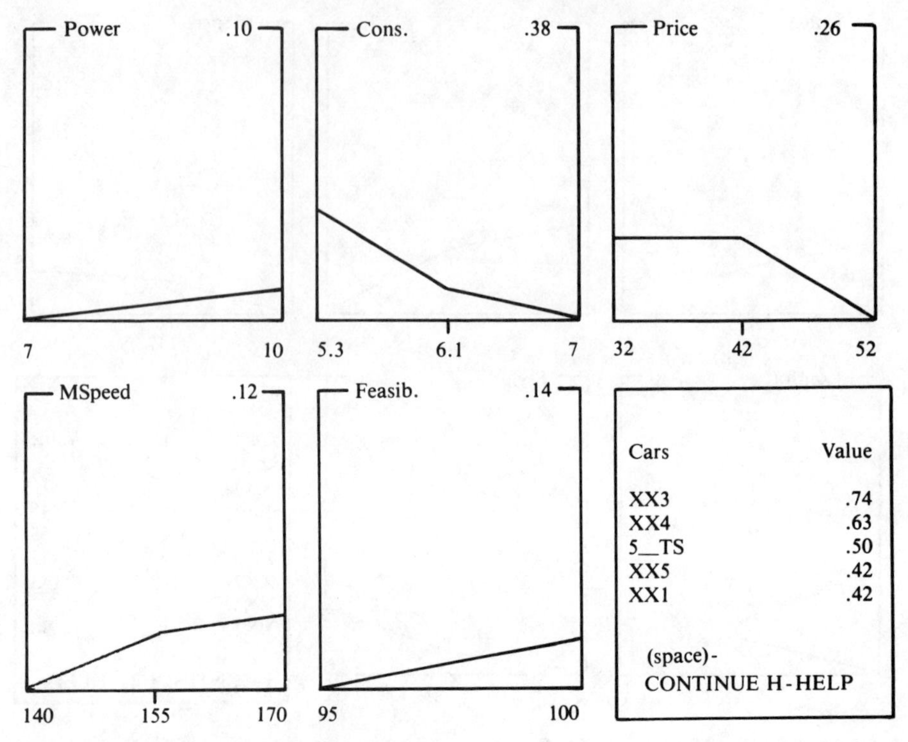

***Figure 11.6.* The Engineering Utility Function**

Table 11.2 Spreadsheet with Utilities

Cars	*Power*	*Cons.*	*Price*	*Speed*	*Comf.*	*Style*	*Feas.*	*MKG*	*ENG*
NB. L	1	−2	−2	2	2	2	1	1	1
MAX	10.71	8.8	52	170	8.5	8.7	100	1	1
MIN	5.32	5.3	32.9	126	7.1	7.3	90	−0.20	−0.20
RI65	8.71	8.6	43.5	150	7.7	8.1	100	0.30	−0.20
205R	8.58	6	48.9	154	8.2	8.5	100	0.52	0.48
2053	8.59	6	46.8	154	8	8.5	100	0.49	0.54
5_TS	8.91	6.6	45	154	8.4	7.8	100	0.56	0.50
5_TX	8.91	6.6	51.8	154	8.3	7.7	100	−0.20	0.31
XX1	7	7	38	145	7.5	8	100	0.36	0.42
XX2	7	7.9	40	145	7.8	8.3	100	0.43	−0.20
XX3	7	5.5	40	145	7.5	8	100	0.37	0.73
XX4	8	5.5	43	150	8	8	95	0.45	0.63
XX5	9	6.5	47	160	8.5	8.7	98	0.75	0.42
XX6	10	6.5	52	170	8.5	8.7	90	−0.20	−0.20

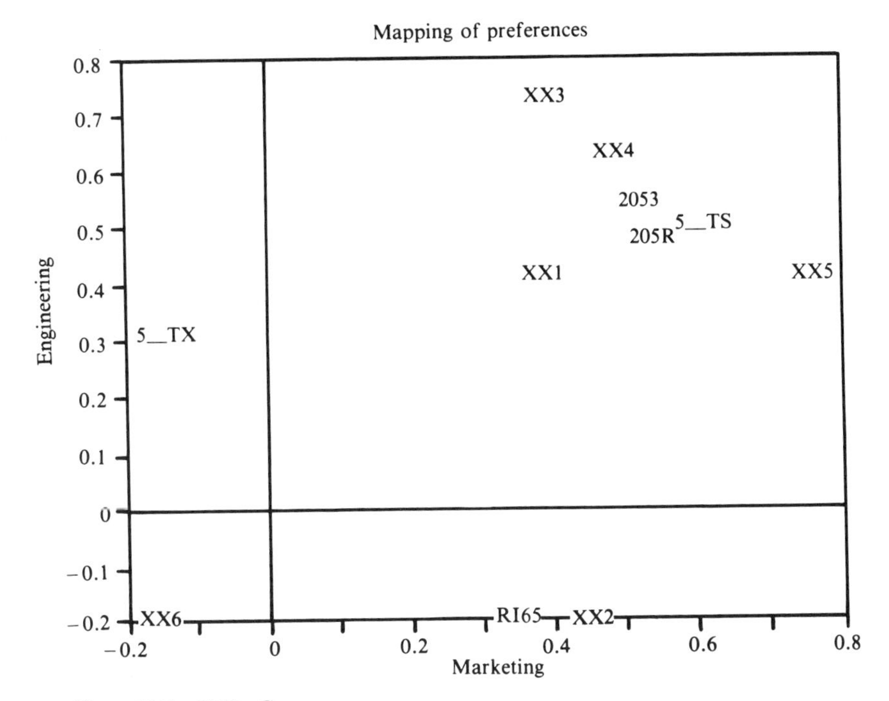

Figure 11.7. **Utility Space**

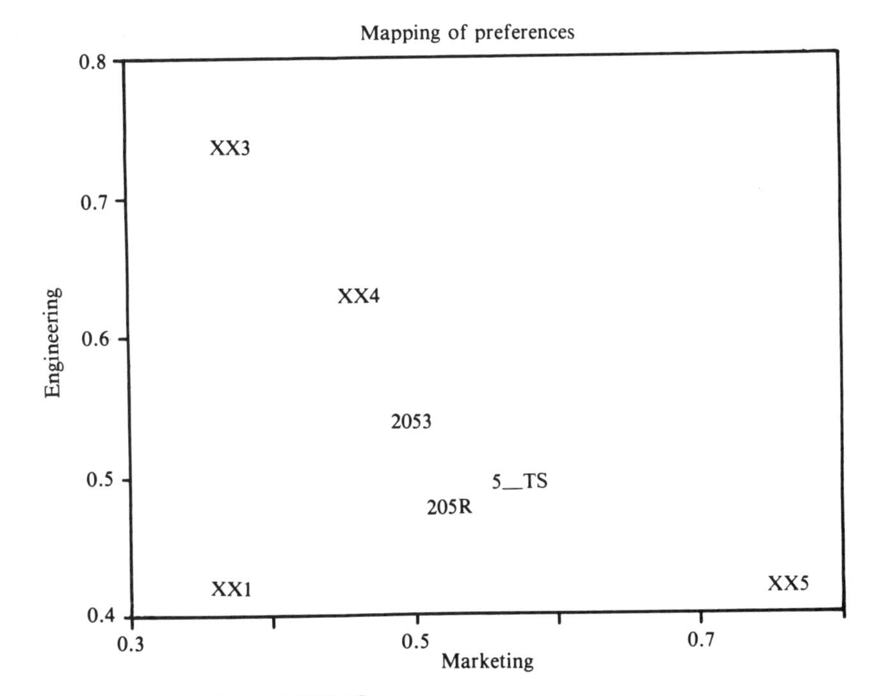

Figure 11.8. **Zoomed Utility Space**

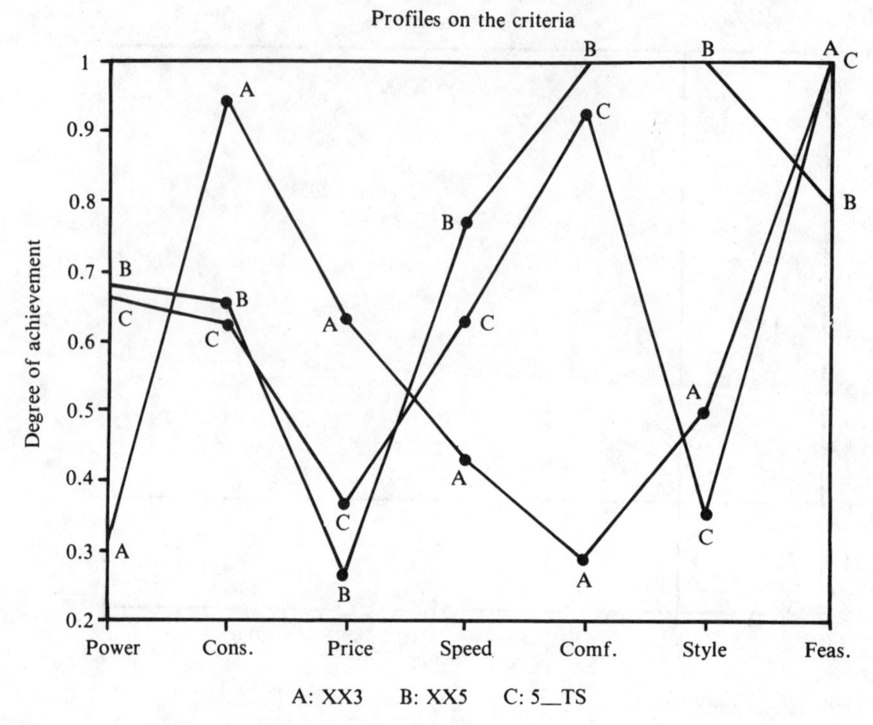

Figure 11.9. **Profiles of Alternatives**

tion under 123. Each criterion is scaled as a degree of achievement (g-gmin)/(gmax-gmin) for a "benefit" criterion.

Of course at any stage of the process the participants could share more preference information, such as their marginal utility functions. Then we could, for instance, superimpose on the same graph their marginal utilities from Figures 11.5 and 11.6 and reveal clearly the criteria chosen and, more generally, their preference structures.

11.5 Concluding Remarks

The decision process involved in new product design is complex and subtle. Traditional operations research tools such as optimization methods are not directly relevant. Even the NSS approach developed here could appear too formalized in some practical situations. However, we believe that this approach is more relevant because participants keep control of the process. They can always rely on their own intuition and nonformalized view of the problem, but they can also, individually or in a group, confront their view with the problem representation they formalize using the NSS. Doing so, they will probably modify their initial view.

Considering process complexity, in actual practice before trying to use the NSS for new car design, we are implementing it on more limited problems, such as choosing the noise level for a new car.

REFERENCES

Giordano, J. L. 1982 "Construction de function de préférence de clients à partir de notes de satisfaction." Working Paper, Regie RENAULT.

Jacquet-Lagreze, E. 1981. "Systèmes de décision et acteurs multiples—Contribution à une theorie de la décision pour les sciences des organizations." Thèse d'état en Gestion, Université de Paris-Dauphine.

______. 1985. PREFCALC, Version 2.0. EURODECISION, BPS7, 78530 Buc, France.

______, and Siskos, J. 1982. "Assessing a Set of Additive Utility Functions for Multicriteria Decision-Making, the UTA Method." *European Journal of Operational Research*, 10, No. 2, June.

______, and Shakun, M. F. 1984. "Decision Support Systems for Semi-Structured Buying Decisions." *European Journal of Operational Research*, 16, No. 1, pp. 48–58. Also in Shakun, M. F. 1988. *Evolutionary Systems Design: Policy Making Under Complexity and Group Decision Support Systems*, Holden-Day, Oakland, CA.

Jarke, M., Jelassi, M. T., and Stohr, E. A. 1984. "A Data Driven User Interface for a Generalized Multiple Criteria Decision Support System." *Proceedings of the IEEE Workshop on Languages for Automation*, New Orleans, pp. 127–133.

Jarke, M., Jelassi, M. T., and Shakun, M. F. 1985. "MEDIATOR: Toward a Negotiation Support System." Graduate School of Business Administration, New York University. Also in Shakun, M. F. 1988. *Evolutionary Systems Design: Policy Making Under Complexity and Group Decision Support Systems*. Holden-Day, Oakland, CA.

Jelassi, M. T., Jarke, M., and Stohr, E. A. 1985. "Designing a Generalized Multiple Criteria Decision Support System." *Journal of Management Information Systems*, 1, No. 4, Spring.

Shakun, M. F. 1985. "Decision Support Systems for Negotiations." *Proceedings of the IEEE International Conference on Systems, Man, and Cybernetics*, Tuscon, AZ. Revised version in Shakun, M. F. 1988. *Evolutionary Systems Design: Policy Making Under Complexity and Group Decision Support Systems*. Holden-Day, Oakland, CA.

______. 1988. *Evolutionary Systems Design: Policy Making Under Complexity and Group Decision Support Systems*. Holden-Day, Oakland, CA.

Sprague, H. M., and Carlson, E. D. 1982. *Building Effective Decision Support Systems*. Prentice Hall, Englewood Cliffs, NJ.

CHAPTER 12

Using MEDIATOR for New Product Design and Negotiation

Eric Jacquet-Lagreze* and Melvin F. Shakun**

New product decisions represent important strategic choices for companies. The design and decision process frequently involves several departments, e.g., marketing, engineering, finance, and the company president. The various players often have different views, which may be partly in conflict. In this chapter we use MEDIATOR—a negotiation on support system (NSS) based on evolutionary systems design (ESD)—to support new product design and negotiations, drawing on Giordano, Jacquet-Lagreze, and Shakun (1985), who operationalize the ESD group problem representation for the new product problem. We further operationalize the evolution of group problem representation by invoking the controls/goals referral process. MEDIATOR supports this evolution, which involves creating additional new designs, concensus seeking, compromise, and decision.

12.1 Introduction

Group decision support systems (GDSS)—also referred to as negotiation support systems (NSS)—provide decision support for problem definition and solution (design) in complex contexts involving multiplayer, multicriteria, ill-structured, dynamic problems. Shakun (1981a, 1981b, 1988) develops evolutionary systems design (ESD) as an artificial intelligence (AI) framework for

*LAMSADE, University of Paris-Dauphine, France.

**New York University, New York, U.S.A.

GDSS. The ESD methodology is background for two complementary papers. In one, Jarke, Jelassi, and Shakun (1985), applying the ESD framework, develop MEDIATOR, a data base–centered GDSS. An application to group car-buying decisions is presented, and system architecture is developed. In the other paper Giordano, Jacquet-Lagreze, and Shakun (1985) operationalize the ESD group problem representation for the design and negotiation of new products. This chapter illustrates the use of MEDIATOR for new product design and negotiation, drawing on these two previous papers. Although the research was undertaken with the cooperation of a major multinational company, the scenario developed here employs student-faculty players and a fictitious product.

THE DESIGN / NEGOTIATION SETTING

A group of three players ($j = 1, 2, 3$), departments in a company, is involved in designing a new product for the ultralight airplane market. Player $j = 1$ is marketing, player $j = 2$ is engineering, and player $j = 3$ is finance. The group will meet with a fourth player ($j = 4$), the company president, after a preliminary design phase.

Players agree to use the group decision support system MEDIATOR (see Jarke, Jelassi, and Shakun 1985) to support design and negotiation of the new product based on the ideas developed by Giordano, Jacquet-Lagreze, and Shakun (1985). Using MEDIATOR, players can together build a common (group) joint problem representation (multiuser case)—in effect, joint mappings from control (decision) space to goal (criteria) space to preference (utility) space. At each stage in the design/negotiation process the common joint problem representation shows the acknowledged degree of consensus (or conflict) among the players, i.e., at each stage players may show different individual problem representations. The evolution of problem representation can be described as a process of consensus seeking–through sharing of views, which constitutes exchange of information–within which compromise is possible. Thus at any stage of consensus seeking MEDIATOR can support compromise through the use of axiomatic solution concepts and/or concession-making procedures in the MEDIATOR model base. Computer display of the evolving problem representation can be used to support continued consensus seeking.

In the design phases MEDIATOR supports players 1, 2, 3. In a later meeting the president (player $j = 4$) acts as a human mediator who makes the final new product decision supported by MEDIATOR. The group interpersonal settings are basically cooperative.

12.2 Initial Design Phase

In the initial design phase players 1, 2, 3, consider four ultralight plane designs —A, B, C, D. Assume each player has worked alone with his individual decision support system (IDSS). Using PREFCALC—a microcomputer program for assessing piecewise linear, additive utility functions (Jacquet-Lagreze 1985)—he has established his initial individual problem representation, i.e., his own mappings from control space to goal space (and through marginal utility functions) to utility space—see Jacquet-Lagreze and Shakun (1984) and Jarke, Jelassi, and Shakun (1985). Then for the group meeting MEDIATOR in Table 12.1 shows the initial group problem representation—group mappings from control to goal (criteria) to utility spaces—based on the union of the individual problem representations established previously.

As shown in Table 12.1, the controls are $x1$ = maximum speed, km/hr; $x2$ = fuel consumption, km/liter, at 120 km/hr; and $x3$ = selling price, thousands of French francs (FF). The goals (criteria) are as follows:

$y1j$ = annual sales volume as estimated by player j, thousands of planes sold, where $j = 1, 2, 3$ (In Table 12.1 player 3 has used player 1's sales estimates, i.e., $y13 = y11$.)

$e1j$ = estimated percent error in $y1j$, where $j = 1, 3$ (Player 2 does not use $e12$ as a criterion and so does not estimate it.)

$y2$ = unit manufacturing cost, thousands of FF

$y3j$ = annual profit, billions of French francs, based on player j's sales estimate $y1j$ (We have $y3j = (x3 - y2)y1j$ where $j = 1, 2, 3$.)

$y4$ = investment, billions of French francs

$y5j$ = annual percent return on investment based on player j's profit estimate $y3j$ where $j = 2, 3$ (Player 1 does not use $y51$ as a criterion and so does not compute it. We have $y5j = 100\ y3j/y4$ for $j = 2, 3$.)

$y6$ = $x1$ = maximum speed, km/hr

$y7$ = $x2$ = fuel consumption, km/liter. (In Europe the unit liters/100 km is normally used, but to facilitate understanding for English readers the reciprocal km/liter is used here in analogy to miles/gallon.)

As mentioned previously, using PREFCALC the three players had determined their respective utilities, $u1$, $u2$, $u3$, shown in Table 12.1 by developing their individual marginal utility functions now plotted by MEDIATOR on the same graphical axes in Figure 12.1 to facilitate comparison.*

In establishing their individual problem representations prior to the group meeting all players knew the control values ($x1$, $x2$, $x3$) for the four planes. Marketing (player 1) estimated annual sales $y11$ and its percent error $e11$ based on market research. This could include use of a mathematical model of

*In Table 12.1 for each player PREFCALC also lists the utility of the IDEAL (ANTI-IDEAL)—a conceptual plane design having the most (least) preferred value in the feasible set of plane designs considered for each criterion used by that player, but a design that, in general, is not feasible—as one (zero).

Table 12.1 **Initial Group Problem Representation**

	Controls			Criteria														Utilities		
Plane design	*Speed* $x1$	*Fuel* $x2$	*Price* $x3$	*Sales* $y11$	$y12$	$y13$	*% errors* $e11 = e13$	*Cost* $y2$	*Profit* $y31$	$y32$	$y33$	*Invest* $y4$	*% invest return* $y52$	$y53$	*Speed* $y6$	*Fuel* $y7$	$u1$	$u2$	$u3$
A	260	8.6	125	90	160	90	8.6	95	2.7	4.8	2.7	19	25.3	14.2	260	8.6	.00	.67	.00
B	140	14.3	60	450	470	450	6.7	50	4.5	4.7	4.5	16	29.4	28.1	140	14.3	.62	.60	.95
C	120	11.8	50	575	550	575	5.9	43	4.0	3.8	4.0	18	21 4	22.4	120	11.8	.77	.21	.51
D	220	11.5	100	220	240	220	4.8	79	4.6	5.0	4.6	17	29 7	27.2	220	11.5	.51	.63	.85
Ideal				575	550	—	4.8	—	4.6	4.8	4.6	16	29.7	28.1	260	14.3	1	1	1
Anti-ideal				90	160	—	8.6	—	2.7	3.8	2.7	19	21.4	14.2	120	8.6	0	0	0

Table 12.2 **Initial Plane Rank Orders and Utilities**

Player 1	*Player 2*	*Player 3*
C .77	A .67	B .95
B .62	D .63	D .85
D .51	B .60	C .51
A 0	C .21	A 0

Table 12.3 **Initial Return on Investment Data**

Plane design	*Percent return on investment* $y52$	$y53$
A	25.3	14.2
B	29.4	28.1
C	21.4	22.4
D	29.7	27.2

the form:

$$y11 = f(x1, x2, x3, \text{marketing mix})$$

i.e., sales is a function of the controls (product characteristics—speed $x1$, fuel $x2$, and price $x3$) and the marketing mix (advertising expenditure, warranty terms, etc.). Based on $y11$ and estimated unit manufacturing cost $y2$ obtained from engineering, marketing estimated profit $y31$. Marketing used sales, percent error in sales, and profit criteria ($y11$, $e11$, $y31$) in establishing its utility $u1$. Note that percent error in sales $e11$ is also reflected as percent error in profit.

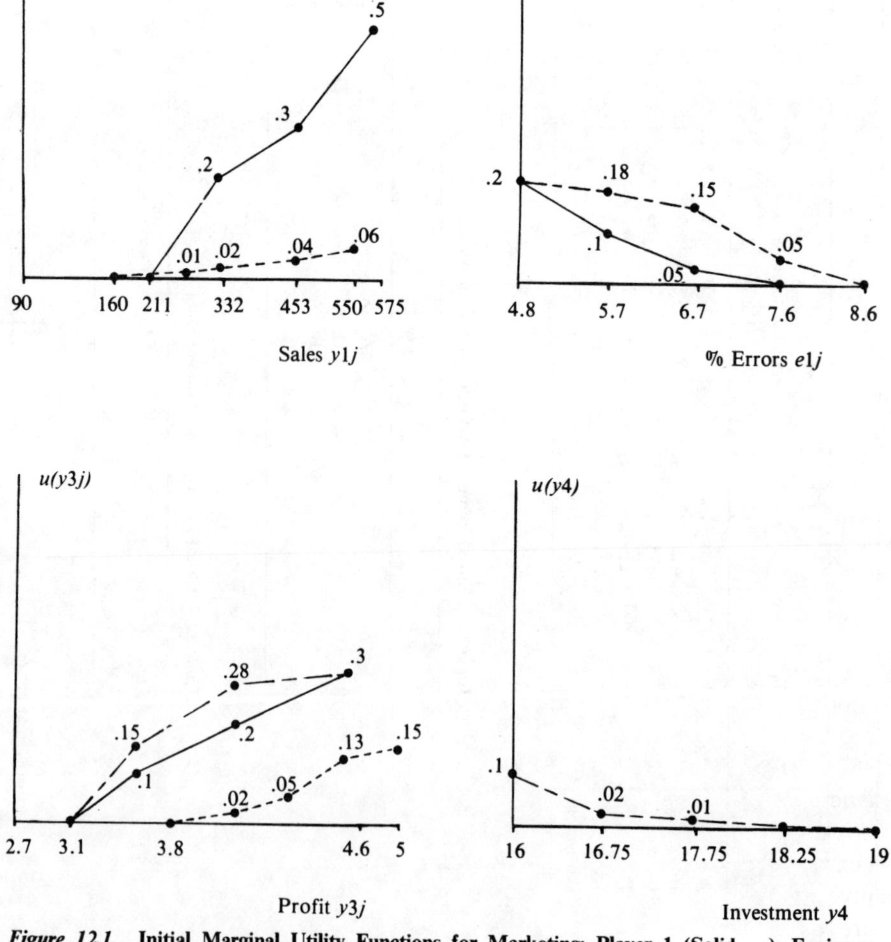

Figure 12.1. **Initial Marginal Utility Functions for Marketing: Player 1 (Solid —); Engineer: Player 2 (Dotted •••••); Finance: Player 3 (Dash-Dotted –•–•–).**

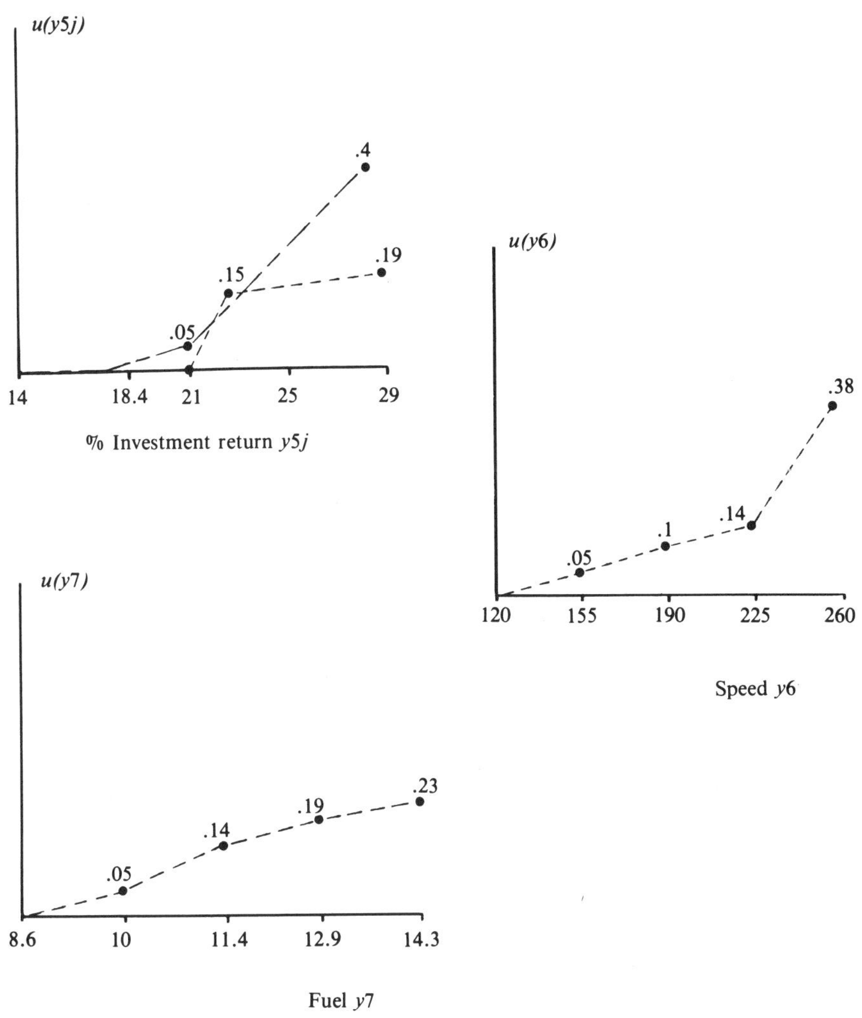

Figure 12.1. **Continued.**

Engineering (player 2) made its own estimate of sales $y12$ considering the product characteristics ($x1$, $x2$, $x3$). Based on $y12$ and unit cost $y2$, engineering estimated profit $y32$ and percent return on investment $y52$ (based on $y32$). Engineering used sales, profit, investment return, and product characteristic criteria ($y12$, $y32$, $y52$, $y6$, and $y7$) in establishing its utility $u2$.

Finance (player 3) accepted marketing's estimate of sales and sales error, i.e., $y13 = y11$ and $e13 = e11$. Finance used profit, investment, and percent investment return criteria: profit estimate $y33$ (which equals $y31$ because they are based on $y13$ and $y11$, respectively, and $y13 = y11$), investment $y4$ (obtained from engineering), percent return on investment $y53$ (based on

$y33 = y31$), and percent error in sales $e13 = e11$ (which is also reflected as percent error in profit and percent return on investment) in establishing its utility $u3$.

A criterion not used by a player carries a weight of zero. In such a case, for simplicity in Figure 12.1, his utility graph for this criterion, which would be zero throughout, is not shown. Although used in estimating profit (and hence return on investment), unit cost $y2$ was not used directly by any player as a criterion, as such. Although listed for information in Table 12.1, $y2$ is not shown at all in Figure 12.1 because it carries a criterion weight of zero for all players. In Table 12.1 the dash for Ideal and Anti-ideal means the criterion was not used.

At the group meeting MEDIATOR displays Table 12.1. From the latter MEDIATOR is then requested to show the plane rank orders and utilities shown in Table 12.2.

Players clearly disagree as to their plane preference. The conflict arises from differences in their individual problem representations—in the underlying marginal utility functions shown in Figure 12.1 and (returning to Table 12.1) in differences in estimated sales and, consequently, profit and return on investment. As players are about to discuss these differences—engage in consensus seeking through exchange of information—player 3, finance, asks for the display shown in Table 12.3, i.e., the projection of Table 12.1 onto plane design, $y52$ and $y53$.

12.3 Design Evolution

Player 3 notes that, although percent return on investment differs among plane designs and players, what strikes him is that none of them is over 30 percent. He asks whether it would be possible to obtain a yield of over 30 percent on investment return. The other players reply that they don't know. Player 3 asks MEDIATOR to insert an a priori goal target constraint, $y53 \geq 30$. He then asks MEDIATOR to display the intersection of the technologically feasible set as shown in Table 12.1 and the a priori goal target. As expected, MEDIATOR replies that the intersection is empty (because no planes have an investment return over 30 percent). Player 3 asks for help. MEDIATOR offers three menu choices:

1. Expand feasible set.
2. Expand a priori goal target.
3. Change dimensions of control or criteria space (or values—not discussed here, see Shakun 1981a, 1988).

Player 3 selects option 1, expand feasible set. MEDIATOR offers two alternatives:

1. Review investment return data.
2. Add new plane design alternatives.

Player 3 selects option 1, review investment return data. MEDIATOR suggests player 3 review all data arising in the calculation of investment return $y53$:

1. investment return $y53 = 100$ profit $y33$/investment $y4$
2. profit $y33 =$ (price $x3$ − cost $y2$) sales $y13$
3. review $x3$, $y2$, $y13$, $y4$

Player 3 asks MEDIATOR for the projection of Table 12.1 onto plane design, $x3$, $y2$, $y13$, $y4$. Player 3 decides these data are correct and does not change them. He has not expanded the feasible set by this option. He then selects menu option 2, add new plane design alternatives. MEDIATOR suggests player 3 try to generate new plane design alternatives having $y53 \geq 30$. Player 3 asks player 2, engineering, who replies that three designs—A, D, B—are on the speed-fuel consumption technological frontier, and all four designs are on the (Pareto optimal) frontier if price is also considered. Therefore engineering says other designs would not yield an investment return over 30 percent. It seems at this point that the feasible set cannot be expanded by adding new design alternatives that give a $y53 \geq 30$.

Player 3 returns to the previous "help" menu and selects option 2, expand a priori goal target. Here MEDIATOR suggests lowering aspiration level $y53$ below 30 percent. Player 3 does not want to do this, at least not yet. Because he does not at this point wish to expand the a priori goal target, he chooses to return to the "help" menu and selects option 3, change dimensions of control or criteria space. MEDIATOR replies as follows:

1. Change control or criteria dimensions using controls/goals referral process (see Shakun 1988).
2. Change criteria dimensions or values using goals/values referral process (see Shakun 1981a, 1988). This option is not discussed here.

Player 3 chooses menu option 1, change control or criteria dimensions using controls/goals referral process. MEDIATOR replies:

> Use controls/goals referral process: Goals are given as rows and controls as columns. Use the following heuristic: Given a particular goal dimension (row) and looking at the control dimensions (columns), ask whether there is any other new control dimension which also delivers the goal. Insert:
>
> Goals: $\underline{y53}$
>
> Controls: $\overline{\underline{x1, x2, x3}}$

Player 3 has inserted $y53$ for goals and $x1, x2, x3$ for controls, so MEDIATOR displays the following controls/goals relation (trimatrix):

	speed	*fuel*	*price*
	$x1$	$x2$	$x3$
invest. return, $y53$	*, *, 1	*, *, 1	*, *, 1

Symbol 1 indicates that at present $y53$ is being delivered by $x1$, $x2$, and $x3$ for player 3. Symbol * means that players 1 and 2 are not using $y53$.
MEDIATOR asks:

> Given the goal dimension return on investment $y53$, is there any other new control dimension which delivers this goal dimension? While player 3 is thinking about this, engineering says: "How about a 2 seater ultraplane?" A new control dimension, seating capacity, has been generated. Since neither marketing nor engineering is against considering a control dimension $x4$, seating capacity, the following evolved controls/goals tri-matrix is displayed:

	speed	*fuel*	*price*	*capacity*
	$x1$	$x2$	$x3$	$x4$
invest. return $y53$	*,*,1	*,*,1	*,*,1	*,*,1

MEDIATOR asks if player 3 wants to consider additional change in control or criteria dimensions. When player 3 says "yes," MEDIATOR suggests considering a heuristic to see if new criteria are now suggested by the introduction of control dimension $x4$. The heuristic is, "Given the control dimension seating capacity $x4$, is there any other new criterion which is delivered by this control dimension?" Player 2 tells player 3 that for him seating capacity itself is also a criterion, i.e., $y8 = x4$. Player 3 says that a new criterion $y9$, feasibility or probability of realizing the plane design, is suggested by control dimension $x4$ because planes having $x4 = 2$ are radically new designs not made before. Thus an evolved group problem representation will have new control dimension $x4$ and new criteria dimensions $y8$ and $y9$. The generation of these new dimensions is an example of cybernetic self-organization (see Shakun 1988).

The evolved group problem representation is shown in Table 12.4. Engineering has developed two additional plane designs, E and F, each having a two-seat capacity. Of course new dimensions $x4$, $y8$, $y9$ are shown. Figure 12.2 presents the $x1$, $x2$ projection of Table 12.4 to exhibit the technological frontiers.

In Table 12.4 players have estimated sales, $y11$, $y12$, $y13$, for the new designs E and F. By exchanging information, they have evolved their sales estimates, $y11$, $y12$, $y13$, for planes A, B, C, D to those shown in Table 12.4 from the initial ones in Table 12.1. Player 3 no longer is simply using $y13 = y11$. After participating in group discussions on sales estimates, he comes up with his own $y13$ estimates. Of course prior to MEDIATOR displaying Table 12.4 players had used their IDSS and PREFCALC to develop their evolved marginal utility functions, which MEDIATOR displays as Figure 12.3. Based on Table 12.4, MEDIATOR is asked to give planes on

Table 12.4 **Evolved Group Problem Representation**

	Controls				Criteria															Utilities		
Plane design	*Speed x1*	*Fuel x2*	*Price x3*	*Capa-city x4*	*Sales y11*	*y12*	*y13*	*% errors e11 = e13*	*Cost y2*	*Profit y31*	*y32*	*y33*	*Invest y4*	*% invest return y52*	*y53*	*Speed y6*	*Fuel y7*	*Capacity y8*	*Feasi-bility y9*	*u1*	*u2*	*u3*
A	260	8.6	125	1	100	150	120	8.6	95	3.0	4.5	3.6	19	23.7	18.9	260	8.6	1	1	.02	.43	.37
B	140	14.3	60	1	450	460	450	6.7	50	4.5	4.6	4.5	16	28.8	28.1	140	14.3	1	1	.44	.35	.65
C	120	11.8	50	1	560	560	560	5.9	43	3.9	3.9	3.9	18	21.8	21.8	120	11.8	1	1	.53	.29	.47
D	220	11.5	100	1	230	240	240	4.8	79	4.8	5.0	5.0	17	29.7	29.7	220	11.5	1	1	.47	.42	.77
E	160	12.4	80	2	470	400	440	9.0	65	7.0	6.0	6.6	21	28.6	31.4	160	12.4	2	.95	.77	.58	.5
F	210	9.9	120	2	250	290	270	9.0	95	6.2	7.2	6.7	22	33	30.7	210	9.9	2	.95	.69	.67	.48
Ideal					560	560	—	4.8	—	7.0	7.2	6.7	22	33	31.4	260	14.3	2	1	1	1	1
Anti-ideal					100	150	—	9.0	—	3.0	3.9	3.6	16	21.8	21.8	120	8.6	1	.95	0	0	0

Table 12.5 **Evolved Plane Rank Orders and Utilities**

Player 1	*Player 2*	*Player 3*
E .77	F .67	D .77
F .69	E .58	B .65
C .53	A .43	E .50
D .47	D .42	F .48
B .44	B .35	C .47
A .02	C .29	A .37

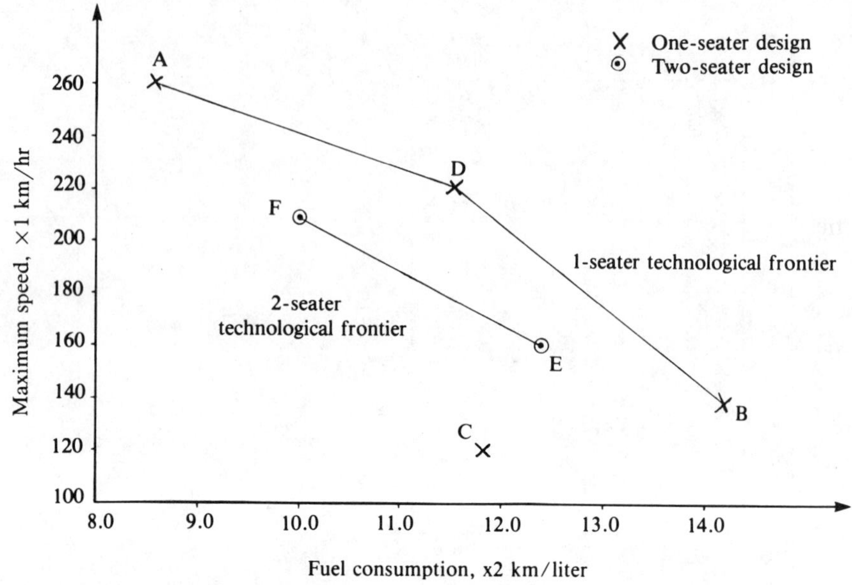

Figure 12.2. **Technological Frontiers.**

the Pareto optimal frontier in utility space (reply: D, E, F) and to show the plane rank orders and utilities, Table 12.5.

New designs E and F occupy the first two ranks for players 1 and 2, although in different order. For player 3 initial designs D and B continue to occupy the top two ranks, although their order is now reversed (compare with Table 12.2). Clearly, players differ in their plane preferences. Players decide to discuss differences in evolved marginal utility functions (Figure 12.3). Engineering feels that finance has put a very large weight, .33, on criterion $y9$, feasibility. "After all," engineering says, "we judge that the feasibility only drops from 1 to .95 in going from a one-seater plane design to a new two-seater design. To put a large weight of .33 on this seems excessive. We've used a weight of .12, as you can see." Finance, influenced by engineering's argument, decides to drop its weight to .25, still double that of engineering. Reusing PREFCALC, finance comes up with second evolved marginal utility functions. Players 1 and 2 do not make any changes in theirs, so their second evolved marginal utility functions are the same as in Figure 12.3. MEDIATOR displays the second evolved marginal functions, Figure 12.4, the second evolved group problem representation, Table 12.6 (only the last column $u3$ is changed from Table 12.4), and the second evolved plane rank orders and utilities, Table 12.7. When asked for Table 12.6, MEDIATOR again replies that planes D, E, F are on the Pareto optimal frontier in utility space.

Table 12.7 shows that, although plane D is still first for player 3, its utility has dropped to .71 from .77. Planes B and E are now tied for second place with a utility of .57 each, and F is close behind with a utility of .55. The rank orders and utilities for players 1 and 2 remain the same.

Player 1 asks MEDIATOR to compute the maximin compromise solution using the current (second) evolved group problem representation, Table 12.6. First MEDIATOR normalizes each player's utilities between 0 (for his last-choice plane) and 1 (for his first-choice plane). For each plane the normalized utility for each player and minimum utility comparing normalized utilities among players are computed and shown in Table 12.8.

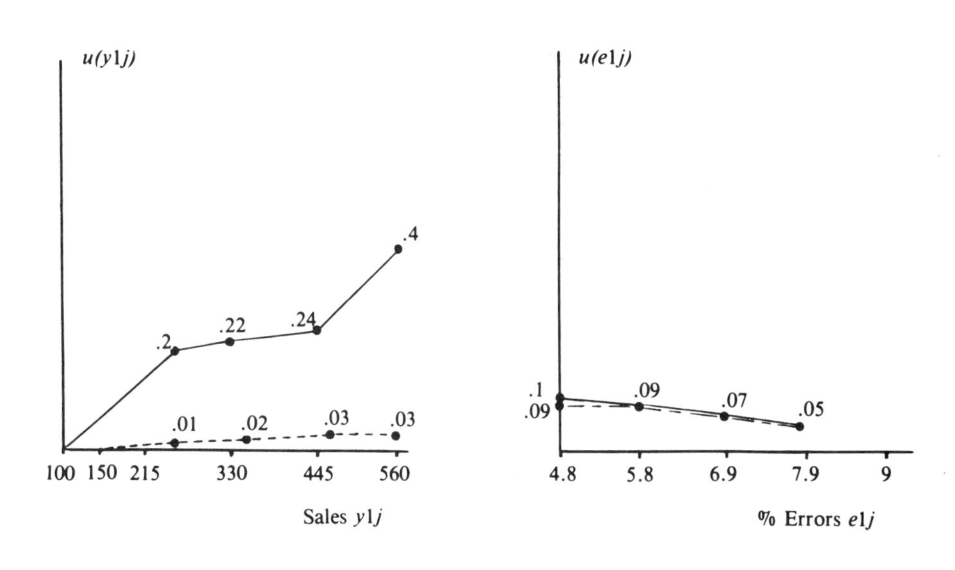

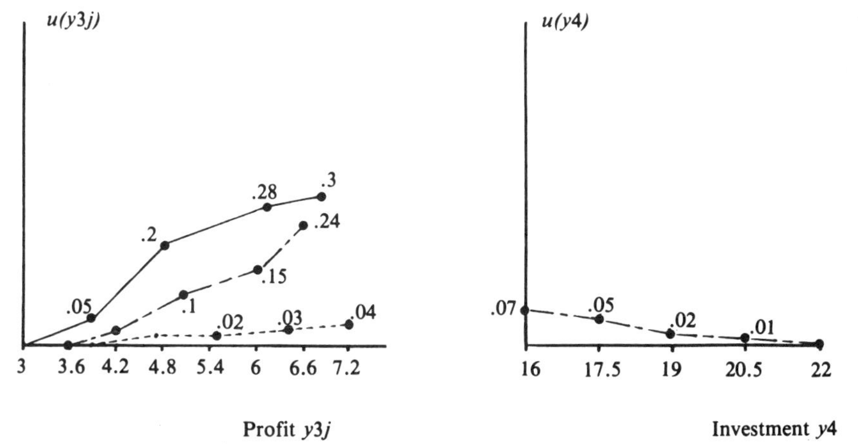

Figure 12.3. **Evolved Marginal Utility Functions for Marketing: Player 1 (Solid —); Engineer: Player 2 (Dotted •••••); Finance: Player 3 (Dash-Dotted –•–•–).**

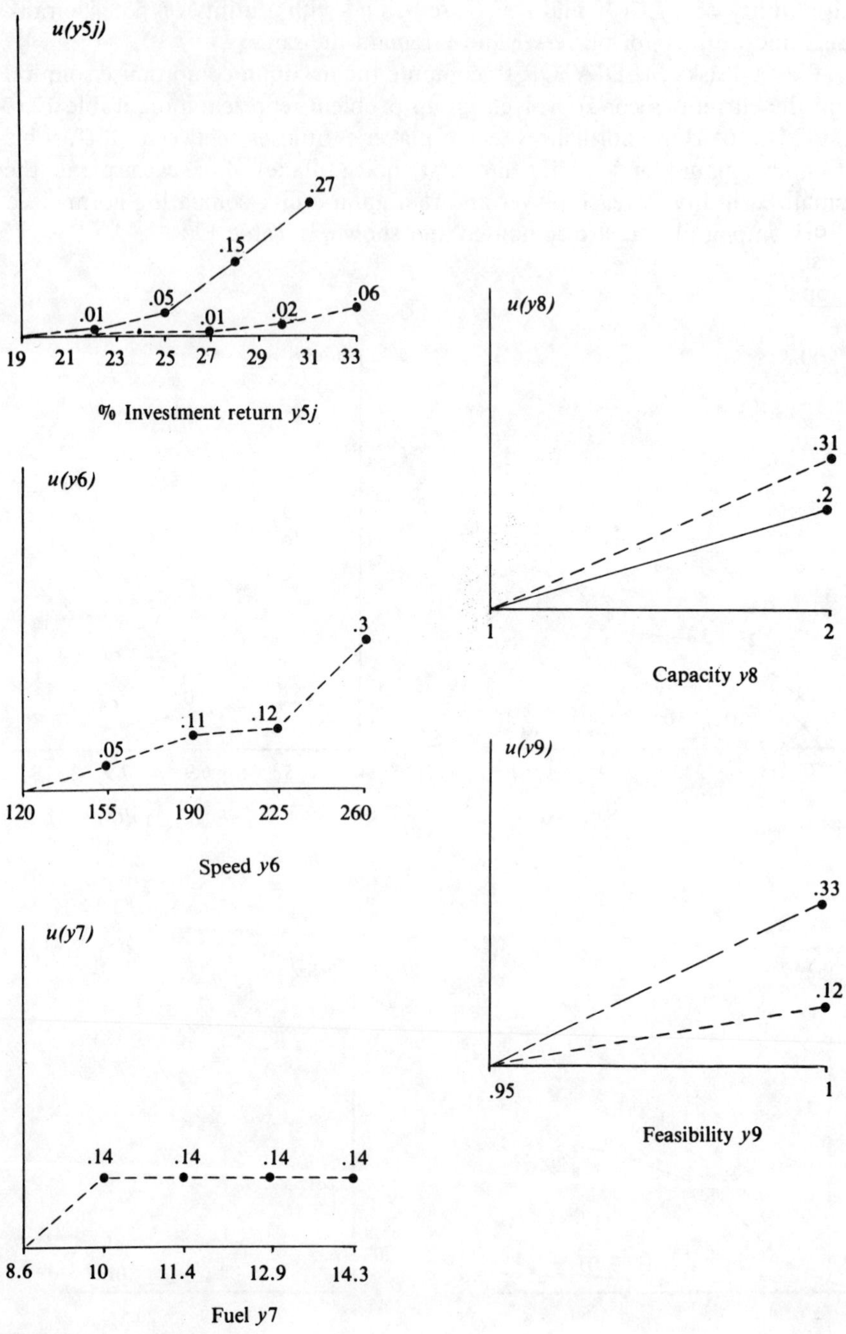

Figure 12.3. **Continued.**

The plane that maximizes the minimum utility, giving a maximin utility of .67, is design E, which is the maximin compromise solution. Of course MEDIATOR can support compromise through use of other axiomatic solution concepts, e.g. Nash, Kalai-Smorodinsky, as well as concession-making procedures, e.g. Rao-Shakun and conditional control target expansion (see Shakun 1985 and Jarke, Jelassi, and Shakun 1985).

Finance is not convinced to compromise on plane E. He still wants D. The three players are scheduled to have a group meeting with the company president, who will make the final decision in any case. "Let's see what happens there," finance says. Engineering agrees; he still wants plane F.

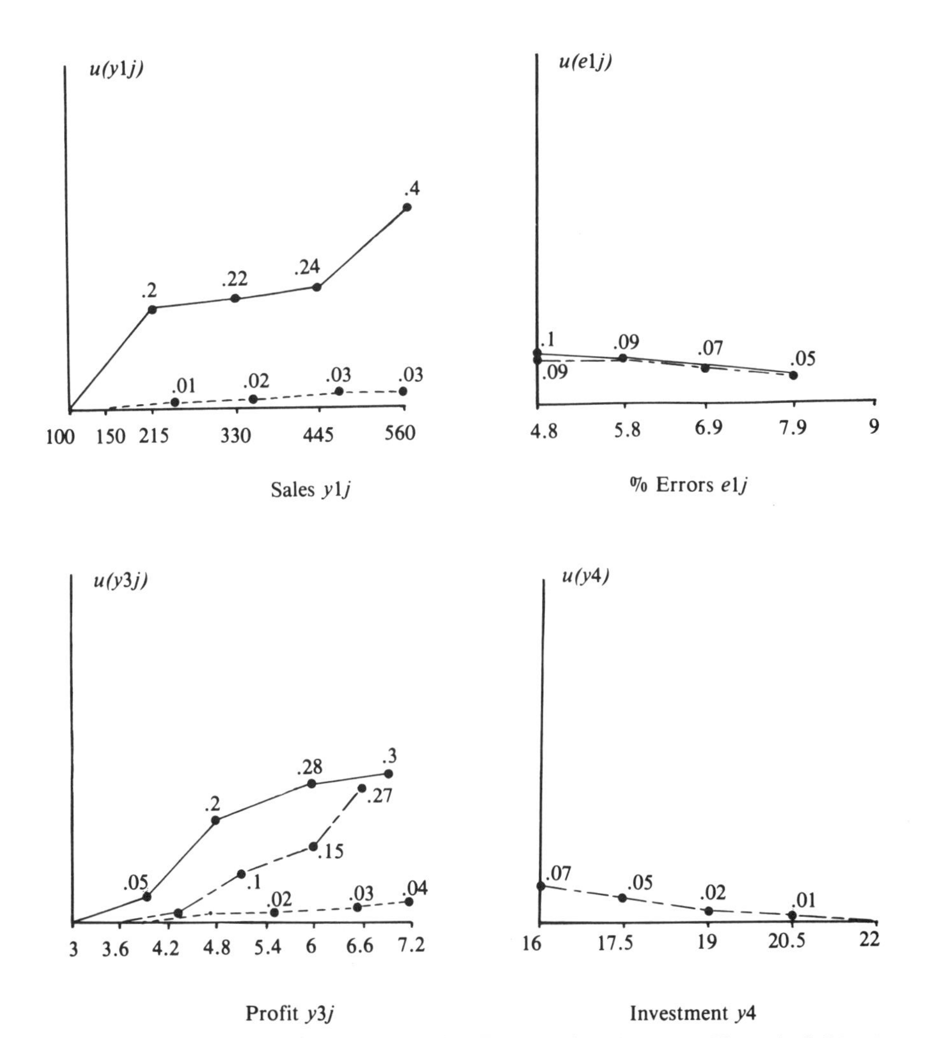

Figure 12.4. **Second Evolved Marginal Utility Functions for Marketing: Player 1 (Solid —); Engineer: Player 2 (Dotted ·····); Finance: Player 3 (Dash-Dotted –·–·–).**

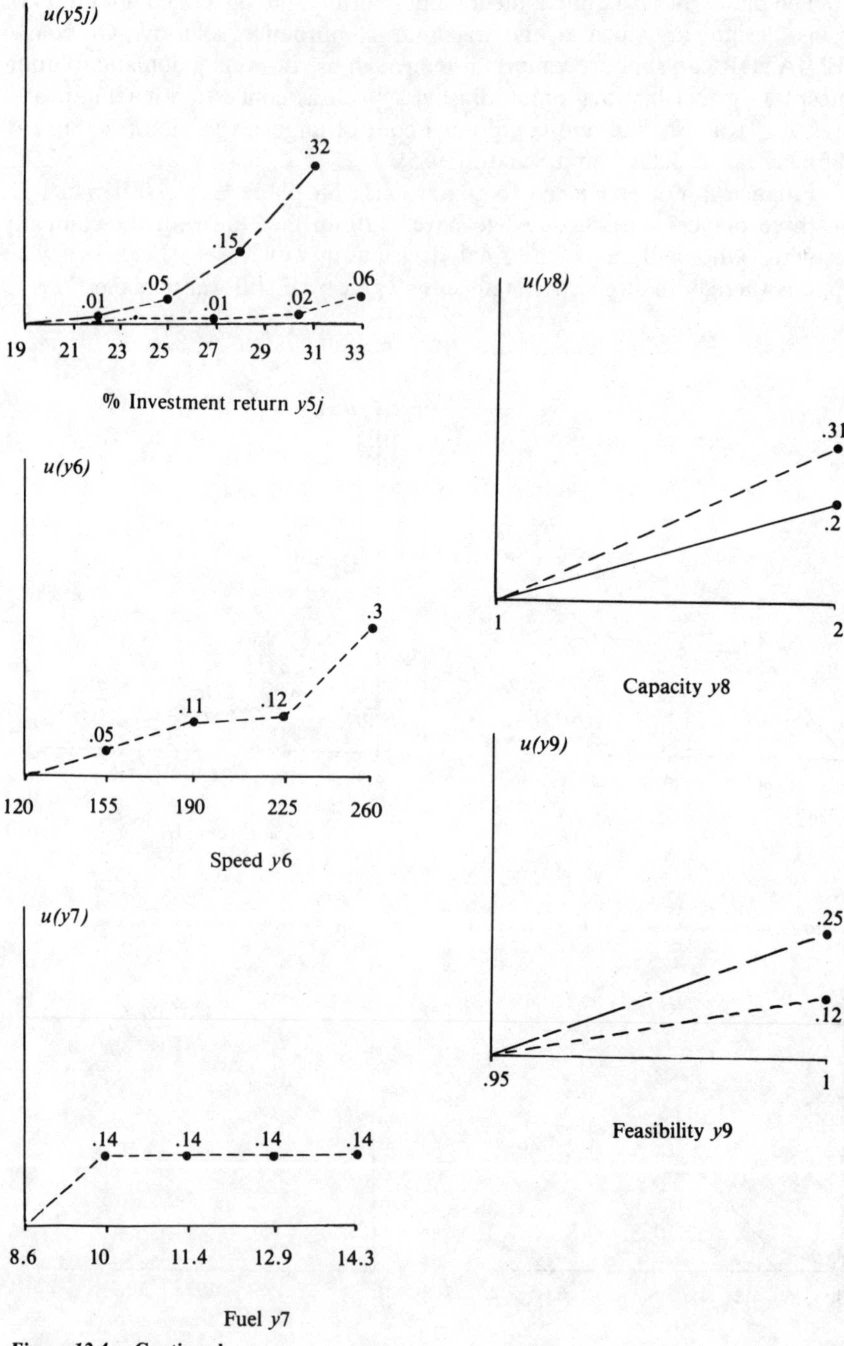

Figure 12.4. **Continued.**

Table 12.6 Second Evolved Group Problem Representation

	Controls				Criteria																Utilities		
Plane design	Speed $x1$	Fuel $x2$	Price $x3$	Capacity $x4$	Sales $y11$	$y12$	$y13$	% errors $e11 = e13$	Cost $y2$	Profit $y31$	$y32$	$y33$	Invest $y4$	% invest return $y52$	$y53$	Speed $y6$	Fuel $y7$	Capacity $y8$	Feasibility $y9$	$u1$	$u2$	$u3$
A	260	8.6	125	1	100	150	120	8.6	95	3.0	4.5	3.6	19	23.7	18.9	260	8.6	1	1	.02	.43	.29
B	140	14.3	60	1	450	460	450	6.7	50	4.5	4.6	4.5	16	28.8	28.1	140	14.3	1	1	.44	.35	.57
C	120	11.8	50	1	560	560	560	5.9	43	3.9	3.9	3.9	18	21.8	21.8	120	11.8	1	1	.53	.29	.39
D	220	11.5	100	1	230	240	240	4.8	79	4.8	5.0	5.0	17	29.7	29.7	220	11.5	1	1	.47	.42	.71
E	160	12.4	80	2	470	400	440	9.0	65	7.0	6.0	6.6	21	28.6	31.4	160	12.4	2	.95	.77	.58	.57
F	210	9.9	120	2	250	290	270	9.0	95	6.2	7.2	6.7	22	33	30.7	210	9.9	2	.95	.69	.67	.55
Ideal					560	560	—	4.8	—	7.0	7.2	6.7	22	33	31.4	260	14.3	2	1	1	1	1
Anti-ideal					100	150	—	9.0	—	3.0	3.9	3.6	16	21.8	21.8	120	8.6	1	.95	0	0	0

Table 12.7 Second Evolved Plane Rank Orders and Utilities

Rank order	Player 1	Player 2	Player 3
1	E .77	F .67	D .71
2	F .69	E .58	B .57
3	C .53	A .43	E .57
4	D .47	D .42	F .55
5	B .44	B .35	C .39
6	A .02	C .29	A .29

Table 12.8 **Second Evolved Normalized Utilities**

Plane	*Player 1*	*Player 2*	*Player 3*	*Minimum utility*
A	0	.37	0	0
B	.56	.16	.67	.16
C	.68	0	.24	0
D	.60	.34	1	.34
E	1	.76	.67	.67
F	.89	1	.62	.62

Maximin utility = .67.

Maximin compromise solution is plane E.

12.4 Meeting with the President: Decision

At the group meeting the president acts as a mediator supported by MEDIATOR. The president is also player 4 who, in fact, will make the final decision. The other three players tell the president that they have been working closely together. Supported by MEDIATOR, they have introduced the two-seater designs, E and F, representing a major evolution of the problem representation. However, players 1, 2, and 3 remain divided in that they prefer planes E, F, and D, respectively. The president asks MEDIATOR to display the current (second) evolved group problem representation of players 1, 2, 3 (Table 12.6). He observes differences among players as shown there and in Table 12.7. He notes the maximin solution is plane E, as shown in Table 12.8. Then the president looks at Figure 12.4, the current (second) evolved marginal utility functions. Considering that his colleagues represent different departments, he says that the marginal utility functions look reasonable, except he thinks the .25 weight on feasibility assigned by finance is high—too conservative. The president feels finance is a little too concerned about the 5 percent reduction in feasibility that engineering associates with going to the new two-seater design concept. Finance replies, "Who knows if it is only a 5 percent reduction? Maybe it is more, and engineering is optimistic in placing the feasibility of the two-seater designs at .95."

Returning to Table 12.6, the president listens thoughtfully as the other three players discuss differences in their sales estimates $y11$, $y12$, and $y13$. They do not change them, however. They tell him that some changes in sales estimate were made in previous meetings, but this is where they stand now. Of course these differences in sales estimates are reflected in variations in profit and return on investment, as everyone notes.

The president decides to develop his own individual problem representation on his IDSS. He adjourns the meeting until the next day. For his sales estimate, $y14$, the president decides to use the average of the estimates made by the other three players, i.e., $y14 = (y11 + y12 + y13)/3$. However for plane F the sales estimate $y14 = 270$ computed by averaging is too low in the president's judgment. He believes engineering's estimate of 290 is better, and

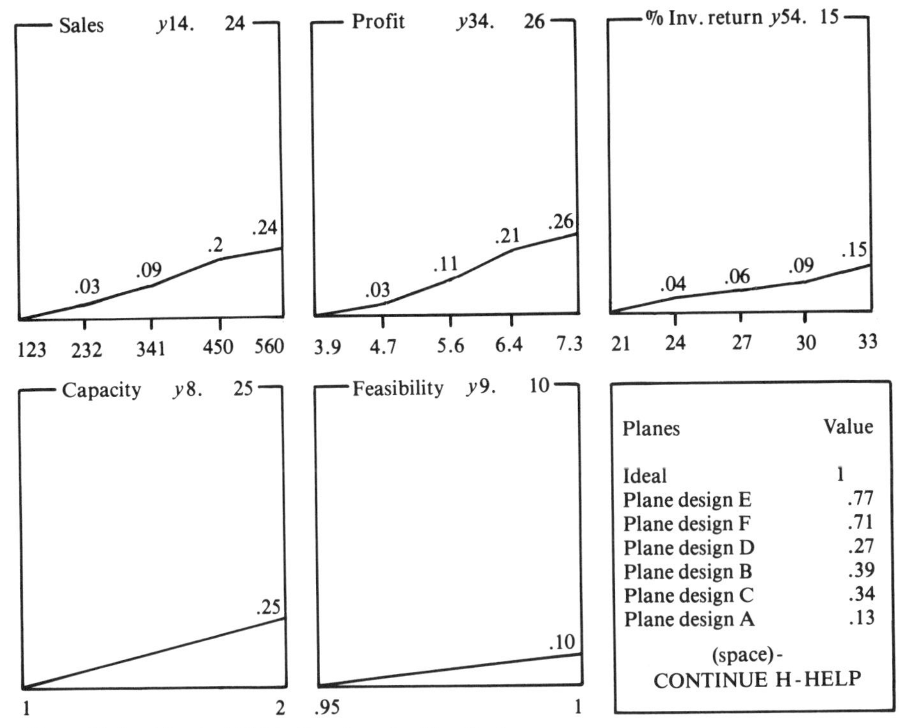

Figure 12.5. **Marginal Utility Functions for President.**

decides to use it. Then his IDSS computes profit as $y34 = (x3 - y2)y14$ and percent return on investment as $y54 = 100y34/y4$. The president works with PREFCALC to develop his marginal utility functions. He uses sales $y14$, profit $y34$, percent investment return $y54$, capacity $y8$, and feasibility $y9$ as criteria.

Figure 12.5 is the computer printout from PREFCALC of the president's marginal utility functions. The abscissas are the criteria $y14$, $y34$, $y54$, $y8$, $y9$ (for which the weights are shown as .24, .26, .15, .25, and .10, respectively), and the ordinates are the respective marginal utilities. The values of .77, .71, .27, .39, .34, .13 in the box are the president's total utilities $u4$ for planes E, F, D, B, C, A respectively.

Table 12.9 shows the president's individual problem representation. Plane E has the highest utility, .77, for the president, with plane F second, with a utility of .71. Plane D has a utility of only .27. The president thinks to himself that we wants one of the two-seater designs, E or F, but which? Plane E has the higher utility, .77, but F is close at .71. Marketing favors E and engineering, F. He decides to ask a "what-if" question, in effect a sensitivity analysis on sales. For plane F what if he uses sales $y14 = 270$, the average of the other three players' sales estimates, instead of the higher figure of 290?

Table 12.9 **President's Individual Problem Representation**

	Controls				*Criteria*					*Utility*
Plane design	*Speed* $x1$	*Fuel* $x2$	*Price* $x3$	*Capacity* $x4$	*Sales* $y14$	*Profit* $y34$	*% invest return* $y54$	*Capacity* $y8$	*Feasibility* $y9$	$u4$
A	260	8.6	125	1	123	4.3	22.6	1	1	.13
B	140	14.3	60	1	453	4.5	28.1	1	1	.39
C	120	11.8	50	1	560	3.9	21.7	1	1	.34
D	220	11.5	100	1	237	5.0	29.4	1	1	.27
E	160	12.4	80	2	437	6.6	31.4	2	.95	.77
F	210	9.9	120	2	290	7.3	33.2	2	.95	.71
Ideal					560	7.3	33.2	2	1	1
Anti-ideal					123	3.9	21.7	1	.95	0

To obtain his "what-if" individual problem representation, in Table 12.9 for plane F the president changes $y14$ to 270 instead of 290, which gives a profit $y34 = 6.7$ and a percent investment return $y54 = 30.7$. With this new data and using PREFCALC he gets Figure 12.6 as the computer printout of his "what-if" marginal utility functions. The values in the box on Figure 12.6 are the president's utilities $u4$ for the "what-if" run. Table 12.10 shows the president's "what-if" individual problem representation with the "what-if" utility values $u4$. In comparing Figures 12.5 and 12.6 and Tables 12.9 and 12.10, note that the changes for F in $y14$, $y34$, and $y54$ have resulted in changes in the ideal values for $y34$ and $y54$. Although the weights associated with criteria $y34$ and $y54$ are the same in Figures 12.5 and 12.6, the change in the $y34$ and $y54$ ideal values at which these weights are the ordinates of the marginal utility curves means that the president has necessarily had to modify these curves somewhat because the maximum ordinates (equal to the weights) have been shifted to the left, i.e., to occur at the lowered values of the $y34$ and $y54$ ideal points. Modification of these utility curves means that the utilities $u4$ of all planes are subject to change to the values shown in Figure 12.6 and Table 12.10.

From Tables 12.9 and 12.10 MEDIATOR displays the president's plane rank order/utility information as Table 12.11, which shows that the plane rank orders from Tables 12.9 and 12.10 are the same. However the utility spread between E and F in Table 12.9—.77 to .71—increases to .84 to .69 in Table 12.10. In other words under the "what-if" assumption that sales estimate $y14$ is 270 instead of 290, the gap in utility between first-ranked plane E and second-ranked plane F increases from .06 to .15. This convinces the president that between planes E and F he wants E, but before definitely deciding, he asks MEDIATOR to compute the maximin compromise solution for the four players using the second evolved normalized utilities of players 1, 2, 3 from Table 12.8 and his own normalized utilities based on his Table

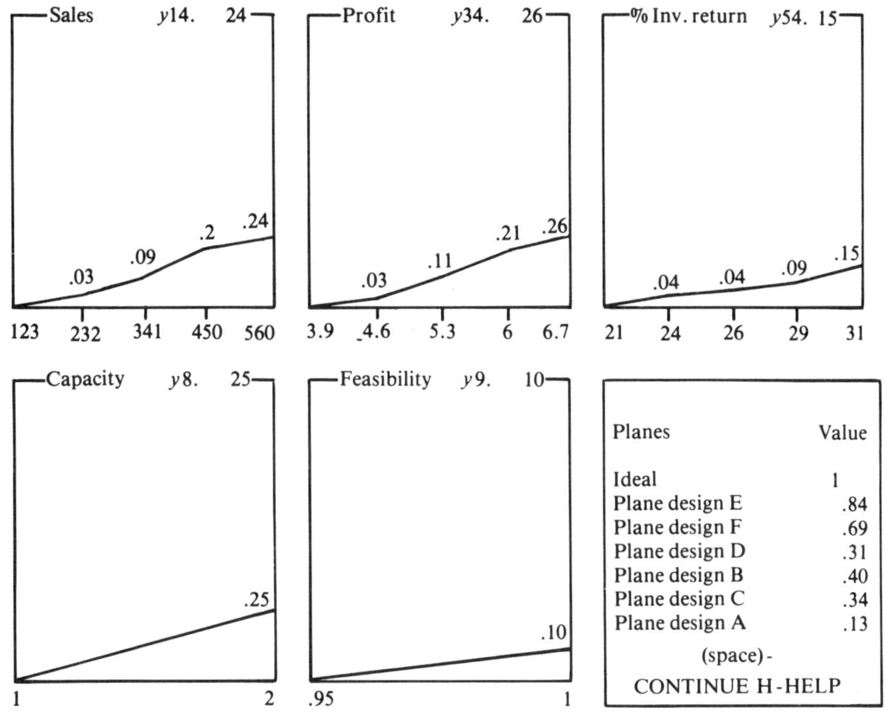

Figure 12.6. **What-if Marginal Utility Functions for President.**

Table 12.10 **President's "What-If" Individual Problem Representation**

	Controls				*Criteria*					*Utility*
Plane design	*Speed x1*	*Fuel x2*	*Price x3*	*Capacity x4*	*Sales y14*	*Profit y34*	*% invest return y54*	*Capacity y8*	*Feasibility y9*	*u4*
A	260	8.6	125	1	123	4.3	22.6	1	1	.13
B	140	14.3	60	1	453	4.5	28.1	1	1	.40
C	120	11.8	50	1	560	3.9	21.7	1	1	.34
D	220	11.5	100	1	237	5.0	29.4	1	1	.31
E	160	12.4	80	2	437	6.6	31.4	2	.95	.84
F	210	9.9	120	2	270	6.7	30.7	2	.95	.69
Ideal					560	6.7	31.4	2	1	1
Anti-ideal					123	3.9	21.7	1	.95	0

Table 12.11 **President's (Player 4) Plane Rank Orders and Utilities**

Rank Order	*Player 4**	*Player 4***
1	E .77	E .84
2	F .71	F .69
3	B .39	B .40
4	C .34	C .34
5	D .27	D .31
6	A .13	A .13

*From Table 12.9.

**From Table 12.10 ("what-if").

Table 12.12 **Normalized Utilities and Maximin Solution for Four Players**

Plane	*Player 1*	*Player 2*	*Player 3*	*Player 4*	*Minimum utility*
A	0	.37	0	0	0
B	.56	.16	.67	.41	.16
C	.68	0	.24	.33	0
D	.60	.34	1	.22	.22
E	1	.76	.67	1	.67
F	.89	1	.62	.91	.62

Maximin utility = .67.

Maximin compromise solution is plane E.

Table 12.13 **Normalized Utilities and Maximin Solution for Four Players Using Player 4's "What-If" Utilities**

Plane	*Player 1*	*Player 2*	*Player 3*	*Player 4*	*Minimum utility*
A	0	.37	0	0	0
B	.56	.16	.67	.38	.16
C	.68	0	.24	.30	0
D	.60	.34	1	.25	.25
E	1	.76	.67	1	.67
F	.89	1	.62	.79	.62

Maximin utility = .67.

Maximin compromise solution is plane E.

Table 12.14 **Final Summary of Plane Rank Orders and Utilities**

Rank order	*Player 1*	*Player 2*	*Player 3*	*Player 4**	*Player 4***
1	E .77	F .67	D .71	E .77	E .84
2	F .69	E .58	B .57	F .71	F .69
3	C .53	A .43	E .57	B .39	B .40
4	D .47	D .42	F .55	C .34	C .34
5	B .44	B .35	C .39	D .27	D .31
6	A .02	C .29	A .29	A .13	A .13

*From Table 12.9.

**From Table 12.10 ("what-if").

12.9. This is shown as Table 12.12. The maximin utility is .67, and the maximin compromise solution is plane E. The president then asks MEDIATOR to compute the maximin solution using normalized utilities of players 1, 2, 3 from Table 12.8 and his own normalized utilities based on his "what-if" representation, Table 12.10. This is shown as Table 12.13.

Again the maximin compromise solution is plane E with maximin utility of .67. As a final summary table the president requests MEDIATOR to join Tables 12.7 and 12.11 to produce Table 12.14. For the latter, upon request MEDIATOR indicates D, E, and F are on the four-player utility Pareto optimal frontier for both sets of the president's utilities.

At the meeting the next day the president decides to have MEDIATOR display his work—Figure 12.5, Table 12.9, Figure 12.6, Table 12.10, Tables 12.11, 12.12, 12.13, 12.14—to the other three players. He discusses it with them. On the basis of this he decides on plane design E as the new product.

12.5 Concluding Remarks

The use of MEDIATOR for new product design and negotiation shows possibilities for decision support in group design processes. The use of the controls/goals referral process to change control and/or criteria dimensions represents support for creative evolution of the ESD group problem representation (as does use of the goals/values referral process—see Shakun 1981a, 1988—to change criteria dimensions and/or values). By supporting consensus seeking and compromise as discussed in this chapter, MEDIATOR helps players work through conflicting views leading to the new product decision. This is facilitated by extensive use of tabular (spread sheet) and graphical displays.

The authors wish to thank Dr. Jean-Louis Giordano for conceptual and practical discussions, and Ms. Carine Ziol for research assistance.

REFERENCES

Giordano, J. L., Jacquet-Lagreze, E., and Shakun, M. F. 1985. "A Decision Support System for Design and Negotiation of New Products." Graduate School of Business Administration, New York University, New York. Also in Shakun, M. F. 1988. *Evolutionary Systems Design: Policy Making Under Complexity and Group Decision Support Systems*. Holden-Day, Oakland, CA.

Jacquet-Lagreze, E. 1985. PREFCALC, Version 2.0. EURO-DECISION, BP 57, 78530 Buc, France.

_____, and Shakun, M. F. 1984. "Decision Support Systems for Semi-Structured Buying Decisions." *European Journal of Operational Research*, 16, No. 1, pp. 48–58. Also in Shakun, M. F., 1988. *Evolutionary Systems Design: Policy Making Under Complexity and Group Decision Support Systems*, Holden-Day, Oakland, CA.

Jarke, M., Jelassi, M. T., and Shakun, M. F., 1985. "MEDIATOR: Toward a Negotiation Support System." Graduate School of Business Administration, New York University, New York. Also in Shakun, M. F. 1988. *Evolutionary Systems Design: Policy Making Under Complexity and Group Decision Support Systems*. Holden-Day, Oakland, CA.

Shakun, M. F., 1981a. "Formalizing Conflict Resolution in Policy Making." *International Journal of General Systems*, 7, No. 3, pp. 207–215.

_____. 1981b. "Policy Making and Meaning as Design of Purposeful Systems." *International Journal of General Systems*, 7, No. 4, pp. 235–251.

_____. 1985. "Decision Support Systems for Negotiations." *Proceedings of the IEEE International Conference on Systems, Man and Cybernetics*. Tucson, AZ, November. Also in Shakun, M. F. 1988. *Evolutionary Systems Design: Policy Making Under Complexity and Group Decision Support Systems*. Holden-Day, Oakland, CA.

_____. 1988. *Evolutionary Systems Design: Policy Making Under Complexity and Group Decision Support Systems*. Holden-Day, Oakland, CA.

CHAPTER 13

Negotiating to Free Hostages: A Challenge for Negotiation Support Systems

Guy O. Faure* and Melvin F. Shakun**

Negotiating to free hostages in international contexts can concern not only governments but also companies operating abroad, particularly in developing countries. We first present a real negotiation case in which a mediator tries to help hostile opposing parties—an international company and a guerilla movement that has kidnapped eight company employees—reach an agreement that would free the hostages. An analysis of the case follows. Evolutionary systems design (ESD)—an artificial intelligence framework for negotiation support systems (NSS)—as implemented in the NSS MEDIATOR is applied to the case. The mediator is supported by MEDIATOR in building his group problem representation. The focus here is on the goals/values relation that shows at any time which values are delivered by which goal dimensions for the parties, thus defining the negotiation goal space at that time. The evolution of the goals/values relation attendant with possibilities for an earlier and more satisfying agreement is supported by MEDIATOR. Although the development of support here is centered on methodology using specified heuristics, the MEDIATOR knowledge base can contain information on past hostage negotiations, hostage taker types, their values, goals, etc., as discussed, so that negotiation support is both methodological and substantive.

*University of Paris V, Sorbonne, France.
**New York University, New York, U.S.A.

13.1 Introduction

The taking of hostages poses a difficult problem. In international contexts it can concern not only governments but also companies operating abroad, particularly in developing countries. Negotiating to free them becomes more and more a necessary skill. If one considers the complexity of the situation—including parties who belong to different cultures, who express antagonistic ideologies, and in general the dramatic issue that has to be dealt with—this type of negotiation is one of the most difficult to perform. From a research viewpoint work on hostage cases, which involve a large number of variables related to the negotiation process, can be of considerable use in contributing to a general theory of negotiation.

Following this introduction, in Section 13.2 we present a negotiation to free hostages. It is a real case, a recent negotiation for which we have complete data often not available to researchers. The hostage taking occurred in a developing country. The parties involved in the negotiation were the top executives of an international company and the leaders of a guerilla movement that had kidnapped eight company employees.

In Section 13.3 we analyze the negotiation, focusing on some specific topics such as the hostage takers' motivations, the conflict of legitimacy, the parties' values and perceptions, the structure of the negotiation process, the problem of uncertainty, face-saving concerns, and the mediator's function and strategy.

In Section 13.4 we discuss a framework—evolutionary systems design (ESD)—for negotiation support systems (NSS), which we apply to the case in Section 13.5.

In Section 13.6 we present concluding remarks.

13.2 Case Presentation

All names and places have been changed in order to make the case impossible to identify. This was a condition of using the data collected for research purposes. We shall call the country "Minaland," the company "ANSAT," and the guerillas "MPR" (mountain people representatives).

Minaland is an underdeveloped country with few inhabitants in a very large land area. The country has many mineral resources, recently discovered and almost unexploited. Most of the population suffers from malnutrition. The political situation is rather unstable, with one military coup d'état following another.

The country is divided into two parts: flatlands and mountains. The flatlands, occupied by the dominating ethnic group, are the site of the capital. The head of state belongs to this dominating group, which monopolizes almost all the civil servant jobs. The mountains, inhabited by the minority group, were the poorest part of the country until mineral resources were discovered. The mountain people have a very strong resentment toward the other ethnic groups. Fifteen years ago they launched a guerilla war to obtain their independence. This war lasted ten years and ceased only after an agreement giving them some autonomy was reached.

Then foreign companies, including ANSAT, started to undertake new projects. Contractor of the central government of Minaland, ANSAT is working on mining projects. In July 1984 seven ANSAT employees were captured by the guerillas. After two days they were sent back with the following message: "Stop building the new plant." ANSAT does not know what to do. The government of Minaland declares that the situation is under control and that ANSAT has nothing to fear.

By October 8 rumors ran rampant about very strong guerilla activity. Natives working for the company vanish. ANSAT decides to "wait and see." The next morning the mountain people launch a general attack against company headquarters. Eight employees, all Westerners, among them two women, arc kidnapped. The guerillas do not leave any demand.

ANSAT is in a very difficult situation. The company is responsible for the fact that the eight employees were still working in the plant and thus exposed to being taken hostage. ANSAT has to do something but considers the mountain people mere gangsters and does not want to lose face by starting negotiations with them. The company uses another strategy—getting in touch with a third party, a kind of mediator, in order to obtain more information about the hostages' health and the MPR's demands. This mediator has been working on welfare projects in the mountains for a long time and knows both parties well. They seem to trust him.

After several meetings with the two parties the mediator determines their respective positions:

- For the MPR: They are representatives of a real nation that has suffered from constant spoliation. They are fighting for their rights; their action is legitimate. ANSAT has been warned to leave and has not taken any notice. The mountain people had to take some action. ANSAT has to bear full responsibility for what has happened. The eight people held should not be considered hostages but "temporary guests." ANSAT's behavior has been seriously injurious to the MPR, and this injury has to be compensated for.
- For ANSAT: The MPR behave as gangsters. They obey no laws, so they do not have to be treated with any respect. Their statements have no basis whatsoever. The only authority ANSAT has to deal with is the legal government of Minaland. ANSAT owes nothing to the mountain people. All the MPR can do to straighten out the situation is free the hostages.

In the meantime the government of Minaland decides to solve the problem in its own way, which is to bomb the place where the hostages are supposedly being kept in the mountains. Some mountain people are killed, but the hostages survive.

The MPR do not make any exact demand, but inform the mediator that their estimated total loss is $24 million ($24m), including the cost of the operations made necessary to get ANSAT out of the mountains, the loss of men, equipment, herds, and pastures due to the bombing from the regular army, and the support of the 30 men assigned to protect the hostages.

When the mediator informs ANSAT, the company declares this $24m amount to be totally absurd. ANSAT is willing to pay no more than $50,000. The mediator does not transmit this offer to the MPR, fearing it might provoke a reaction against the hostages. Basically, ANSAT decides not to make any move, feeling that time is on its side because the guerilla war has flared up—escalating day after day—and that the MPR are badly in need of men, money, and equipment. However the MPR do not show any sign of being in a hurry to reach an agreement.

ANSAT strives to control its own environment in two ways: (1) keeping the hostages' families quiet by pointing out (threatening) the risks involved to the hostages in any unwise move and (2) giving very little information to the mass media.

Now the two parties meet directly. At each meeting they stick to a similar strategy. Each side tries futilely to make the other admit that it is wrong in its way of looking at the situation. ANSAT is surprised when the leader of the MPR negotiating team is killed. The murder was evidently ordered by his own people at home, who were suspicious that he was cooperating with the company. This was not true. After several months the deadlock becomes so obvious that the MPR interrupt a meeting and leave, breaking off negotiations.

Later on, with the help of the mediator, negotiations resume. The MPR decide to put aside their complaints and not to speak any longer about values and principles. ANSAT sees this as a sign of weakness and does not reciprocate. It does not drop its own demands that the MPR look at the situation from ANSAT's viewpoint, i.e., accept ANSAT's values. ANSAT decides to wait until the MPR give up all their demands.

There is an escalation in threats from both sides. ANSAT tells the MPR that if anything happens to the hostages, the MPR would be fully responsible, and this would ruin their reputation at the international level. ANSAT threatens that if things do not quickly take a more positive orientation, it might stop the blackout on the mass media, and the MPR would then be considered as no more than mere gangsters. ANSAT also suggests that it could interfere in the guerilla warfare by convincing the village chiefs to stop supporting the MPR. Thus by lowering the MPR's conflict payoff through threats, ANSAT hopes to induce the MPR to agree to a lower payoff.

The MPR also elicits quite a few threats. Among them is the risk of death to the hostages through accident or illness. ANSAT would have to bear the entire responsibility for this because of its unwillingness to make the negotiations progress. There could also be new bombings from the central government, resulting in potential harm to the hostages. A few days ago there had been a huge forest fire, and the hostages were at risk. The MPR again lost some men while protecting them.

In the meantime ANSAT learns that the hostages have been transferred to the other side of the border, to the neighboring country that supports the rebels. The mediator keeps going from one side to the other trying to lessen the demands and improve the negotiating atmosphere.

Other third parties start to increase the pressures on both sides. The holding company controlling ANSAT strongly urges it to reach an agreement. The situation has to be resolved rapidly in order to avoid loss of image and credibility and the bad consequences for other contracts with the central government. Some Western countries push ANSAT strongly to compromise quickly to end the problem, hoping other minorities struggling for more autonomy will not latch on to the idea of hostage taking.

At the same time the neighboring country that supports the MPR puts pressure on them to come to a solution because it seeks economic help from the country in which ANSAT has its headquarters.

Gradually the level of MPR demands has decreased from $24 to $2 million. Things seem to be in better shape for reaching an agreement. However, ANSAT sticks to its former position, seeing in this drop in demands yet another sign of weakness. ANSAT does not make a monetary offer, feeling it just needs to wait a little longer for the last stage, which should be for the MPR to release the hostages, if possible with apologies. ANSAT also tries to get rid of the mediator, whom it suspects of not standing firm enough on its behalf. With ANSAT not making an offer a second deadlock occurs. The MPR break off negotiations.

The mediator comes back into the negotiations, working separately with each side. He tries to make ANSAT adopt a more realistic attitude, striving to point out to the MPR the positive side of past ANSAT actions in the mountains, where the company had provided free aid to the natives—medicines, tools, roads. He emphasizes the stress and suffering of the hostages' families and the unfair situation of the hostages, who are in any case innocent victims. He also points out the necessity of not jeopardizing future cooperation with the West.

The pressure from the holding company increases, and the company president decides to lead the negotiation himself in place of his vice-president. He pays more attention to not hurting the mountain people's feelings. ANSAT now leaves aside all principles and values. Step by step the parties' positions come closer, and an agreement is outlined. A "financial compensation" will be given to the MPR for the "care and protection" of the hostages: $750,000 plus a "humanitarian donation" consisting of medicine and equipment, which will be delivered first as a sign of goodwill.

The medicine and equipment are sent to the MPR, but they refuse the whole cargo, declaring it to be defective—not what was agreed upon. They retaliate by raising the amount to be paid from $750,000 to $1 million, not negotiable, after delivery of good medicine and equipment.

After one more month ANSAT gives in. Several ways of implementing the agreement are suggested to the MPR.

1. Half the total sum, release of the hostages, then the other half.
2. Half the total sum, release of half the hostages, then the other half of the money, then the other hostages.

The MPR rejects both propositions, and the final formula is payment of the full amount of the money, then release of the hostages. The agreement is implemented without any major setback, although it is done in a paranoid atmosphere. The hostages are eventually released.

The whole negotiation lasted one year, requiring ten meetings between the parties. The MPR consisted of six people around the table. Two people negotiated successively for ANSAT. The mediator participated in the negotiation until the very last moment.

After the release of the hostages the company made a brief statement to the press, saying that the hostages had been freed without any counterpart, that is to say, no ransom of money, medicine, and equipment. The company added that there had never been any negotiation with the rebels, just some interventions. The MPR, through one of their representatives in a Western capital, stated that the eight people had been released without any counterpart and only for "humanitarian reasons." This is a summary account of the negotiation. More background information and detail are given in Faure (1987).

13.3 Analysis

THE HOSTAGE TAKERS' MOTIVATIONS

We note three types of hostage takers, each one corresponding to a different set of motivations:

1. the extortionist who is usually called a gangster and is motivated by money (A bank robber surrounded by police who takes hostages in order to negotiate his freedom may also be considered an extortionist.)
2. the political militant who is motivated by power, influence, fame, political recognition, and/or money
3. the mentally ill person who has motivation oriented toward psychological satisfaction and can be a psychopath or paranoic

Type 1 has a simpler rationale than type 2, who is far more difficult to deal with. Type 3 is even more complex. Types 1, 2, and 3 are ideal types; the three profiles are often found combined in the same person.

The MPR are first of all political militants. They are challenging the central government, showing that they can establish a separate power that cannot be removed. But their action is not only political; they practice extortion, making money by kidnapping hostages. What they call "financial compensation" is nothing more than a ransom to release the hostages.

Are the MPR's financial and political gains more than counterbalancing the risk of moral loss attached to the fact of taking hostages? Can we postulate a paranoid feature in their personality that helps them forget about the least acceptable aspects of their behavior?

THE CONFLICT OF LEGITIMACY

A company buying from or selling to another company does not usually wonder if the other party is morally qualified to enter into that type of relation. But when the deal does not concern goods but human beings, questions arise. Negotiating with terrorists has in itself something of an illegitimate nature because no government officially sanctions terrorism, hijacking, and extortion. This is one reason there was no progress in the negotiation for many months. There could not be any efficient interaction because only negative arguments were exchanged. The company's only aim was to prove to the MPR that their behavior was wrong, unacceptable, and that they had to apologize for it and send back the hostages. The company viewed terrorism as illegitimate.

For the company this illegitimacy morally justified its behavior in lying to the press after the negotiation, saying there was never any ransom given nor even any negotiation carried out, just some interventions.

Another interesting moral question: With a negotiation considered as dishonest, illegitimate, do we have to be honest and stick to our words? ANSAT's answer is obviously no. This can be observed through its attempt to cheat the MPR with the medicine and equipment. Because the other party does not consider itself to be behaving in an illegitimate or dishonest way, its reactions may carry severe consequences, especially if it feels offended and retaliates.

EACH PARTY'S VALUES AND PERCEPTIONS

ANSAT belongs to the industrialized West. The MPR are part of a traditional society that belongs to the third world. Each side carries its own total vision of the context in which the negotiation takes place.

For ANSAT the conflict is between a civilized company coming from a civilized country—respecting the law and contributing to the development of a backward country—and an unruly party that has rather uncivilized ways. In fact the MPR are quite violent, having killed one of their own negotiators between two meetings.

For the MPR the conflict is between a people struggling for freedom on its own land and a wealthy company supporting imperialistic views and looting the third world.

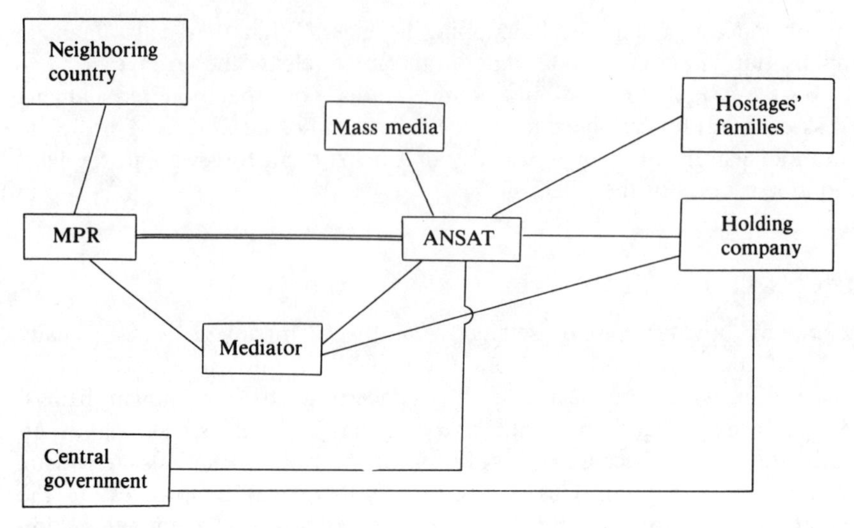

Figure 13.1. **System of Interactions.**

The discrepancy between these two ways of looking at the situation—the antagonism between value systems and the impossibility of meeting on some common principles—add a lot to the difficulties of reaching an agreement.

Zartman (1976) develops a theory about international negotiations according to which the encounter between parties who belong to very different cultures leads to few attempts to persuade the other but, instead, to many maneuvers. This case does not verify the theory, for 90 percent of the negotiation time is spent in attempts to modify the other party's general perceptions and values—the way he has to look at the situation. However we have to consider that this type of hostile negotiation is not the most common or representative of what can be seen in the domain of international negotiations.

THE STRUCTURE OF THE NEGOTIATION PROCESS

First we consider the system of interactions (Figure 13.1). There are two types of relations: negotiation (= =) and influence (−). We note four significant nonrelations:

1. MPR/central government shows the weakness of the Minaland government.
2. MPR/hostages' families emphasizes that ANSAT controls communication to these families.
3. MPR/mass media reflects the image concern of the MPR.
4. MPR/holding company reflects the concern of the holding company not to be directly involved in the negotiation.

Researchers like Ikle (1964), Zartman (1976), Stevens (1963), and Pruitt (1981) divide the negotiation process into two stages:

1. outlining a general formula that agrees on main principles, i.e., consensus-seeking phase
2. bargaining on details to reach the final balance, i.e., concession-making phase

The first stage is mostly competitive; the second stage is mostly cooperative (Stevens 1963) or coordinative (Pruitt 1981).

For Pruitt (1981) the core of negotiation is the passage from stage 1 to 2. If we consider the case under study, 90 percent of the negotiators' time and energy was devoted to the first stage, with no result—the more discussions, the more disagreements. The negotiation appears as a mere confrontation, a substitute for war. No mutual adjustments on values, principles, or issues to negotiate are performed, no real moves are made, the situation remains frozen.

The inability to reach a kind of consensus on the overall situation—severe values conflict—is probably one of the main characteristics of negotiations concerning hostages and, in general, negotiations under hostility. It also gives a better understanding of the reasons why such negotiations last so long (in the case of a favorable environment for the terrorists), e.g., the U.S. hostages in Teheran were kept for more than 400 days.

In the case under study no agreement could be foreseen. The process just led to a deadlock. After some time to get out of this deadlock, the MPR decided to put aside any discussion of principles. But ANSAT did not reciprocate.

Pruitt (1981) suggests a model of "strategic choice" to explain how negotiations may switch from competitive to coordinative behavior. He states that, in a deadlock, competitive actions become decreasingly attractive because of the other party's firmness. Eventually all possible threats and commitments have been made, and the opponent is unwilling to make further concessions. In addition the danger of reaching a deadline without agreement is becoming increasingly apparent. Coordination is the only approach left.

The negotiation between ANSAT and the MPR corresponds roughly to what Pruitt (1981) suggests except that no coordinative action is taken. The final impetus to adopt coordinative behavior was given by the environment—the holding company and the neighboring country. This impetus never came from the dynamics of the negotiation itself. This is not only realistic but rather usual in this type of negotiation because very seldom are negotiators in a position that allows them to ignore their environment.

Such a fact tends to show that agreements can be obtained without previously reaching any consensus on the general situation. We can further say that here the agreement was reached almost without going through any coordinative stage, except for some of the very last moments. The briefness of that last stage reveals a sort of unconscious wish to speed up the process in order to avoid the painful reality of the means that have to be used and the compromise that has to be made to get the hostages freed.

UNCERTAINTY

There is no negotiation that does not carry a certain amount of uncertainty. Uncertainty involves more or less all the components of the situation: the other party's goals, values, and psychological profile; external factors; the probability of reaching an agreement, etc.

If we consider the issues at stake in a negotiation to free hostages, uncertainty adds something quite dramatic. For instance it is not unusual to see terrorists killing some of the hostages. In our case no one could really be sure that such a thing would not happen.

However one still has to carry on, relying on two hypotheses:

1. MPR have not killed the hostages.
2. MPR will prefer to let them free once they have what they want.

To confirm the first hypothesis, it is possible to try to get some evidence by asking, for instance, for letters from the hostages. Regarding the second hypothesis, there is no evidence proving that the MPR really intend to free the hostages after receiving the ransom. However, if we assume that they are rational, the MPR should free the hostages in order not to spoil their image. This is a major difference between political militants and mere extortionists, who are not usually worried about their reputation.

Usually in negotiations there is a minimum trust between parties. This minimum does not appear in this negotiation.

- ANSAT thinks that, because the mountain people dared to take hostages, they are capable of anything.
- The MPR think that, because ANSAT did not leave their land when it was asked to, they cannot rely on any goodwill from the company.

If negotiation can be defined as a learning process (Cross 1977), in this particular case what was learned more than anything else was mistrust. As has been shown in laboratory experiments (Brehmer and Hammond 1977), the higher the level of uncertainty, the higher the level of conflict.

THE FACE-SAVING CONCERN

Deutsch (1973) concludes that more face-saving actions arise if the negotiation atmosphere is conflicting. The observations taken from the case under study verify such a conclusion. Moreover the problem of the legitimacy of the other side, of its recognition as a negotiating party, reinforces the face-saving concern.

The fear of losing face is related to the desire to maintain an image of capability and strength (Brown 1977). For ANSAT several targets (or audiences) have to be considered: itself, the MPR, the holding company, and the mass media. ANSAT shows its sensitiveness toward face evaluations in refusing to admit that there was any negotiation and in denying the payment of any ransom. In order not to appear weak ANSAT, in the very limited information it gives to the mass media, states that the interruption of the mining work and plant construction was not connected with the taking of the hostages. One of the main reasons why the negotiation did not show any significant progress for a long period was the care taken by ANSAT not to appear weak in the eyes of the holding company.

Face-saving attempts are made not only for oneself, but also for the other party. Not only does one have to save one's own face; one also must maintain the opponent's face. Such a concern could be noticed during the negotiation through the use of verbal devices. For instance the word "hostages" was never pronounced during the meetings between the two parties.

While addressing the mass media, MPR's reference to "humanitarian reasons" for releasing the hostages can also be taken as a device to save face. It is just a blanket phrase used to avoid showing any weakness. The donation of medicine and equipment for those so-called "humanitarian reasons" reveals the same concern.

It seems that during the whole negotiation what was mainly shared by both parties was not a common desire to compromise but the fear of losing face in front of the other party or in front of audiences like the press, the holding company, or inhabitants of the mountain area.

THE MEDIATOR

Mediation is a very common and efficient practice in highly conflicting negotiations. For instance the negotiation between the U.S. and Iran about the hostages in Teheran eventually came to a conclusion through the mediation of a third country, Algeria.

The mediator is not an arbitrator; he does not have any power to decide instead of the parties. However he is trusted by them, and this trust gives him the possibility to influence both parties and work to get their respective positions closer and closer.

ANSAT finds definite advantages to operating through a mediator:

- There will be lower cost for the company if the mediator is killed or taken as an additional hostage compared to the cost it would have to face if an ANSAT representative were targeted for such action.
- The mediator has good knowledge of the inhabitants of the mountains, of their culture, traditions, values.
- He enjoys a good image among them.
- Use of the mediator provides a way to minimize direct contacts with the other party. This serves a face-saving purpose (ANSAT does not admit openly to be negotiating) and avoids difficulties with the Minaland government (a private company is not supposed to deal with rebels).

In a deadlock situation the mediator fulfills an essential task: maintaining indirect contact between the two parties when nothing else can be done. To serve this purpose the mediator here used several devices. Chiefly he manipulated the information both ways, emphasizing the positive words, ideas, or moves; understating verbal aggression; and hiding part of the negative attitudes.

For instance the mediator did not transmit the first offer from ANSAT to the other party. (This offer was 1/500 of what was asked: $50,000 for the $24m asked.) He considered that such a low offer would definitely spoil the whole negotiation from the beginning.

A mediator's position is sometimes very difficult to handle. If he is not seen by the side with which he is interacting as favoring its own views, he may lose its confidence and sometimes may merely become a substitute target (Rubin 1980). ANSAT started to see the mediator as a third party who could not be trusted any longer and tried to get him out of the game. But the MPR realized that without him it would be extremely difficult to start negotiating again, so they managed to bring him back.

ADDITIONAL THOUGHTS

Focusing on case studies of negotiations concerning hostages—especially at an international level—is of special interest regarding research on negotiations because they involve a very large number of interacting variables. These variables operate on each of the four dimensions that structure this type of situation: economic, ideological, psychological, and cultural.

Regarding the ANSAT/MPR negotiation, we can identify background factors and conditions that influence the process, drawing on the Sawyer and Guetzkow (1965) model (see Figure 13.2). Time has not been mentioned previously, but one may observe several facts concerning this type of negotiation:

- Each day that passes has a very high cost because it is a day taken away from the life of the hostages—a day that can never be given back.
- The point for the terrorists (except if they want purposely to humiliate the other party, which happens sometimes) is usually not to keep hostages forever but to get, as quickly as possible, something in exchange.
- The company has no real interest in making the negotiations last indefinitely because this would show a lack of capability to solve the problem. It could have harmful consequences for its image. (Recall the case of President Carter and the hostages in Teheran.) Despite these points the negotiation lasts for a year or more. This shows the influence of some other variables, like the values of the negotiators, the face-saving concern, and the level of uncertainty, on the process.

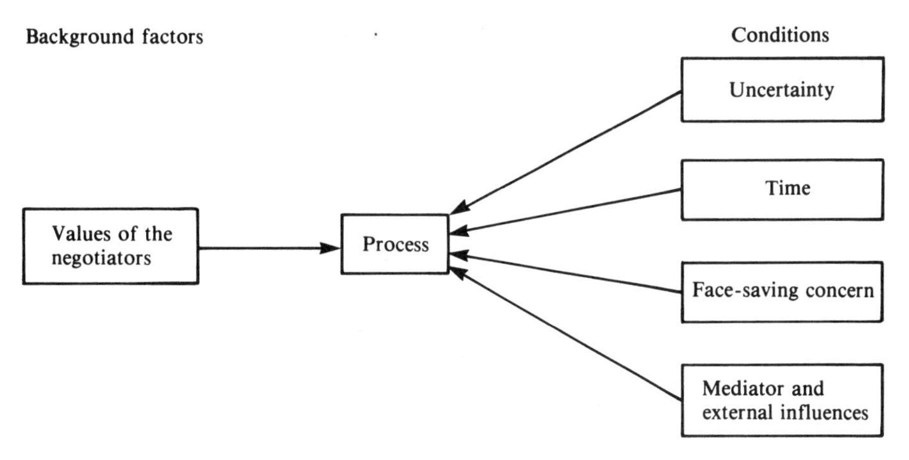

Figure 13.2. **Background Factors and Conditions Influencing the Negotiation Process.**

There are two totally different ways to look at a negotiation to free hostages according to the party concerned:

- The company sees it as an exchange—"How much money do we need to give to get the hostages back?" In other words "How much are they worth?" They are seen in this way as an "exchange value."
- The MPR see the hostages as a means of pressure used in order to implement a negotiation related to another subject. ANSAT should be made to pay for the role—relating, contracting, being friendly—it has played with the government. The valuation put on this other subject is not connected with that of the hostages.

This discrepancy in ways of looking at the situation is one of the major causes of deadlock during the negotiation.

13.4 A Framework for Negotiation Support Systems: ESD

Evolutionary systems design (ESD) (see Shakun 1988) is an artificial intelligence framework for negotiation support systems (NSS). The ESD methodology is characterized by an evolving group problem representation involving relations among values, goals, preferences, and controls. ESD is a unified approach to evolving problems that at least initially may have no feasible solution or may have multiple feasible solutions requiring a choice. Of course, if there is group conflict regarding that choice, then from a group decision viewpoint there is also initially no feasible solution.

MEDIATOR (Jarke, Jelassi, and Shakun 1985) is a negotiation support system based on ESD and data base–centered implementation. In the basic scenario a group of players is involved in negotiations. A human mediator supports these negotiations, and he in turn is supported by MEDIATOR. The (human) mediator supports negotiation by assisting the players in a process of consensus seeking within which compromise is possible. In less hostile settings the mediator, using MEDIATOR, aids consensus seeking by helping the players build a common (group) joint problem representation of the negotiations. Under conditions of strong interpersonal hostility (highly noncooperative), as in the ANSAT/MPR hostage negotiation, the mediator would use MEDIATOR to build his own private group problem respresentation—the chances would be small for a common joint problem representation built and shared by players because of their hostility.

Then assume in the ANSAT/MPR negotiation that the mediator is supported by MEDIATOR in building his own private group problem representation. The negotiation support framework will be simultaneously descriptive in that what actually did happen will be one possible outcome of the support process.

Following ESD (Shakun 1988), at any time t the group problem representation consists of (1) a goals/values relation that shows which values are delivered by which goal dimensions for the players, thus defining the goal space at that time, and (2) mappings from control space to this goal space to preference (e.g., utility) space. We note in many negotiations that the control and goal spaces are the same. Because the heart of the ANSAT/MPR negotiation involved values conflict and the attendant difficulty in redefining the operational goal space, we shall focus on (1) here. Although it will be referred to later, for detailed discussion and application of (2) see Shakun (1988).

SHARED VALUES AND OPERATIONAL GOALS

Conflict must finally be resolved at the operational goal level—it is the nonempty intersection of players' operational goal targets that resolves conflict (Shakun 1988). However it is easier to find operational goal dimensions on which players agree to negotiate—i.e., easier to achieve goal dimension agreement (thus defining a negotiation goal space) using the goals/values referral process—if they share common values than if they do not. Further, negotiating a solution (nonempty intersection) in the negotiation goal space is facilitated if underlying values are shared. ANSAT and MPR failure to identify common values makes it difficult for the players to successfully use the goals/values relation (referral process) to redefine the operational goal space, a powerful method for conflict resolution (see Chapters 1 and 2 of Shakun 1988).

THE GOALS / VALUES RELATION

Mathematically the goals/values relation is given by:

$$\lambda^j(t) \subset v(t) \times y(t) \qquad j = 1, 2 \in C$$

where $\times$ represents a Cartesian product and is shown by the goals/values matrix

	$y(t)$
$v(t)$	$(\lambda jki(t))$

where t = present time with $t = 0, 1, 2 \ldots$ representing a moving present time t.

$j = 1, 2$ with $j = 1$ representing MPR and $j = 2$ representing ANSAT. In ESD we write $j \in C$, i.e., j is a member of a (here reluctant) coalition C that must solve the hostage problem. More generally, $j = 1, 2, \ldots, J$ allows us to represent additional players in coalition C.

$v(t) = (vk(t))$ represents values $(k = 1, 2, \ldots, K)$

$y(t) = (yi(t))$ represents operational goal dimensions $(i = 1, 2, \ldots, p)$ comprising the dimensions of coalition C's goal space at time t.

$\lambda jki(t) = 1$ indicates player j is "for" value vk being delivered by goal dimension yi (i.e., he favors both the value vk and the goal dimension yi as an operational expression of this value; $\lambda jki(t) = 0$ indicates player j is against value vk being delivered by goal dimension yi; $\lambda jki(t) = *$ indicates player j is neutral or does not perceive value vk as being delivered by goal dimension yi.

Shakun (1981, 1988) develops the use of the λ relation formalization—described there also as the goals/values referral process—by which the relations $\lambda(t)$ can change, including change in the dimensions of the operational goal space $y(t)$ and values $v(t)$. This can lead to goal dimension agreement—agreement among players $j \in C$ on a common set of goal dimension on which to negotiate. If negotiations on these common goal dimensions become deadlocked so there is no goal target agreement, i.e., no nonempty intersection of players' goal targets, then the goals/values referral process can be used again to redefine the goal space.

13.5 Applying the ESD Framework to the Case

We shall consider the evolution of the goals/values relation in negotiation support for the ANSAT/MPR negotiation. To facilitate this we present a summary of major negotiation events over time (months) in Figure 13.3. We define time $t = 0$ as the point at which the hostages are taken. Shortly thereafter, by $t = .5$, the mediator learns that the MPR has defined its goal target by a demand (monetary payment) of \$24 million, and ANSAT has defined its target by an offer of \$50,000 (\$.05 million). Players have evidently agreed to negotiate on money, but there is no goal target intersection (agreement).

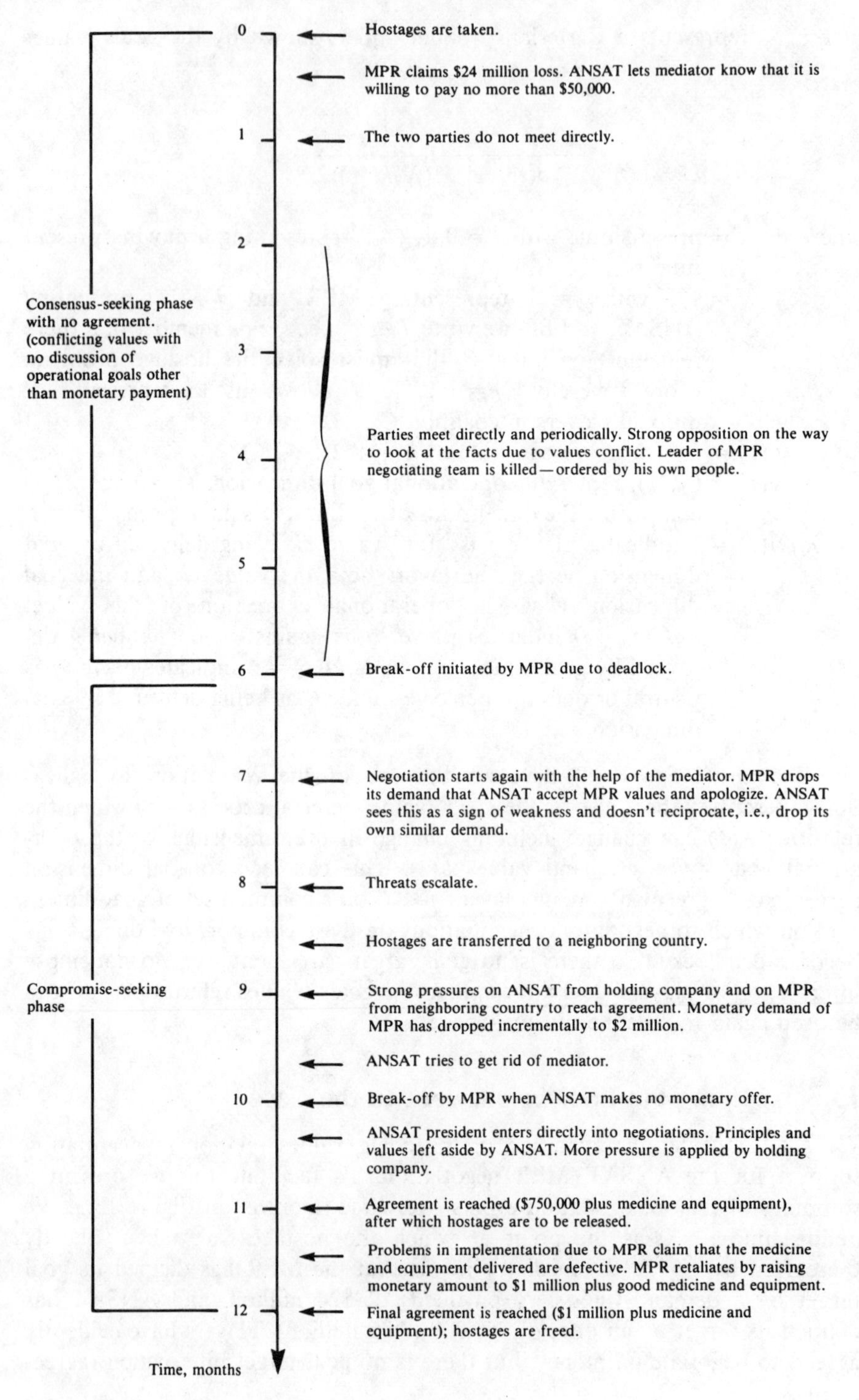

Figure 13.3. **Time Scale of ANSAT / MPR Negotiation.**

The mediator defines MPR's goal target by $y11 > 24$ and ANSAT's by $y12 \leq .05$ and asks MEDIATOR for their intersection. Of course MEDIATOR replies that the intersection is empty. The mediator asks for help. MEDIATOR offers two menu choices:

HELP

1. Expand player goal targets.
2. Change goal dimensions or values using goals/values referral process.

Because the players are extremely far apart, it seems unlikely that option 1 will work, so the mediator chooses option 2:

CHANGE GOAL DIMENSIONS OR VALUES USING GOALS / VALUES REFERRAL PROCESS

Use goals/valucs referral process. Values are given as rows and goals as columns. Use the following heuristics:

1. Given a particular value (row) and looking at the goal dimensions (columns), ask whether there is any other new goal dimension that also delivers the value.
2. Given a particular goal dimension (column) and looking at the values (rows), ask whether there is any other new value that is also delivered by the goal.
3. Given a particular value (row), is there any other new value (more general or less general) that also expresses this value?
4. Is there any other additional value that is important to this negotiation?
5. Given a particular goal dimension (column), is there any other goal dimension that is suggested by this goal?
6. Is there any other additional goal dimension that is important to this negotiation?
7. Is there any other additional player who should now be included in the goals/values relation?

Begin by inserting:

time: .5

values: hostages should be free, $v1$; MPR recovers loss, $v2$

goals: money, $y1$

For time = .5 the mediator has inserted two values—hostages should be free, $v1$, and MPR recovers their loss, $v2$—and one goal—money, $y1$. MPR recovers loss means MPR gets reparations for exploitation of natural resources and costs of holding the hostages. MEDIATOR displays the following goals/values relation (bimatrix):

time = .5	money $y1$
hostages should be free $v1$	1, 1
MPR recovers loss $v2$	1, *(0)

where the mediator has entered 1, 1 indicating that MPR and ANSAT are, representatively, both in favor of value $v1$, hostages should be free, being delivered by goal dimension $y1$, money. His entry 1, *(0) means MPR favors value $v2$ being delivered by goal $y1$, and ANSAT is not against negotiation on money. An entry of 0 by ANSAT would block negotiating on money so that even though ANSAT is against $v2$ being delivered by $y1$, the mediator enters a neutral * because ANSAT is willing to negotiate on money. The entry *(0) symbolizes that a 0 underlies the *. Thus there is goal dimension agreement to negotiate on money. However, as mentioned, players are adamant about their goal targets, so negotiations on this dimension are at the moment deadlocked. MEDIATOR implements the goals/values referral process by asking the mediator to use the above heuristics. Before doing so, the mediator decides to talk further with both sides to obtain better insight into their values and goals.

Afterwards, now at time = 1 month, the mediator tries to apply the above heuristics to the goals/values bimatrix.

Heuristic 1: Given a particular value, does any other new goal dimension deliver this value? Insert:

value: hostages
should be free $v1 \rightarrow$ goal: <u>none</u>

value: MPR
recovers loss $v2 \rightarrow$ goal: <u>none</u>

Because this heuristic applied to $v1$ and $v2$ does not suggest a new goal to him, the mediator inserts "none." MEDIATOR then displays heuristic 2:

Heuristic 2: Given a particular goal dimension, is there any other new value that is also delivered by this goal? Insert:

goal: money $y1 \rightarrow$ value: <u>MPR gets resources</u>

value: <u>MPR punishes ANSAT</u>

The mediator has entered two new values. MEDIATOR displays the current goals/values relation:

time $t = 1$	money $y1$
hostages should be free $v1$	1, 1
MPR recovers loss $v2$	1, *(0)
MPR gets resources $v3$	1, *(0)
MPR punishes ANSAT $v4$	1, *(0)

where the mediator has entered 1, *(0) for ($v3$, $y1$) and ($v4$, $y1$), indicating that MPR favors $v3$ and $v4$ being delivered by $y1$ and ANSAT is not against negotiating on money. Again a 0 by ANSAT would block negotiation on money so that, even though ANSAT is against $v3$ being delivered by $y1$ and against $y4$, the mediator enters a neutral * because ANSAT is willing to negotiate on money. As before, the entry *(0) symbolizes this information.

> Heuristic 3: Given a particular value, is there any other new value (more or less general) that also expresses this value? Insert:

value: hostages should be free	$v1 \rightarrow$ value: none
value: MPR recovers loss	$v2 \rightarrow$ value: no exploitation of natural resources without fair compensation
value: MPR gets resources	$v3 \rightarrow$ value: none
value: MPR punishes ANSAT	$v4 \rightarrow$ value: none

The mediator has entered a new value. MEDIATOR displays the current goals/values relation:

time $t = 1$	money $y1$
hostages should be free $v1$	1, 1
MPR recovers loss $v2$	1, *(0)
MPR gets resources $v3$	1, *(0)
MPR punishes ANSAT $v4$	1, *(0)
no exploitation of natural resources without fair compensation, $v5$	1, *(0)

where the mediator has entered 1, *(0) for ($v5$, $y1$).

> Heuristic 4: Is there any other additional value that is important to this negotiation? Insert:

value: MPR should follow the law and not take hostages

value: MPR wants to show weakness of Minaland government

value: MPR wants favorable international recognition

value: ANSAT saves face

value: MPR saves face

The mediator has entered five new values: MEDIATOR displays the current goals/values relation:

time $t = 1$	money $y1$
hostages should be free $v1$	1, 1
MPR recovers loss $v2$	1, *(0)
MPR gets resources $v3$	1, *(0)
MPR punishes ANSAT $v4$	1, *(0)
No exploitation of natural resources without fair compensation $v5$	1, *(0)
MPR should follow the law and not take hostages $v6$	*(0), *(0)
MPR wants to show weakness of Minaland government $v7$	1, *(0)
MPR wants favorable international recognition $v8$	*, *(0)
ANSAT saves face $v9$	*, *(0)
MPR saves face $v10$	*(0), *

The mediator has entered * where he feels that a player is neutral or does not perceive a particular value being delivered by $y1$. The other entries—1 and *(0)—are interpreted as previously explained.

Heuristic 5: Given a particular goal dimension, is any other goal dimension suggested by this goal? Insert:

goal: none

The mediator inserts "none."

Heuristic 6: Is any other additional goal dimension important to this negotiation? Insert:

goal: none

The mediator inserts "none."

Heuristic 7: Should any other additional player now be included in the goals/values relation? Insert:

player: none

The mediator inserts "none." Of course there are other interested players, e.g., the Minaland government, ANSAT's holding company, the neighboring country supporting the MPR, etc., but the mediator decides not to include them in the formal goals/values representation (at least not now).

The mediator considers the current goals/values relation. Although he has used nonblocking * entries because players are willing to negotiate on money, paired entries are frequently 1, *(0), representing either conflicts in values or conflicts in values being delivered by the goal dimension *y*1, money payment.

In direct meetings between ANSAT and MPR starting at time $t = 2$ months (Figure 13.3) these value conflicts pervade the discussions and persist. Players do not identify common values. Neither side makes a concession on monetary payment. From time to time during private discussions the mediator asks the MPR to check the calculation of its "losses," suggesting that perhaps the losses are not as high as the MPR originally thought. At time $t = 6$ months negotiations break off, initiated by the MPR due to deadlock.

This is what actually happened from time $t = 2$ to 6 months. Even if the mediator were supported by MEDIATOR, this is one possible outcome—thus our negotiation support framework serves as a descriptive model. However, continuing our support scenario, suppose at $t = 2$ months the mediator supported by MEDIATOR applies heuristic 1 to the current (above) goals/values (bi)matrix having ten values (rows) and one goal dimension (column).

> Heuristic 1: Given a particular value, does any other new goal dimension deliver this value? Insert: (We omit listing values where no new goals are generated.)
> value: no exploitation of natural resources without fair compensation
> *v*5 → goal: <u>ANSAT shares profits with MPR</u>
> value: MPR should follow the law and not take hostages
> *v*6 → goal: <u>MPR frees hostages</u>

The mediator has entered two new goal dimensions. MEDIATOR displays the current goals/values relation in which the mediator, using the symbols, has entered his judgments:

time $t = 2$	money *y*1	ANSAT shares profits with MPR *y*2	MPR frees hostages *y*3
Hostages should be free *v*1	1, 1	*, *	1, 1
MPR recovers loss *v*2	1, *(0)	1, 1	*, *
MPR gets resources *v*3	1, *(0)	1, 1	*, *
MPR punishes ANSAT *v*4	1, *(0)	*, *(0)	*, *
No exploitation of natural resources without fair compensation *v*5	1, *(0)	1, 1	*, *

time $t = 2$	money $y1$	ANSAT shares profits with MPR $y2$	MPR frees hostages $y3$
MPR should follow the law and not take hostages $v6$	*(0), *(0)	*(0), *	*(0), 1
MPR wants to show weakness of Minaland government $v7$	1, *(0)	*, *	*, *
MPR wants favorable international recognition $v8$	*, *(0)	*, *	*, *
ANSAT saves face $v9$	*, *(0)	1, 1	1, 1
MPR saves face $v10$	*(0), *	1, 1	*, *

None of the goal dimension columns has a direct entry 0. The negotiation goal space has been redefined—it now has three goal dimensions, $y1$, $y2$, $y3$. Actually, although entered now explicitly, $y3$ was implicit in the $t = .5$ and 1 goals/values representations, but goal dimension $y2$ is new (an example of cybernetic self-organization—see Shakun 1988). Column $y2$ has a considerable number of (1, 1) entries, showing that players could possibly identify common values. The realization that there are shared values is frequently obscured by the inability to find operational goal dimensions to express them. The generation of goal $y2$ (profit sharing) makes the mediator realize that players might share values $v2$, $v3$, $v5$, $v9$, $v10$ as signaled by the (1, 1) entries in column $y2$. Many of the zeros in the *(0) entries in column $y1$ (whether in the $t = 2$ or previous $t = 1$ goals/values representations) result from ANSAT's opposition to paying ransom money rather than values disagreement. Thus in our scenario the mediator could now suggest that players consider goal dimension $y2$ (profit sharing) while not focusing on goal dimension $y1$ (ransom money) for the time being. Later, if things go well, the MPR might be willing to drop entirely the $y1$ dimension of ransom money, which would, of course, be readily acceptable to ANSAT. The MPR *(0) entry for ($v6$, $y3$) might eventually even be changed to 1 and for ($v6$, $y2$) to * (although this is not required because *(0) does not block goals $y3$ and $y2$). MPR might also drop value $v4$, to which ANSAT would readily agree. At this point both players would agree on all entry pairs. The central government might not be against ANSAT and MPR negotiating on $y2$, seeing profit sharing as a way to pacify the mountain people and deescalate guerilla activity. It might agree to allow ANSAT to share profits with MPR if the level of profit sharing is not too high. In any case ANSAT is sure to leave Minaland, which will hurt the government economically, unless the MPR accepts ANSAT's presence.

Of course this is only a scenario illustrating how MEDIATOR, through the goals/values referral process, could have supported the mediator early in the negotiation (say at $t = 2$ months) in redefining the negotiation goal space, thus opening up possibilities for conflict resolution. Because of the very hostile situation, there is no guarantee that such support would have led to an earlier and more satisfying agreement. Nevertheless, MEDIATOR-type support seems worthwhile when one considers that (1) negotiations lasted a whole year and (2) although the hostages were finally freed at a ransom the company could afford, they lost a year of their normal lives, and ANSAT necessarily had to abandon its operations in Minaland.

THE LAST FIVE MONTHS OF THE NEGOTIATION

We now discuss the use of our negotiation support framework starting at time $t = 7$ months in Figure 13.3. At the same time it will serve as a descriptive model of what actually happened.

Assume that the generation of goal $y2$ (profit sharing) has not occurred. Eliminating the $y2$ column from the goals/values relation for time $t = 2$ and remembering that no evolution of this matrix occurred between $t = 2$ and $t = 7$, we have the following goals/values relation for the mediator at $t = 7$:

time $t = 7$	money $y1$	MPR frees hostages $y3$
Hostages should be free $v1$	1, 1	1, 1
MPR recovers loss $v2$	1, *(0)	*, *
MPR gets resources $v3$	1, *(0)	*, *
MPR punishes ANSAT $v4$	1, *(0)	*, *
No exploitation of natural resources without fair compensation $v5$	1, *(0)	*, *
MPR should follow the law and not take hostages $v6$	*(0), *(0)	*(0), 1
MPR wants to show weakness of Minaland government $v7$	1, *(0)	*, *
MPR wants favorable international recognition $v8$	*, *(0)	*, *
ANSAT saves face $v9$	*, *(0)	1, 1
MPR saves face $v10$	*(0), *	*, *

When at time $t = 7$ months (Figure 13.3) the MPR drops its demand that ANSAT accept MPR values, formally in the goals/values relation at $t = 7$ this means that MPR is no longer demanding ANSAT change its *(0) entries in the $y1$ column to * or 1. (ANSAT's entries in the $y3$ column are already all * or 1.) When ANSAT doesn't reciprocate, formally it means that the company continues to demand that MPR change its *(0) entry for ($v6$, $y1$) and ($v6$, $y3$) to * or, better, to 1. The escalation of threats at $t = 8$ (Figure 13.3) formally means that, for given goal dimensions, threats—by lowering his perceived conflict payoff—can influence an opponent to expand his goal target (make concessions).

At time $t = 9$ (Figure 13.3) there is strong pressure on ANSAT from the holding company and on MPR from the neighboring country to reach agreement. MPR's monetary demand has dropped incrementally to $2 million. The mediator, using heuristic 4, adds two values to the $t = 7$ goals/values relation: ANSAT wants to satisfy holding company, $v11$, and MPR wants to satisfy neighboring country, $v12$. MEDIATOR shows the following goals/values relation for $t = 9$ in which the mediator, using the symbols, has added his entries for ($v11$, $y1$), ($v11$, $y3$), ($v12$, $y1$), and ($v12$, $y3$):

time $t = 9$	money $y1$	MPR frees hostages $y3$
Hostages should be free $v1$	1, 1	1, 1
MPR recovers loss $v2$	1, *(0)	*, *
MPR gets resources $v3$	1, *(0)	*, *
MPR punishes ANSAT $v4$	1, *(0)	*, *
No exploitation of natural resources without fair compensation $v5$	1, *(0)	*, *
MPR should follow the law and not take hostages $v6$	*(0), *(0)	*(0), 1
MPR wants to show weakness of Minaland government $v7$	1, *(0)	*, *
MPR wants favorable international recognition $v8$	*, *(0)	*, *
ANSAT saves face $v9$	*, *(0)	1, 1
MPR saves face $v10$	*(0), *	*, *
ANSAT wants to satisfy holding company $v11$	*, 1	*, 1
MPR wants to satisfy neighboring country $v12$	1, *	1, *

When ANSAT makes no monetary offer, the MPR breaks off negotiations at $t = 10$ (Figure 13.3). At $t = 10.5$ ANSAT drops its demand that MPR accept the company's values. Formally this means that ANSAT is no longer demanding that MPR change its *(0) entry for ($v6$, $y1$) and ($v6$, $y3$) to * or, better, to 1. The holding company applies more pressure on ANSAT. At $t = 10.5$ the mediator applies heuristic 1 to the last ($t = 9$) goals/values relation.

Heuristic 1: Given a particular value, does any other new goal dimension deliver this value? Insert:

value: hostages should be free	$v1 \rightarrow$	goal: medicine
		goal: equipment

The mediator has entered two new goal dimensions generated by referral to value $v1$. He does not continue with heuristic 1 but asks MEDIATOR to display the updated goals/values relation ($t = 10.5$) in which he enters as judgments in the new goal columns, $y4$ and $y5$, the same entries as for $y1$:

time $t = 10.5$	money	MPR frees hostages	medi-cine	equip-ment
	$y1$	$y3$	$y4$	$y5$
Hostages should be free $v1$	1, 1	1, 1	same entries as for $y1$	
MPR recovers loss $v2$	1, *(0)	*, *		
MPR gets resources $v3$	1, *(0)	*, *		
MPR punishes ANSAT $v4$	1, *(0)	*, *		
No exploitation of natural resources without fair compensation $v5$	1, *(0)	*, *		
MPR should follow the law and not take hostages $v6$	*(0), *(0)	*(0), 1		
MPR wants to show weakness of Minaland government $v7$	1, *(0)	*, *		
MPR wants favorable international recognition $v8$	*, *(0)	*, *		
ANSAT saves face $v9$	*, *(0)	1, 1		
MPR saves face $v10$	*(0), *	*, *		
ANSAT wants to satisfy holding company $v11$	*, 1	*, 1		
MPR wants to satisfy neighboring country $v12$	1, *	1, *		

The mediator thinks as follows: There are now four operational goals that define a ($y1$, $y3$, $y4$, $y5$) goal space. There is now intense pressure on ANSAT by the holding company and on MPR by the neighboring country to reach agreement. This pressure has generated new values $v11$ and $v12$. The identification of these new values means that players might newly sense expanded goal targets within the ($y1$, $y3$, $y4$, $y5$) goal space. In particular both sides are now attaching a lot of weight to goal $y3$, MPR frees hostages, to deliver values $v11$ and $v12$ and much less (little) weight on counterpart—money $y1$, medicine $y4$, and equipment $y5$ goals. The introduction of new goals $y4$ and $y5$ also facilitates goal target expansions by both sides. In other words both players currently have high preference (utility) for goal $y3$, releasing the hostages. In ESD (Section 13.4), if the group problem representation mapping from goal space to utility space were formally established using PREFCALC—a microcomputer program for utility function assessment (Jacquet-Lagreze 1985)—both players would have very high utilities if hostages are released, and the utility difference to a player from paying or receiving more or less counterpart would be small. Thus almost any reasonable counterpart proposal in the middle (after all the MPR's money demand is down to $2 million), such as $750,000 plus medicine and equipment—the actual agreement at $t = 11$—is likely to work. Hostages are to be released after the counterpart is delivered.

However there are implementation problems due to an MPR claim that medicine and equipment delivered before the monetary payment are defective. MPR retaliates (punishes ANSAT—implements value $v4$) by raising the monetary amount to $1 million (contracting its goal target) plus good medicine and equipment. ANSAT gives in, with final agreement coming at $t = 12$ months (Figure 13.3).

13.6 Concluding Remarks

In the actual case each player was preoccupied for a long time (roughly seven months for MPR and ten months for ANSAT) in demanding that the other side accept its values. Without external pressures from the holding company and the neighboring country, this phase would have lasted even longer. Once this phase was over, players reacting to satisfy these pressures placed high weight on goal $y3$, freeing the hostages, and rather readily expanded their goal targets on counterpart—money $y1$, medicine $y4$, and equipment $y5$ goals—leading to agreement.

If MEDIATOR were available—as in the scenario presented—the ESD framework for negotiation support might have aided the mediator early in the negotiation ($t = 2$) in generating (through the goals/values referral process) a new operational goal, e.g., $y2$, profit sharing, perhaps with the realization that the players might indeed share some values. Recognition of shared values can facilitate negotiating a solution in the redefined goal space (defined by money $y1$, ANSAT shares profits with MPR $y2$, and MPR frees hostages $y3$).

If in fact even supported by MEDIATOR the mediator is unable to bring about an early agreement (at around $t = 2$) because of the extreme hostility, the mere attempt at redefining the goal space might break each player's preoccupation with demanding that the other side accept its values. Clearly, negotiations remained stalled until this phase (10.5 months) was over, i.e., until both sides dropped their demands that the other accept its values.

Even without the generation of a new goal (such as $y2$) leading to redefinition of the operational goal space and possible agreement in the redefined space, MEDIATOR can support the mediator in the last five months of the negotiation by supporting (1) the generation of values $v11$ (ANSAT wants to satisfy holding company) and $v12$ (MPR wants to satisfy neighboring country) in response to pressures on ANSAT by the holding company and on MPR by the neighboring country, and (2) the generation of goals $y4$ (medicine) and $y5$ (equipment). At the same time we have shown that the ESD negotiation framework serves as a descriptive model of what actually happened.

Our discussion of the support MEDIATOR can give the mediator has been centered on ESD as a methodology for negotiation support. The MEDIATOR knowledge base can contain information on past hostage negotiations, hostage taker types, their values, goals, etc., as discussed in Section 13.3, so negotiation support can be substantive as well as methodological.

REFERENCES

Brehmer, B., and Hammond, K. 1977. "Cognitive Factors in Interpersonal Conflict." In Druckmann, D. (ed.). *Negotiations, Social-Psychological Perspectives*. Sage Publications, Beverly Hills, CA.

Brown, B. 1977. "Face-Saving and Face-Restoration in Negotiation." In Druckmann, D. (ed.). *Negotiations, Social-Psychological Perspectives*. Sage Publications, Beverly Hills, CA.

Cross, J. 1977. "Negotiation as a Learning Process." In Zartman, W. (ed.). *The Negotiation Process*. Sage Publications, Beverly Hills, CA.

Deutsch, M. 1973. *The Resolution of Conflict*. Yale University Press, New Haven, CN.

Faure, G. 1987. *La négotiation: Conditions externes et logiques internes*. Thèse d'Etat, Sorbonne, Paris.

Ikle, F. 1964. *How Nations Negotiate*. Harper and Row, New York.

Jacquet-Lagreze, E. 1985. PREFCALC, Version 2.0. EURO-DECISION, BP 57, 78530 Buc, France.

Jarke, M., Jelassi, M. T., and Shakun, M. F. 1985. "MEDIATOR: Towards a Negotiations Support System." Graduate School of Business Administration, New York University, New York. Also in Shakun, M. F. 1988. *Evolutionary Systems Design: Policy Making Under Complexity and Group Decision Support Systems*. Holden-Day, Oakland, CA.

Pruitt, D. 1981. *Negotiation Behavior*. Academic Press, New York.

Rubin, J. Z. (ed.). 1980. *Dynamics of Third Party Intervention*. Praeger, New York.

Sawyer, J., and Guetzkow, H. 1965. "Bargaining and Negotiations in International Relations." In Kelman, H. (ed.). *International Behavior: A Social-Psychological Analysis*. Holt, Rinehart and Winston, New York.

Shakun, M. F. 1981. "Formalizing Conflict Resolution in Policy Making." *International Journal of General Systems*, 7, No. 3.

______. 1988. *Evolutionary Systems Design: Policy Making Under Complexity and Group Decision Support Systems*. Holden-Day, Oakland, CA.

Stevens, C. M. 1963. *Strategy and Collective Bargaining Negotiations*. McGraw-Hill, New York.

Zartman, W. 1976. *The Fifty Percent Solution*. Doubleday, Garden City, NY.

CHAPTER 14

Effectiveness, Productivity, and Design of Purposeful Systems: The Profit-Making Case*

Melvin F. Shakun† and Ephraim F. Sudit††

14.1 Introduction

In this chapter we examine the interactive role of effectiveness and productivity in the design of purposeful systems. Managerial designs have traditionally emphasized financial objectives, occasionally supplemented by economic efficiency goals. Little attention has been accorded to the relationship of effectiveness and efficiency in systems design. Our proposed model integrates productivity (efficiency) and effectiveness standards into goal/technology conflict resolution processes designed to the extent possible to generate feasible and acceptable solutions for goal attainment.

In Section 14.2 relationships between values, goals, controls, and technology characterizing systems designs as a dynamic difference game are formulated. Section 14.3 presents our generalized goal attainment model of effectiveness and develops a measure of effectiveness that measures how close technology comes to delivering desired goal targets at any given hierarchical level. Section 14.4 reviews the concept of total factor productivity as a measure of overall physical efficiency. In Section 14.5 efficiency (productivity) and cost standards are incorporated as control variables in assessing and monitoring technologically feasible sets at the production level of a profit-making system. Sections 14.6 and 14.7 deal with two illustrations of the methodology as applied to automobile production. Section 14.8 presents concluding remarks.

*This chapter appeared in the *International Journal of General Systems*, Vol. 9, 1983, pp. 205–215.

†New York University, New York

††Rutgers—The State University of New Jersey, Newark, New Jersey

Table 14.1 **Goals / Values Incidence Matrix at Time *t***

	$g(t)$
$v(t)$	$\lambda_{kij}(t)$

14.2 Design of Purposeful Systems

Shakun (1981a, 1981b) and in Chapters 1 and 2 considers the design of purposeful systems to deliver K values $v(t) = (v_k(t))$ at time $t(k = 1, 2, \ldots, K;\ t = 0, 1, \ldots, T)$ to N players $j = 1, 2, \ldots, N$. Values are nonoperational goals that are delivered in the form of operational goals $g(t) = (g_i(t))$ for $i = 1, 2, \ldots, p$. The values and goals are related by an incidence matrix shown in Table 14.1 where $\lambda_{kij}(t) = 1$ when player j is for value v_k being delivered by goal g_i at time t (player j favors both v_k and g_i as an operational expression of v_k); $\lambda_{kij}(t) = 0$ when a player j is against v_k being delivered by g_i at time t; $\lambda_{kij}(t) = X$, when player j is neutral or does not perceive v_k as being delivered by g_i at time t.

The operational goals are expressed mathematically in a dynamic difference game:

$$x(t+1) = f(x(t), u(t), t), \qquad x(0) = x_0 \tag{1}$$

$$y(t) = g(x(t), u(t), t) \in Y(t) \tag{2}$$

$x(t) = (x_1(t), \ldots, x_n(t))$ gives the state of the system at time t;

$y(t) = (y_1(t), \ldots, y_p(t))$ are systems outputs (goals);

$Y(t)$ = a set of admissible outputs;

$u(t) = u_1(t), \ldots, u_m(t)$ represents system inputs (controls).

With N players we write:

$$u(t) = (u_1(t), \ldots, u_m(t)) = (u^1(t), \ldots, u^N(t))$$

which is a partition of m dimensional input $u(t)$ among controls $u^1(t), \ldots, u^N(t)$ assigned to the N players denoted by the superscripts. Equation (1) expresses the state transition function. Equation (2) constrains outputs $y(t)$ to a set of admissible outputs $Y(t)$. Normally $Y(t) \subset R^p$, i.e., $Y(t)$ is a subset of R^p, the p-dimensional real vector space. We can also have constraints $x(t) \in X(t)$, $u(t) \in U(t)$ where $X(t) \subset R^n$, $U(t) \subset R^m$.

Let $U^C(t)$ be a set of c dimensional admissible controls available to a coalition C of the set η of N players ($C \subset \eta$). We note that coalition C may be the grand coalition of all N players. Let $Y^C(t)$ be a set of p-dimensional admissible outputs (goals) defined by the coalition C as the intersection of admissible outputs $Y^j(t)$ for all players $j \in C$. The nonemptiness of $Y^C(t)$ is a necessary condition called "goal target agreement" for problem definition and solution by coalition C. If $Y^C(t)$ is initially empty (i.e., there is conflict), it may become nonempty by expansion of the $Y^j(t)$ through negotiation among coalition members within a given operational goal space. However, an important approach to conflict resolution is to redefine the dimensions of the operational goal space itself, e.g., by using a goals/values referral process (Shakun 1981a, 1981b). Within the new goal space, either originally or after negotiation, to expand individual coalition members' sets of admissible goals $Y^j(t)$, $Y^C(t)$ may be nonempty.

Let $\mathcal{R}^C(t)$ be the set of states $x(t)$ in R^n reachable from x_0 at time t by application of $U^C(t)$ and $U^{\bar{C}}(t)$ where $U^{\bar{C}}(t)$ is a set of $(m - c)$ dimensional admissible controls available to players not in coalition C who can form one or more coalitions $\bar{C}$. Find the intersection $\mathcal{A}(t)$ of $\mathcal{R}^C(t)$ with $X(t)$, the set of admissible states. $\mathcal{A}(t)$ is the set of admissible reachable states. Then, using (2), $g(\mathcal{A}(t), U^C(t), U^{\bar{C}}(t), t)$ representing technological feasible performance is the set of outputs reachable by coalition C regardless of the controls $U^{\bar{C}}(t)$ exercised by $\bar{C}$. If the intersection of target $Y^C(t)$ and performance $g(\mathcal{A}(t), U^c(t)), U^{\bar{C}}(t), t)$ is not empty, then coalition C has defined and solved its problem; otherwise target-performance conflict remains. Thus conflict resolution (problem definition and solution) for coalition C requires both goal target agreement by its members and target-performance conflict resolution. The methodology is called evolutionary systems design (ESD). See Section 1.3 for a more complete formulation.

14.3 Effectiveness

Our model views effectiveness in terms of generalized goal attainment. It includes the goal attainment, systems, and ecological models of effectiveness described in the literature (Miles 1980). In our generalized goal attainment model nominal organizational goals—profit, market share, physical output, efficiency (output/input ratio), etc.—incorporated in the standard goal attainment model are expanded to include, in goal form, structures and processes that are the focus of the systems model (Miles 1980). The latter concentrates on the means—structures and processes—needed for system survival. However, because means themselves are ends, they can also be formulated as goals and incorporated in coalition C's goal target. Measures of such means include measures of employee strain, ability to acquire resources, flexibility, employee self-development, balance in resource allocations to maintenance, adaptive and transformation processes—in short, measures of structures and processes conferring survival.

In the generalized goal attainment approach we also consider goals of various strategic constituencies described in the ecological model of effectiveness (Miles 1980). Our coalition C includes various constituencies who desire to attain goals for themselves. The ecological model treats structures and processes as determinants of effectiveness on an ongoing basis. Our approach includes this but in addition permits incorporation into coalition C's goal target of measures of these structures and processes. Our generalized goal attainment model is quantitatively formulated and provides a comprehensive framework for effectiveness. We begin with some definitions.

A goal set $g^C(t)$ is effective with respect to a value set $v^C(t)$ for a coalition C if there is a goal target agreement, i.e., $Y^C(t)$ is nonempty. For a coalition C a technology $g(\mathscr{A}(t), U^C(t), U^{\bar{C}}(t), t)$ is effective with respect to a nonempty goal target $Y^C(t)$ if their intersection is nonempty.

A technology $g(\mathscr{A}(t), U^C(t), U^{\bar{C}}(t), t)$ is effective with respect to a values set $v^C(t)$ for a coalition C if (1) there is goal target agreement, i.e., $Y^C(t)$ is nonempty, and (2) there is nonempty target/performance intersection, i.e., the intersection of target $Y^C(t)$ and technological performance $g(\mathscr{A}(t), U^C(t), U^{\bar{C}}(t), t)$ is nonempty for each t. If this target/performance intersection consists of a single point, then it is the solution. A single point may result from an adaptation process (Shakun 1981a, 1981b) in which sets $Y^C(t)$ and $g(\mathscr{A}(t), U^C(t), U^{\bar{C}}(t), t)$ initially have an empty intersection and then expand, or overexpand and then contract. If this intersection consists of more than one point, one of them can be chosen as the solution if coalition members agree (e.g., they could choose by optimizing a criterion). If they cannot agree, then through adjustment processes—geometrically speaking, set contraction—they could arrive at a single intersection point (Shakun 1981a, 1981b), or due to disagreement an empty intersection could result. Thus finally there is either an empty or single point intersection.

Note that effectiveness for coalition C necessitates definition and agreement on goal dimensions in relation to values, and then agreement on goal target $Y^C(t)$ within the goal space defined. In other words first goal dimension agreement and then goal target agreement is required. This is undertaken internally by members of coalition C. However the coalition is influenced by external referents, e.g., outside organizations and higher levels in the hierarchy of the same organization. Behavioral theory (Cyert and March 1963) suggests that the set of admissible outputs $Y^C(t)$ if expressed by coalition C as aspiration levels will depend on coalition C's past aspiration levels, past performance, and past reference group performance. Fundamentally, effectiveness depends on the observer or evaluator who from his viewpoint can define a goal target $Y^C(t)$ to be pursued by coalition C. This evaluator can be coalition C itself or some other outside coalition.

In a hierarchical system with levels $s = 1, 2, \ldots, S$, a subvector $\tilde{U}_s^C(t)$ of the coalition control vector $U_s^C(t)$ producing goals $y_s^C(t) = g_s(\mathscr{A}_s(t), U_s^C(t), U_s^{\bar{C}}(t), t)$ at a higher goal level s becomes the goal target $Y_{s+1}^C(t) = \tilde{U}_s^C(t)$ to be produced by lower goal level $(s + 1)$ controls, $U_{s+1}^C(t)$ operating with technology

$$y_{s+1}^C(t) = g_{s+1}\left(\mathscr{A}_{s+1}(t), U_{s+1}^C(t), U_{s+1}^{\bar{C}}(t), t\right).$$

In other words all lower level output goals are fed up the hierarchy and contribute to delivering higher goals and eventually values.

In terms of the hierarchical notation the preceding discussion has involved the highest goal level, $s = 1$. At this level the goal target/technology adaptation process has resulted in a single point intersection, a subvector $\tilde{U}_1^C(t)$ of whose controls become the goal point target $Y_2^C(t)$ at level $s = 2$, i.e., $Y_2^C(t) = \tilde{U}_1^C(t)$. If the intersection of goal point target $Y_2^C(t)$ and technology $g_2(\mathscr{A}_2(t), U_2^C(t), U_2^{\bar{C}}(t)$ is not empty (i.e., point $Y_2^C \in g_2$), then the subset $\tilde{U}_2^C(t)$ of the controls $U_2^C(t)$ becomes the goal point target at level 3, $Y_3^C(t) = \tilde{U}_2^C(t)$. If the intersection of goal point target $Y_2^C(t)$ and g_2 is empty, then the set g_2 could expand until there is an intersection. This process continues until we reach goal level $s = S - 1$, at which level we have goal point target $Y_{S-1}^C(t)$. If the intersection of $Y_{S-1}^C(t)$ and technological performance g_{S-1} is nonempty, we have said the technology g_{S-1} is effective with respect to the goal $Y_{S-1}^C(t)$.

The above defines a situation where at each level s $(s = 1, 2, \ldots, S - 1)$ the technology g_s is effective with respect to the goal point target $Y_s^C(t)$, i.e., Y_s and g_s have a nonempty intersection. This is not always the case—sometimes at level s the adjustment of the system does not result in a nonempty intersection of Y_s, and g_s is not effective with respect to Y_s.

To define effectiveness at level s on a numerical scale we say that the effectiveness E_s is one if the intersection of goal point target $Y_s^C(t)$ and technological performance $y_s^C(t) = g_s(\mathscr{A}_s, U_s^C(t), U_s^{\bar{C}}(t), t)$ is nonempty. Our measure of effectiveness E_s measures how close at level s the technology comes to delivering the goal compared to a zero output. If the technology g_s gives a vector of zero output, the effectiveness is zero. Thus at any level s for time period t we define a measure of effectiveness $E_s(t)$:

$$E_s(t) = 1 - \frac{\| y_s^C(t) - Y_s^C(t) \|}{\| 0 - Y_s^C(t) \|} \tag{3A}$$

and over a time sequence $t = 0, 1, \ldots, T$ a measure of effectiveness E_s:

$$E_s = 1 - \frac{\sum_{t=0}^{T} \| y_s^C(t) - Y_s^C(t) \|}{\sum_{t=0}^{T} \| 0 - Y_s^C(t) \|} \tag{3B}$$

where $\|y_s^C(t) - Y_s^C(t)\| = d_s(y_s^C(t) - Y_s^C(t))$ = distance between $y_s^C(t)$ and $Y_s^C(t)$. Negative effectiveness indicates one or more of the outputs in $Y_s^C(t)$ is even further from the goal target $Y_s^C(t)$ than is the origin. In Euclidean space we take

$$\| y_s^C(t) - Y_s^C(t) \| = \sqrt{\sum_{i=1}^{p} [y_i^C(t) - Y_i^C(t)]^2} \tag{4}$$

If, for all t, $Y_s^C(t) \in g_s$, i.e., we have the case of nonempty intersection, this distance is zero. In fact if all the constraints from level 1 and all lower level constraints are included in the level 1 design problem, and if, after adjustment, the intersection of $Y_1^C(t)$ and g_1 is a single point, then $Y_s^C(t) \in g_s$ for all s; so the effectiveness $E_s = 1$ for all s. But if after adjustment the intersection between $Y_1^C(t)$ and g_1 is empty for at least one t, then $Y_s^C(t) \notin g_s$ for at least one t for all other levels s as well. For any such t, for $y_1^C(t)$ the coalition chooses a preferred point as a feasible target in the technologically feasible set g_1. Then in Euclidean space:

$$\| y_1^C(t) - Y_1^C(t) \| = \sqrt{\sum_{i=1}^{p} [y_i^C(t) - y_i^*(t)]^2} \tag{5}$$

where $y^*(t) \in Y_1^C(t)$ is the closest point to preferred point $y_1^C(t)$ and where subscript i denotes the ith component in the y and y^* vectors.

Choice of a preferred point suggests that if the coalition reluctantly had (as it does because there is no nonempty intersection) to expand the goal target set $Y_1^C(t)$ to give a point intersection with g_1, that expansion would result in an intersection at the preferred point. One particular preferred point in (3A) is obtained by choosing $y_1^C(t)$ through choice of $U^C(\tau)$ for $\tau = 0, 1, \ldots, t$ as close as possible to the set $Y_1^C(t)$ and this is obtained by

$$\min_{U^C(\tau)} \max_{U^{\bar{C}}(\tau)} Z(t) = \| y_1^C(t) - Y_1^C(t) \| \tag{6}$$

where the right-hand distance in (6) is defined, as before, by (5). For (3B) we use $\Sigma Z(t)$ in (6) where the sum is taken over $t = 0, 1, \ldots, T$ and $\tau = 0, 1, \ldots, T$ as well.

This procedure gives a designed (planned) effectiveness, as for a corporate production plan or a national planned economy. Alternatively, for $y_s^C(t)$ in (3) we may use the output actually produced. This gives an actual production effectiveness.

Finally we note that at any level s the effectiveness measures (3A), (3B) apply not only to the full p_s dimension goal space, but may also be computed for various subspaces as may be of interest. Thus we may compute effectiveness on individual goal dimensions (including physical efficiency defined in Section 14.4), as well as on goal subspaces dealing with nominal goal attainment (e.g., profit-market share), management of human resources, adaptation, integration, etc. Of course the overall effectiveness of the system is that computed for the full p_1 dimensional goal space at level $s = 1$, the highest goal level.

14.4 Physical Efficiency: Total Factor Productivity Analysis

Total factor productivity analysis is designed to measure the degree of overall physical efficiency and changes thereof in attaining a physical set of outputs (goals). Let $U_s = F_s(U_{i,s})$ and $U_{s+1} = F_{s+1}(U_{h,s+1})$ be aggregator functions of p outputs ($i = 1, \ldots, p$) and m inputs ($h = 1, \ldots, m$) occurring at hierarchical levels s and $s + 1$ respectively. The level of the Total Factor Productivity (TFP) index at any point in time is defined as a ratio of an aggregate of the outputs to an aggregate of the inputs (Diewert 1976; Jorgenson and Griliches

1967; Solow 1957):

$$\mathrm{TFP} = \frac{U_s}{U_{s+1}}. \tag{7}$$

Taking logs and differentiating with respect to time, the percent change in TFP over time is therefore:

$$\frac{\dot{\mathrm{TFP}}}{\mathrm{TFP}} = \frac{\dot{U}_s}{U_s} - \frac{\dot{U}_{s+1}}{U_{s+1}}$$

where the dot denotes the first derivative with respect to time.

Jorgenson and Griliches (1967) propose Divisia (1926) indexes for aggregating outputs (goals) and inputs (control variables) for the purpose of constructing a Divisia index for TFP change:

$$\frac{\dot{\mathrm{TFP}}}{\mathrm{TFP}} = \sum_{i=1}^{p} \frac{\dot{U}_{i,s}}{U_{i,s}} \alpha_{i,s} - \sum_{h=1}^{m} \frac{\dot{U}_{h,s+1}}{U_{h,s+1}} \alpha_{h,s+1} \tag{8}$$

where

$$\alpha_{i,s} = \frac{P_{i,s} U_{i,s}}{\sum_{i=1}^{p} P_{i,s} U_{i,s}}$$

$$\alpha_{h,s+1} = \frac{P_{h,s+1} U_{h,s+1}}{\sum_{h=1}^{m} P_{h,s+1} U_{h,s+1}}$$

and $P_{i,s}$ and $P_{h,s+1}$ are the unit prices of the outputs at level s and the inputs at level $s + 1$, respectively. Economically, the change in TFP over time could be interpreted as the difference between the weighted sum of outputs (goals) and the weighted sum of inputs (controls).

For a business enterprise at a certain level of systems design, outputs could be defined as physical quantities or commodities, and inputs could be defined as physical quantities of factors of production. In this context the weights $\alpha_{i,s}$ and $\alpha_{h,s+1}$ are respectively the share of the revenue generated by the ith commodity in total revenue and the share of the costs incurred by the hth factor of production in total costs.

The Divisia TFP index approximation for discrete period changes, say from period t to period $t + 1$, following Tornquist (1936) could be stated as:

$$\frac{\mathrm{TFP}_{t+1}}{\mathrm{TFP}_t} = \frac{\prod_{i=1}^{p} \left[\frac{U_{i,s}(t+1)}{U_{i,s,}(t)} \right]^{1/2[\alpha_{i,s}(t) + \alpha_{i,s}(t+1)]}}{\prod_{h=1}^{m} \left[\frac{U_{h,s+1,}(t+1)}{U_{h,s+1,}(t)} \right]^{1/2[\alpha_{h,s+1,}(t) + \alpha_{h,s+1,}(t+1)]}} \tag{9}$$

We note that the TFP calculation can be made between any lower levels controls (inputs) and any higher level goals (outputs) (with input and output levels not necessarily adjacent), provided that we know unit prices of outputs and unit costs of inputs. Thus, for example, lower level $s+1$ can be replaced by $s+2$ in (9).

14.5 Cost and Efficiency Controls

TFP is a measure of overall physical efficiency in resource allocation and utilization. To supplement physical efficiency controls, cost efficiency controls are needed to monitor unit cost controls against standards. Consequently, at level s the subsystem design problem can be stated as:

$$\text{Attain a set of goals } (u_{i,s}) \qquad i = 1, \ldots, p_s \tag{10}$$

which include

Physical efficiency controls:

$$\text{TFP} \geqq \text{TFP(min)} \tag{10a}$$

$$\Delta(\text{TFP}) \geqq \Delta(\text{TFP(min)}) \tag{10b}$$

Unit cost controls:

$$P_{h,s+1} \leqq P_{h,s+1}(\text{max}) \tag{10c}$$

$$h = 1, \ldots, m_s.$$

The minimum TFP performance standards for level s and changes over time (denoted by Δ), TFP(min) and Δ(TFP min), can be predetermined relative to historical performance of the system, related industry performance, or estimated performance by successful competitors plus desired improvement margins. Maximum unit costs standards P(max) can be set relative to market unit prices for materials and capital equipment and market wage rates for labor. To the extent that the output goals do not incorporate quality, separate minimum quality standards can be specified as constraints in (10). The levels of the cost and efficiency controls in (10) affect higher level goals (e.g., profitability growth, market share, etc.).

Failure of the actual output to fall within the goal target may result from actual total factor productivity and/or unit cost levels falling short of their respective standards. Consequently, improvements in productivity and/or cost performance may generate an output level consistent with the goal target. In the design phase, low cost and productivity standards may lead to effectiveness of less than unity. It may be desirable in these cases to search for higher productivity and/or lower unit cost standards to increase design effectiveness as an alternative to further goals/values adaptation processes.

14.6 A Numerical Example: Producing Cadillacs and Chevrolets

We shall consider a two time period production problem for a system producing Cadillacs and Chevrolet automobiles. The values/goals/control hierarchy is shown in Figure 14.1. The higher level goals at level $s = 1$ are

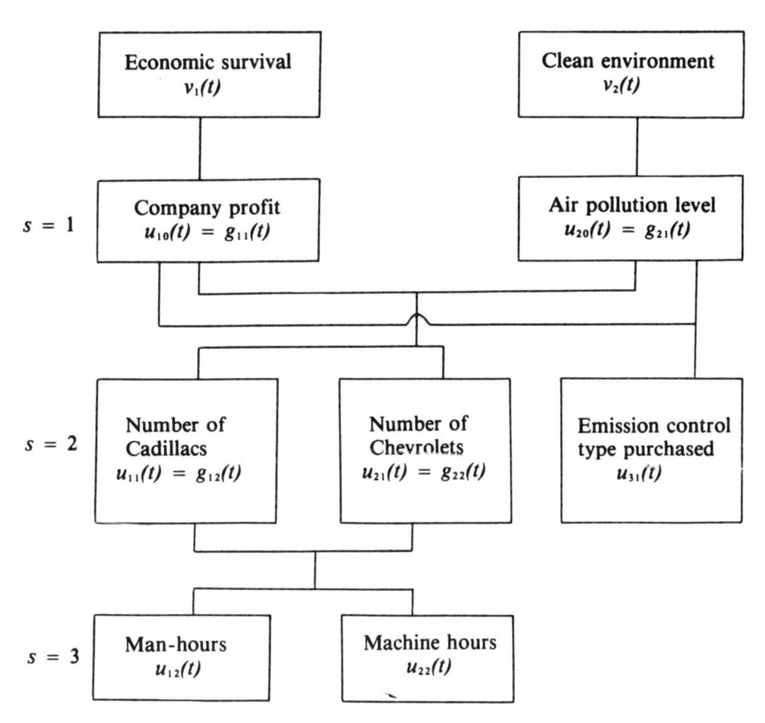

Figure 14.1. **Values / Goals / Controls Hierarchy in Cadillac–Chevrolet Production Problem. $U_{hs}(t) = h^{th}$ Control for Hierarchical Goal Level s for Time Period t. $U_{hs}(t)$ is Drawn at Level $s + 1$. $h = 1, 2, \ldots, m_s$; $s = 1, 2, \ldots, S$; $t = 0, 1, \ldots, T$; here $m_1 = 3$, $m_2 = 2$, $S = 3$, $T = 1$. $g_{is}(t) = i^{th}$ Goal at Goal Level s for Time t, $i = 1, 2, \ldots, p_s$; here $p_1 = p_2 = 2$. $v_1(t)$ and $v_2(t)$ are Values for Time Period t.**

profits and air pollution level; the goals at level $s = 2$ are the numbers, say in hundreds of thousands, of Cadillacs and Chevrolets. The controls for producing the profits and the air pollution goals are the number of Cadillacs, number of Chevrolets, and the emission control type drawn at level $s = 2$. The controls for producing the level 2 goals—number of Cadillacs and number of Chevrolets—are man-hours and machine hours drawn at level $s = 3$. We have the following relationships:

Profit,

$$g_{11}(t) = g_{11}(U_{11}(t), U_{21}(t), U_{31}(t))$$
$$= [a_{11}(U_{31}(t))]U_{11}(t) + [a_{12}(U_{31}(t))]U_{21}(t). \tag{11}$$

Emission Control Type $U_{31}(t)$ for t = 0, 1

	$U_{31} = 1$	$U_{31} = 2$	$U_{31} = 3$
a_{11}	3	2	1
a_{12}	5	4	3

where the numerical values are in thousands of dollars.

Air pollution level,

$$g_{21}(t) = g_{21}(U_{11}(t), U_{21}(t), U_{31}(t))$$
$$= [a_{21}(U_{31}(t))]U_{11}(t) + [a_{22}(U_{31}(t))]U_{21}(t) \quad (12)$$

	$U_{31} = 1$	$U_{31} = 2$	$U_{31} = 3$
a_{21}	3	2.5	2
a_{22}	2	1.5	1

There are three types of emission controls, $U_{31}(t) = 1, 2, 3$, which are increasingly costly and pollution effective so that profit and pollution coefficients per car go down in the same order, 1, 2, 3. The other admissible controls, U_{11} and U_{21}, are defined by:

$$0 \leqq U_{11}(t) \leqq 4 + t \quad \text{for} \quad t = 0, 1 \quad (13)$$

$$0 \leqq U_{21}(t) \leqq 6 \quad \text{for} \quad t = 0, 1 \quad (14)$$

the upper limits are market size constraints defining the state of the system at level 2 and the lower limits are nonnegativity constraints (integer requirements are ignored).

The goal target $Y_1^C(t)$ is defined by $g_{11}(0) \geq 36$, $g_{21}(0) \leq 15$, $g_{11}(1) \geq 36$, $g_{21}(1) \leq 18$ assuming that for period $t = 1$ the government relaxes the pollution constraint from 15 to 18 when it realizes from the results below for $t = 0$ that the company will not be able to earn a viable profit of 36.

To simplify the problem a bit, suppose the emission control $U_{31}(t)$ is fixed to type 1. Then equations (11), (12) for profit and air pollution, respectively, become (i.e., the technology at level $s = 1$ becomes):

$$g_{11}(t) = 3U_{11}(t) + 5U_{21}(t) \quad (15)$$

$$g_{21}(t) = 3U_{11}(t) + 2U_{21}(t). \quad (16)$$

Further suppose the technology at level $s = 2$ is

$$U_{12}(t) = (6 - 0.9t)U_{11}(t) + (5 - 0.75t)U_{21}(t) \quad (17)$$

$$U_{22}(t) = (2 - 0.2t)U_{11}(t) + (1 - 0.1t)U_{21}(t) \quad (18)$$

where, noting Figure 14.1, $U_{12}(t)$ and $U_{21}(t)$ are man-hours and machine hours, respectively, and $U_{11}(t) = g_{12}(t)$ and $U_{21}(t) = g_{22}(t)$ are the numbers of Cadillacs and Chevrolets, respectively. Note that the production coefficients in (17) and (18) have been assumed to be capable of productivity improvement by 15 percent and 10 percent, respectively, from period $t = 0$ to period $t = 1$.* Suppose the admissible controls for man-hours and machine hours are defined

*In equation (15) the use of constant unit profit margins from period $t = 0$ to $t = 1$ of $3,000 and $5,000 imply these production improvements have been offset, we assume here by a combination of (1) increasing costs per man-hour and machine hour and (2) decreasing selling prices necessitated by increasing market competition from $t = 0$ to $t - 1$.

by

$$0 \leqq U_{12}(t) \leqq 50 \tag{19}$$

$$0 \leqq U_{22}(t) \leqq 8 + t \tag{20}$$

the upper limits defining the state of the system at level 3.

Setting $U_{12}(t) = 50$ and $U_{22}(t) = 8 + t$, we obtain as upper boundary conditions for (17), (18):

$$(6 - 0.9t)U_{11}(t) + (5 - 0.75)U_{21}(t) = 50 \tag{21}$$

$$(2 - 0.2t)U_{11}(t) + (1 - 0.1t)U_{21}(t) = 8 + t. \tag{22}$$

Using (13), (14), (15), (16), (21), (22), we plot the output goal space in Figure 14.2. Equations (13), (14), (15), (16) define a feasible region $ABCD$ when $t = 0$ and $AEFD$ when $t = 1$; adding constraints (21), (22) reduces the feasible regions to $ABGD$ when $t = 0$ and $AEHD$ when $t = 1$. The reduced feasible regions are due to constraint (22); constraint (21) is not binding.

For period $t = 0$, referring to Figure 14.2, it is clear that at level $s = 1$ the company cannot achieve goal target (36, 15) which using (15), (16) translates into goal target (1/3, 7) at $s = 2$, i.e., 1/3 Cadillac, and 7 Chevrolets. Suppose the company chooses (33, 15) as the preferred feasible output point. Then using (15), (16), the corresponding feasible output at level 2 for $t = 0$ is (1, 6). For $t = 1$, from Figure 14.2 it is clear that the company can achieve goal (36, 18), which using (15), (16) translates into a goal target (2, 6) at level $s = 2$. In other words at $t = 1$, goals (36, 18) and (2, 6) at levels $s = 1, 2$ respectively are in fact feasible outputs, i.e., goal targets and technological performances at levels 1 and 2 have nonempty point intersections.

Using (3B) the effectiveness values at levels 1 and 2 are computed as follows:

$$E_1 = 1 - \frac{\sqrt{3^2 + 0^2} + 0}{\sqrt{36^2 + 15^2} + \sqrt{36^2 + 18^2}}$$

$$= 1 - \frac{3}{39 + 40.24} = 0.96$$

$$E_2 = 1 - \frac{\sqrt{(2/3)^2 + 1^2} + 0}{\sqrt{(1/3)^2 + 7^2} + \sqrt{2^2 + 6^2}}$$

$$= 1 - \frac{1.4444}{7.0079 + 6.3246} = 0.89.$$

Using (17), (18) the controls $U_{12}(t)$ man-hours and $U_{22}(t)$ machine hours, which give $y_2^C(0) = (1, 6)$ and $y_2^C(1) = (2, 6)$ are $U_{12}(0) = 36$, $U_{22}(0) = 8$, $U_{12}(1) = 35.7$, $U_{22}(1) = 9$.

For the TFP computation for level $s = 2$ the levels of outputs (numbers of Cadillacs and Chevrolets) at time periods $t = 0, 1$ are $U_{11}(0) = 1$, $U_{21}(0) = 6$,

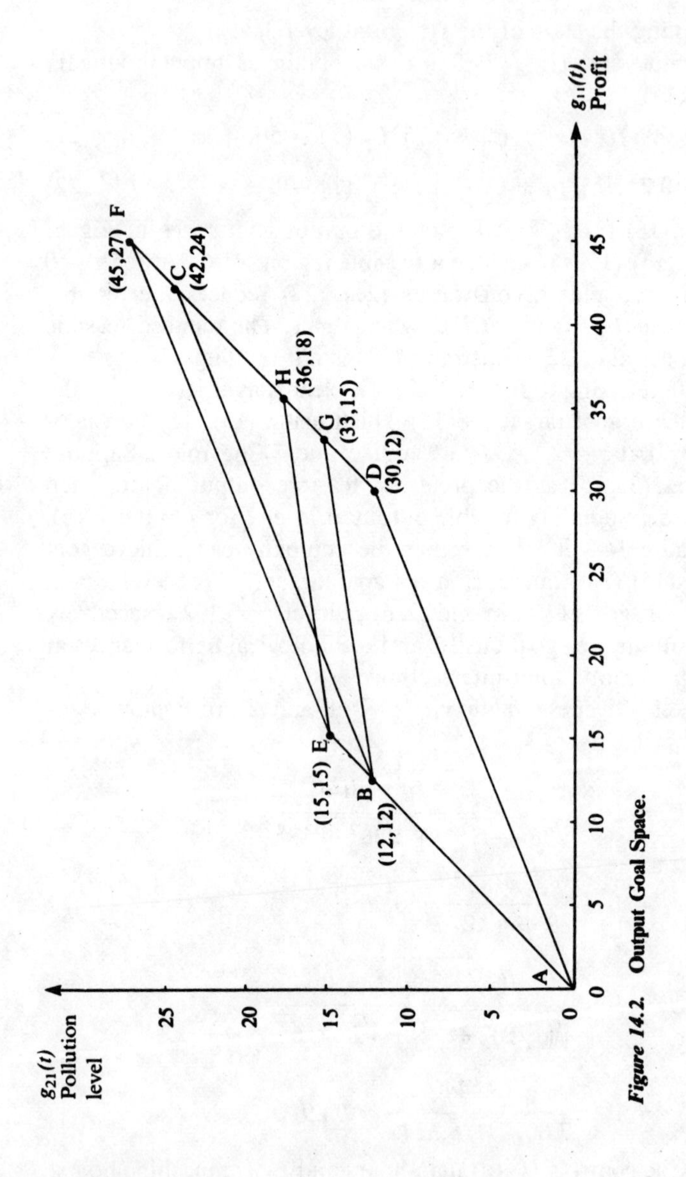

Figure 14.2. **Output Goal Space.**

$U_{11}(1) = 2$, $U_{21}(1) = 6$. The levels of the inputs (man-hours and machine hours) are $U_{12}(0) = 36$, $U_{22}(0) = 8$, $U_{12}(1) = 35.7$, $U_{22}(1) = 9$. Suppose the prices in thousands of dollars for Cadillacs and Chevrolets are $P_{11}(0) = \$20.00$ and $P_{21}(0) = \$13.53$, respectively, for $t = 0$ and $P_{11}(1) = 18.76$, $P_{21}(1) = 12.91$ for $t = 1$. The unit costs in thousands of dollars per man-hour and machine hour are $P_{12}(0) = 0.0150$ and $P_{22}(0) = 8.455$, respectively, for $t = 0$ and $P_{12}(1) = 0.0155$, $P_{22}(1) = 8.710$ for $t = 1$. These numerical values for prices and costs are consistent and satisfy the following relations for $t = 0, 1$*:

$$P_{11}(t) - (6 - 0.9t)P_{12}(t) - (2 - 0.2t)P_{22}(t) = 3 \tag{23A}$$

$$P_{21}(t) - (5 - 0.75t)P_{12}(t) - (1 - 0.1t)P_{22}(t) = 5 \tag{23B}$$

where the coefficients come from (17) and (18). In general the left-hand sides of (23A) and (23B) replace, respectively, the coefficients 3 and 5 in equations (15), (16).

According to (9) the TFP change between period 0 and 1 may be computed as follows:

$$\frac{\mathrm{TFP}_1}{\mathrm{TFP}_0} = \left(\frac{\left(\frac{2}{1}\right)^{1/2(0.198+0.324)} \left(\frac{6}{6}\right)^{1/2(0.802+0.674)}}{\left(\frac{35.7}{36.0}\right)^{1/2(0.0079+0.0071)} \left(\frac{9}{8}\right)^{1/2(0.9921+0.9929)}} \right)$$

$$= 1.11 \tag{23C}$$

In this numerical example TFP has increased by approximately 11 percent from period 0 to 1. Suppose that the expected performance standard was a TFP increase of 25 percent giving only a 0.44 effectiveness on an efficiency increase goal. Then productivity results of 11 percent in the planning process would be unacceptable. Other things being equal, deficient productivity performance will be reflected in lower profitability performance at the higher goal level. With respect to unit cost control the relatively moderate increases in input prices from period 0 to period 1 may be regarded as a favorable performance with respect to minimum unit cost controls (standards). The projected shortfall in productivity performance, however, may lead to redesign of production plans, redefinition of goals, or expansion of the goals set.

As another scenario suppose (36, 18) is the profit-pollution goal target for both $t = 0$ and $t = 1$. The 11 percent increase in TFP results in an expansion of the set of technologically feasible outputs in Figure 14.2 from *ABGD* for $t = 0$ to *ABHD* for $t = 1$. This allows a profit-pollution output change from (33, 15) to (36, 18) associated with a car output increase from (1.6) to (2.6). Thus the company attains its (36, 18) target at $t = 1$ after missing it at $t = 0$. Using (3A), this 11 percent increase in TFP is reflected in an increase of

*Consistent with the preceding footnote.

effectiveness from $t = 0$ to $t = 1$ at level $s = 1$ as follows:

$$E_1(0) = 1 - \frac{\sqrt{3^2 + 3^2}}{\sqrt{36^2 + 18^2}} = 0.89$$

$$E_1(1) = 1$$

and at level $s = 2$

$$E_2(0) = 1 - \frac{\sqrt{1^2 + 0^2}}{\sqrt{2^2 + 6^2}} = 0.84$$

$$E_2(1) = 1.$$

14.7 A Second Example: Illustrating Profit-Pollution Control Trade-Offs

In our previous numerical example we specified in the design of the problem three types of emission controls that were increasingly costly and pollution effective. However, for the sake of computational simplicity we fixed the emission control input to one type only. In the present illustration we focus on the trade-offs between profit and pollution control goals. We therefore allow for a continuous choice among levels of pollution control inputs (e.g., labor, capital, and materials required for the production, purchase, and installation of pollution control devices in cars). The higher the level of those inputs, the lower the level of pollution and the lower the level of profits—hence the profit-pollution control trade-off problems that are of major importance in determining limits to environmental legislation.

In outlining the economic-environmental problem we posit that the pollution level z is generated as an external joint by-product in the production of cars. To simplify the illustration we assume that only one type of car is manufactured, with the number of cars produced denoted by y. If the minimum acceptable profit level at time t is $\hat{\pi}_t$ and the maximum acceptable pollution level at time t is $\hat{z}_t$, then the solution to the economic-environmental problem can be formulated as a search for a feasible and acceptable set of profit and pollution levels $\{\pi, z\}$ satisfying the following conditions:

$$\pi_t = P_t y_t - \sum_i w_{it} u_{it} - \sum_j v_{jt} \geqq \hat{\pi}_t \tag{24a}$$

$$z_t \leqq \hat{z}_t \tag{24b}$$

$$F_t(y_t, z_t, u_t, x_t, T_t/T_{t-1}) = 0 \tag{24c}$$

where π_t is the actual level of profits at time t, P_t is the price of a car, x_{jt} is the level of the quantity of the jth control or input dedicated to pollution abatement, and v_{jt} is its corresponding price. The variables u_{it} and w_{it} are respectively the quantity and price of the ith ordinary input. T_t/T_{t-1} denotes

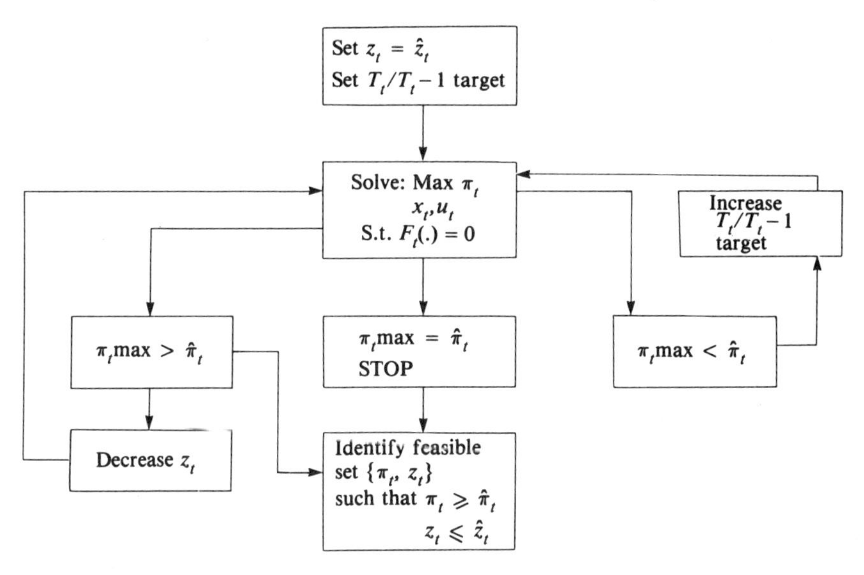

Figure 14.3. **Algorithm for the Identification of Feasible and Acceptable Set** $\{\pi_t, z_t: \pi_t \geqq \hat{\pi}_t, z_t \leqq \hat{z}_t\}$

neutral technical change from time $t - 1$ to time t. Symbols u_t and x_t are respectively the nonpollution and pollution control vectors.

Equations (24a) and (24b) can be viewed as the operational goal constraining equations. Equation (24c) is a state transition function reflecting changes in output and controls (i.e., production feasibilities). The technological change over time T_t/T_{t-1} introduces the dynamic element into the state transition function by making the state of the system dependent on change in technology over time.

Because technical change and antipollution devices are designed to reduce pollution level, $\partial z/\partial x_j < 0$ for all j, and $(\partial z/\partial(T_t/T_{t-1})) < 0$. The levels of antipollution devices are assumed to have no effect on car output. Consequently, $(\partial y/\partial x_j) = 0$ for all j. The level of demand $y_t \leqq \bar{y}_t$ is given (i.e., $\bar{y}_t$ is a market size constraint). Technical change T_t/T_{t-1} is assumed to be Hicks-neutral (i.e., does not affect optimal input combinations).

Figure 14.3 proposes an algorithm for identifying a feasible and acceptable set of operational goals $\{\pi_t, z_t\}$. The set is *feasible* if it stays within the production constraint (24c). It is *acceptable* if $\pi_t \geqq \hat{\pi}_t$ and $z_t \leqq \hat{z}_t$.

Assuming that the planning is done at time $t - 1$, with no external change in y, p, w_i and v_j from time $t - 1$ to t, the only change to be forecast is the technical change (productivity) T_t/T_{t-1}. The SOLVE module in Figure 14.3 is a standard Lagrange solution. $\pi^{\max}$ is the maximum level of profits computed by SOLVE. If $\pi_t^{\max} < \hat{\pi}_t$ implies $\{\pi_t, z_t\} = \phi$, then no feasible solution exists.

The same effectiveness given by equation (3) can be applied to the results of the algorithm. Thus effectiveness of one would indicate the existence of a feasible and acceptable set. Effectiveness of less than unity would indicate that the feasible and acceptable set is empty. If the latter is the case, higher productivity T_t/T_{t-1} targets could be attempted via a process of identifying improved productivity procedures potentially capable of rendering feasible and acceptable solutions through the use of the algorithm in Figure 14.3. This could be done before resorting to the referral process of modifying goals vis-à-vis values.

14.8 Concluding Remarks

Effectiveness and productivity measures play an important and useful role in the successful design of profit-making systems. We have developed in this chapter a measure of effectiveness that indicates the closeness between desired goals (targets) and the technological feasibility of their attainment. Total factor productivity indexes at the production level coupled with cost controls monitor technological performance. Productivity measures the physical efficiency by which inputs are transformed into outputs in the production process. Productivity (efficiency)—itself one effectiveness goal—is related to effectiveness in that higher productivity can result in an expansion of the set of technologically feasible outputs, which in turn can increase effectiveness. In fact the choice of operational goal dimensions in relation to values can be affected by efficiency.

The measure of efficiency used in the chapter, namely, total factor productivity, uses market prices of outputs and inputs and hence is directly applicable to profit-making systems. We note that other measures of efficiency can be used in the absence of market prices and costs (Charnes, Cooper, and Rhodes 1978; Lewin and Morey 1981). These measures of efficiency can also be used in conjunction with the measure of effectiveness developed in this chapter to characterize organizational performance.

REFERENCES

Charnes, A., Cooper, W. W., and Rhodes, E. 1978. "Measuring Efficiency of Decision Making Units." *European Journal of Operations Research*, 2, No. 6, November.

Cyert, R. N., and March, J. G. 1963. *A Behavioral Theory of the Firm*. Prentice Hall, Englewood Cliffs, NJ.

Diewert, E. W. 1976. "Exact and Superlative Index Numbers." *Journal of Econometrics*, 4, No. 2, pp. 115–146.

Divisia, F. 1926. *L'Indice monétaire et la théorie de la monnaie*. Société Anonyme du Recueil, Paris.

Jorgenson, D., and Griliches, Z. 1967. "The Explanation of Productivity Change." *Review of Economic Studies*, 34, No. 99, pp. 249–283.

Lewin, A. Y., and Morey, R. C. 1981. "Measuring the Relative Efficiency and Output Potential of Public Sector Organizations: An Application of Data Envelopment Analysis." Fuqua School of Business, Duke University.

Miles, R. H. 1980. *Macro Organizational Behavior*. Goodyear Publishing Company, Santa Monica, Chap. 12.

Shakun, M. F. 1981a. "Formalizing Conflict Resolution in Policy Making." *International Journal of General Systems*, 7, No. 3.

______. 1981b. "Policy Making and Meaning as Design of Purposeful Systems." *International Journal of General Systems*, 7, No. 4.

Solow, R. M. 1957. "Technical Change and the Aggregate Production Function." *Review of Economics and Statistics*, 39, pp. 312–320.

Tornquist, L. 1936. "The Bank of Finland Consumption Price Index." *Bank of Finland Monthly Bulletin*, 10.

CHAPTER 15

Irrationality and Effectiveness in Public Decision Making

Bertrand R. Munier*, Jean-Louis Rulliere*, and Melvin F. Shakun**

15.1 Introduction

In many cases consideration of the decisions made by public agencies acting in the domains of transportation, urban development, health, and higher education, for example, raises some strange questions. It seems that (1) nobody is ready to take responsibility for the choice of the decision selected and (2) a large number—even all—of the members-users of the organization involved see the chosen decision as unjustifiable through any kind of maximizing type of rationality. Due to considerations of inadequate information flows, the decision can sometimes be dominated by another possible solution. In this latter case the decision can questionably be claimed to be rational from the points of view of the organization *as a whole, even on a simple bounded rationality basis*. Failure to actively define values and goals can be another source of difficulty. In these cases the decision can be regarded again as "irrational." Yet the chosen solution lies within the intersection of the set of "technological" capabilities and of the goal-target of the prevailing coalition of actors. Following Shakun and Sudit (1983), the decision is thus "effective" for this coalition. It works and can indeed turn out to work much better than had been estimated. Nevertheless it is an irrational decision at the moment it is taken.

*GRASCE, Université d'Aix-Marseille III.
**New York University.

How can such "irrational and effective" decisions obtain? Two main lines of arguments can be offered—and eventually combined—to explain this strange result:

1. L. Sfez (1981) insists on "overcoding" (surcode), i.e., on the fact that one single project can have different interpretations for different coalitions within the decision system. Each coalition has its own rationality, and the interplay of these lead—with no *common* meaning—to the "final" decision. Munier (1982) offers a similar argument.
2. We intend to give here a different argument, which gives to the *procedure* chosen the crucial responsibility in obtaining irrational effective decisions. We formalize the model and give a sufficient condition (proposition 2) for the result to obtain.

In Section 15.2 we briefly refer to three empirical cases that show that irrational and effective decisions are important ones. Section 15.3 examines a relevant model of preferences, and Section 15.4 describes, with reference to one of the three cases mentioned in Section 15.2, the process of coalition formation in terms of evolutionary systems analysis. Section 15.5 presents concluding remarks.

15.2 Empirical Examples

The examples mentioned hereunder all belong to the domain of public transportation in France but are relevant to very different types of authorities (local, regional, national). In this section we give some details of the local case and cite references for the other two.

The local problem arose in the city of Lyon in the Rhône-Département, France, in the sixties to develop a subway system. At the same time, however, a cable railway linking the city hall area to the old-time hill quarter called "Croix-Rousse" was to finish its life. For safety reasons the equipment had to be replaced. By what and how was and still is debatable. Although no one dares to say that he made this particular decision—indeed, the council of the Rhône-Département once voted against it explicitly—the facts are as follows:

1. The technique of the rack railway was selected.
2. The line was included in the subway system and extended north to the suburban town of Caluire.

It turns out that this solution is not the least expensive and that the northern extension never was considered a priority decision by a majority of the actors involved in the system.

The main explanation put forward is that a sequence of coalitions was able to form and (1) split the decision into pieces; (2) concentrate the attention of the relevant bureaus on "technical consenses" (i.e., on technicalities upon which agreement was easy to reach), avoiding the possibility that the long-term consequences of these technicalities could fall under the scrutiny of other actors in the system; so that (3) implicit necessities could arise now and then to fill the gaps of the emerging global system.

The final state of the system looks very much like the product of some chance and a great deal of unanticipated necessity, not exactly the same was what we would expect from a "rational" decision maker after 20 years of costly studies and time-consuming debates.

Sfez (1981, pp. 362–363, 371–373) draws fairly similar conclusions concerning the R.E.R. (regional express railway system) in Paris, although the explanation put forward is of type 1 above. Similar situations seem again to have arisen in the decision-making process of the T.G.V. (high-speed train) between Paris and Lyon (Gazier 1983). On the latter see Section 15.4.

15.3 Preferences, Bureaucratic Procedures, and Ambiguity

Empirical evidence shows that at a given time public authorities think about a certain number of perceived possible "moves" or "elementary acts." Let us call E the set of these elementary acts. We define a "project" as a nonempty subset A of E.

A decision-procedure is an ordered set of bureaus that have as an assignment to examine in this order the project A. To "examine" the project, each bureau issues an opinion about it, which will be assumed to take only two values:

1. Either the bureau accepts every aspect of the project: It then "clears" the project.
2. Or the bureau disagrees with a certain number of elements of the project: It then "blocks" the project.

A decision-procedure will thus be modeled as a path in a graph, the vertices of which represent the bureaus, linked by oriented arcs to indicate the order in which they are to examine the project.

We define an index i for each bureau B in the following way: Let I, $I \subseteq N$ be a finite set of indices. $\forall\, i, i' \in I,\ i < i' \Leftrightarrow B_i$ examines the project A before $B_{i'}$.

BEHAVIOR OF A GIVEN BUREAU

We view each bureau assigned to examine a given project as a *filter*. It has to sort project elements in the two categories previously assumed: accepted or rejected. In problems of sorting we feel that preferences should be modeled differently than in problems of maximization or ranking for two reasons:

1. Ranking is a much more demanding mental operation than sorting. It should then be natural to have a more general model of preferences than an order or a preorder on elementary acts.
2. In sorting it is *crucial* not to consider an elementary act in *itself*, but to consider it as a *part of a whole*. Some act is acceptable if we think of

***Table 15.1* Classification of elementary acts**

	$y \in A$	$y \in A^c$
$(x, y) \in U_i$ $(y, x) \in U_i$	Elementary acts x and y in A are fully consistent (straight acceptance).	The project could very well include also y (possible y-extension).
$(x, y) \in U_i$ $(y, x) \notin U_i$	Evoking x leads to refuse y, inconsistent with the former (internal blocking).	The project, containing x, could in no way be extended to include y (no y extension possible).
$(x, y) \notin U_i$ $(y, x) \in U_i$	Evoking y leads to refuse x, inconsistent with the former (internal blocking).	Evoking y outside the project leads to refuse x within the project (external blocking).
$(x, y) \notin U_i$ $(y, x) \notin U_i$	x and y, both within the project, are mutually inconsistent (straight blocking).	Same as above, with the qualification that y should not be combined with the project as it stands (qualified blocking).

some other within (or without) the project. It may not be so if we evoke another act. In other words we do *not* assume here that we have some sort of measuring device or at least a yardstick that would enable us to rank elementary acts We try only to judge every element of a project by trying to give to it a meaning, which makes sense only if we think of *other* acts in and outside the given project. We define: $\forall\ i$, $i \in I, \forall\ A, A \subseteq E,\ A \neq \phi$, the bureau B_i sorts $x, \forall\ x \in A, \forall\ y \in E$ by $(x, y) \in U_i$ or $(x, y) \notin U_i$ with $U_i \subseteq E \times E$.

$(x, y) \in U_i$ can be interpreted as: "Given the capacity of bureau B_i to evoke the elementary act y in E, B_i finds x an acceptable part of A."

$(x, y) \notin U_i$ must not be interpreted as "B_i prefers y over x"—again, we do not assume any *order* of preference here*—but rather as: "Given the capacity of B_i to evoke y, B_i finds x unacceptable as part of A." Table 15.1 gives some further meaning to B_i's possible attitudes when examining an elementary act x of a project A. To do so we distinguish between the case where $y \in A$ and the case where $y \in A^c$, with $A^c = \complement_E A$, the complement of A in E. The elementary act x is always assumed to be part of A.

*To state clearly the difference with the type of "tastes" assumed here, let us remark that the usual preorder of economic theory, say $x \underset{i}{\lesssim} y$ is implied by the restrictive condition:

$\forall\ z, z \in E, (x, z) \notin U_i$ *and* $(y, z) \in U_i$.

With some type of continuity, x and y can be chosen in such a way that the reciprocal implication holds.

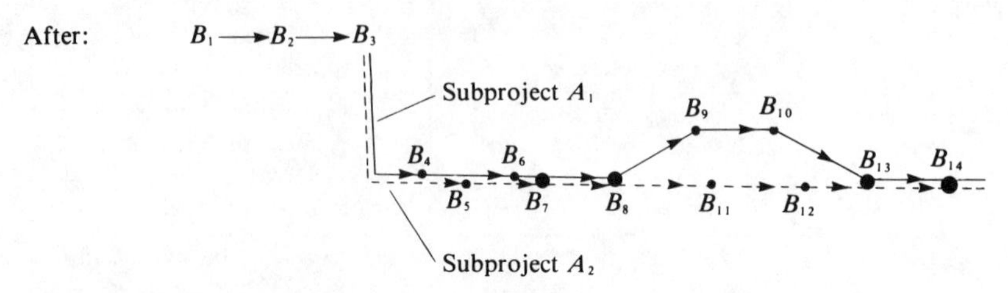

Figure 15.1. **Example of Revision of Procedure.**

RESOLUTION OF BLOCKED SITUATIONS

Define $BL_i(A)$ as

$$BL_i(A) = \{x, x \in A | \exists y \in E, (x, y) \notin U_i\}$$

Then if $BL_i(A) \neq \phi$, project A is blocked by bureau i, the worst case being $BL_i(A) = A$.

If project A is not to be changed, the only way to get out of such a blocked situation is, *if organizationally (technologically) possible*, to *split* the project in two or more subprojects and define therefore new assignments to the set of bureaus. The examination of project A will then follow two (or more) subpaths. On each subpath each bureau will know of only the subproject(s) it has been assigned to. If such a revision of procedure requires the agreement of *all* the bureaus, the procedure will be called *nonmanipulable*. If a single bureau or a coalition of some of the bureaus* in a situation to trigger such a new course, the procedure will be called *manipulable*.

Technically, the subprojects defined must form a partition of project A. Thus given A, $A \subseteq E$, $A \neq \phi$, let H be a finite set of indices h, with $H \subseteq N$, $H \neq \phi$. The subprojects in a revision of procedure will form a family of parts of $A(A_h)_{h \in H}$ such that:

1. $\bigcup_{h \in H} A_h = A$ and:
2. $\forall\, h \in H, \forall\, h' \in H,\ A_h \cap A_{h'} = \phi$

Figure 15.1 shows an example of such a revision of procedure.

PROPOSITION 1

Suppose, given A and i, $BL_i(A) \neq \phi$.

A revision of procedure will unlock a blocked situation if and only if

$$BL_i(A) \cap A_h^i = \phi$$

*Coalitions of some bureaus *and* some external actors interested in the project should also be considered, as empirical evidence shows.

where A_h^i stands for the subproject assigned to bureau i after the revision of procedure.
Remark: If $BL_i(A) = A$, no possible revision of procedure will do the job to unlock the situation, except if $A_h^i = \phi$ for the given i. This is very unlikely to obtain. Thus most of the time the revision of procedure gives rise to another organizational pattern: the pattern of homogeneous examination groups.

HOMOGENEOUS EXAMINATION GROUPS AND LIMITED CONSENSUS

Define the degree of a vertex α as the number of arcs originating at or ending in α. Following Berge (1970) we can write:

$$d(\alpha) = Card(\Gamma_\alpha^+) + Card(\Gamma_\alpha^-)$$

where Γ_α^+ stands for the set of the vertices following α, Γ_α^- for the set of the vertices preceding α, and $d(\alpha)$ for the degree of α. In the preceding example we have $d(B_4) = d(B_6) = d(B_{11}) = 2$ and $d(B_7) = d(B_8) = d(B_{13}) = 4$.

By definition we will say that a set of bureaus is a *homogeneous examination group* G if and only if:

$$G = \{B_j | j \in J,\ J \subseteq I\} \text{ with:}$$

(1) There exists a unique path linking all the vertices of G.
(2) $\forall\ j,\ j \in J\ d(B_j) = k$
where k is a constant natural number.
Remarks:

1. This definition implies that the criterion of homogeneous examination group defines a partition of the bureaus.
2. All bureaus belonging to an H.E.G. examine one and the same subproject, following a unique path in the graph.
3. Two bureaus belonging respectively to two H.E.G., linked by an arc, do not examine the same subproject.

Following Berge (1970) we can construct a hypergraph to clearly show the three above-stated properties. We should make sure (1) that each vertex of the hypergraph corresponds to a set of vertices of the graph with the same degree and (2) that each arc linking two vertices of the hypergraph represents an arc of the initial graph linking two different H.E.G.

Figure 15.2 shows the hypergraph constructed upon the example of Figure 15.1 (after the revision of procedure).

Finally, we define a *limited consensus* for a given H.E.G. as a subproject cleared by every bureau in the H.E.G. More formally: Let G_h be the H.E.G. to which subproject A_h of A is assigned. If, for a given $h \in H$,

$$\forall\ i,\ i \in G_h,\ BL_i(A_h) = \phi$$

A_h is called "limited consensus" for the H.E.G. G_h.

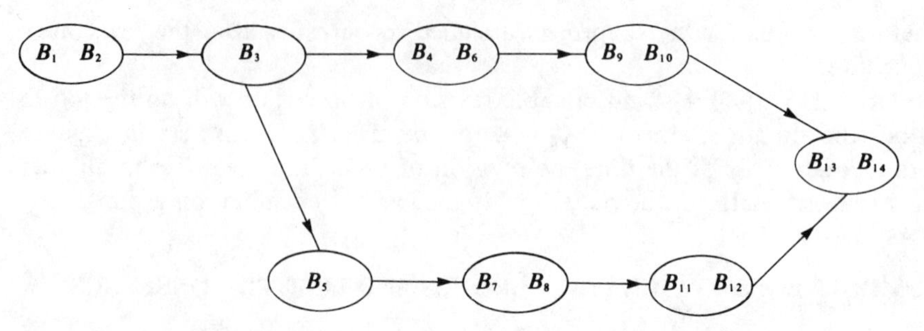

Figure 15.2. **Hyperrgraph.**

AMBIGUITY, EFFECTIVE AND IRRATIONAL DECISIONS

We now can explain why and how, according to the procedure chosen for the decision-making process, irrational decisions may be "accepted," although some (or even all) bureaus did not want them really, had they known that the subproject they have been assigned to were part of the whole "decided upon" project. As Mr. Dreyfus, high-ranking engineer of the French Civil Engineering Department, is reported to have put it: "No political power takes the responsibility of engaging major operations over the long range. The approach is by successive touches, and then, one day, it is realized that the point of no return has been reached" (Gazier 1983).

We are now in a position to give a *necessary and sufficient condition* for a procedure to obtain such an "irrational" decision at the level of a bureau (proposition 2a) or of an H.E.G. (proposition 2b).

Proposition 2a

Let G_h be the H.E.G. associated with the subproject A_h.

Let $BL_i(A - A_h) = \{x \in (A - A_h) | \exists\, y \in E, (x, y) \notin U_i\}$

Then, if:

(1) $(A - A_h) \neq \phi$ (The partition into subprojects is nontrivial.)

(2) A_h is a limited consensus associated with G_h.

(3) $BL_i(A - A_h) \neq \phi$ for some $B_i \in G_h$

B_i clears the subproject A_h, although he would not do so with the whole project A.

Table 15.2 Cases of irrational decision

State of Preferences \ Cases	$x \in A_h$ $y \in A_h$	$x \in A_h$ $y \in A - A_h$	$x \in A_h$ $y \in A^c$	$x \in A - A_h$ $y \in A - A_h$	$x \in A - A_h$ $y \in A_h$
$(x, y) \in U_i$ $(y, x) \in U_i$	Straight Acceptance	←		→	Straight Acceptance
$(x, y) \in U_i$ $(y, x) \notin U_i$	Internal Blocking	(1)	Demand of no extension	(2)	
$(x, y) \notin U_i$ $(y, x) \in U_i$				(3)	(4)
$(x, y) \notin U_i$ $(y, x) \notin U_i$				(5)	

Proposition 2b
Let the notations and the conditions (1) and (2) be, *for every h*, the same as in proposition 2a.

$$\text{Let } \tilde{B}_h(A - A_h) = \bigcup_{i \in G_h} BL_i(A - A_h)$$

If, for every $h \in H$:

(3) $\tilde{B}_h(A - A_h) \neq \phi$

Then each H.E.G. G_h *clears* a limited consensus A_h

but $A = \bigcup_{h \in H} A_h$ is an irrational decision.

The proof of proposition 2a is straightforward.* Clearly, proposition 2b follows immediately from proposition 2a. To find cases in which condition (3) is meaningful and effective, we have only to look for these cases in which, for a given $B_i \in G_h$, we have:

$$\exists\, x \in (A - A_h^i),\ y \in E \text{ such that } (x, y) \notin U_i$$

which is indeed the same as:

$$\exists\, i, \text{ such that } BL_i(A - A_h) \neq \phi.$$

Table 15.2 shows the different possible cases of such irrational decisions. Notations are the same as previously. We exclude only the case $x \in (A - A_h)$

*The proof rests on per absurdum reasoning: Suppose A is a rational decision. Then no bureau i should object to it as a whole. Clearly, condition (3) of Proposition 2a implies that there is at least one bureau objecting to it.

and $y \in A^c$, which hardly makes sense, but this does no harm to the generality of the analysis.*

Proof: Among the five possible cases spotted cases (2) and (3) are symmetric in x and y. So are (1) and (4). So we limit ourselves to three cases, dropping cases (3) and (4).

1. *Cases* (1) *and* (2)

Simply take $x = y$ and $y = x$. We end up with:

$$x \in (A - A_h), \qquad y \in A_h$$

$$\textit{or} \qquad x \in (A - A_h), \qquad y \in (A - A_h)$$

$$\textit{and} \qquad (x, y) \notin U_i \text{ and } (y, x) \in U_i.$$

2. *Case* (5).

Similarly, take element y for x and element x for y. We end up again with:

$$x \in (A - A_h), \qquad y \in (A - A_h)$$

$$\textit{and} \qquad (x, y) \notin U_i \text{ and } (y, x) \notin U_i.$$

In all cases we found: $\exists\ x \in (A - A_h)$, $\exists y \in E$ such that $(x, y) \notin U_i$. Condition (3) of proposition 2a is thus met.

15.4 Coalition Formation and Model Dynamics

If the conditions of proposition 2 obtain, we will have an effective and yet irrational decision. How do these conditions appear?

One can "blame" a bad organizational design showing such "flaws" as conditions of proposition 2 above. But one cannot dismiss the idea that these conditions are met because some coalition of players in the organization (bureaus) or in the environment forms and *imposes* the corresponding revision of the procedure. For explanatory purposes we can use the general terminology of Shakun (1981a, 1981b) developed in his methodology of "evolutionary systems design" (ESD).†

In ESD players define and try to attain goals as operational expressions of underlying values. The players, who may change over time, are viewed as playing a dynamic game in which a coalition C (which may also change over time) of the set of players can form if it can deliver to its members a set of agreed-upon goals, thus defining and solving its policy-making problem. Formally, this means that for each time period t the coalition goal target

*Blank squares of the table are excluded as impossible due to the hypothesis that A_h is a limited consensus for G_h.

†For convenience in representing and modifying goals, ESD uses the p-dimensional real vector space. However we note that a nonempty intersection between $Y^c(t)$ and $y^C(t)$ (see next paragraph) is implied by the hypothesis that a set of points in this intersection is a limited consensus for the members of the coalition C. Note also that the definition of limited consensus in 3.3 does not require use of metric spaces.

$Y^C(t)$ has a nonempty intersection with the technologically (organizationally) feasible performance $y^C(t)$.*

The cases sketched in Section 15.2 can then be modeled as a series of one-period decision-problems with a moving present time period t ($t = 0, 1, 2, \ldots$). Consider a project $A^C(t)$ for coalition C at time t. Let $\mathscr{R}(C; t) = \mathscr{R}^p(C; t)$ be the p-dimensional ($p = p(C; t)$) real vector space that characterizes goals of coalition C at time t. Let g be a function from $\{A^C(t)\}$ to $\mathscr{R}^p$. Then $y^C(t) = g(\{A^C(t)\}, t)$ is the set of possible outputs representing technologically feasible performance for coalition C at time t. In $\mathscr{R}^p$ coalition C's goal target, $Y^C(t) = \bigcap_{j \in C} Y^j(t)$ where $Y^j(t)$ is player j's goal target.

We refer only hereunder to the T.G.V. case (between Paris and Lyon). Although it may not be the best example of the three quoted, the present model certainly embodies part of the truth and contributes to understanding that particular decision-making process, as some empirical studies suggest (Gazier 1983). The process may be divided into five stages.

STAGE 1. $t = 0$

The engineers of S.N.C.F. are player $j = 1$. With $C = j = 1$ and $t = 0$, we have $A^C(t) = A^1(0) = a$, the T.G.V. project. $A^1(0)$ gives output $y^C((t) = y^1(0)$, which intersects goal target $Y^C(t) = Y^1(0)$, a subset of $\mathscr{R}(C; t) = \mathscr{R}(1; 0)$. But it is anticipated that the project would not be cleared on financial grounds mainly.

STAGE 2. $t = 1$

The engineers form a coalition $C = (1, 3)$ with some urban planners ($j = 3$, outside the S.N.C.F.) interested in the decentralization of economic activities, which, they believe at that time, the T.G.V. will bring. Then coalition $(1, 3)$ goes to S.N.C.F. top authorities $j = 2$ and tries to form a coalition $C = (1, 2, 3)$. But the intersection of $Y^C(1)$ and $y^C(1)$ where $C = (1, 2, 3)$ is empty on financial grounds again. The project is not cleared.

STAGE 3. $t = 2$

The engineers try to obtain a *revision of procedure*. They form a coalition with $j = 4$, the track (right-of-way) department within the S.N.C.F. They *split the project* and define two subprojects: the new track project, the T.G.V.-cars and -locomotives project.

Coalition $(1, 4)$, arguing that the new track is needed to expand system capacity,† goes to S.N.C.F. top authorities and tries to form a coalition

*We suggest here that this "intersection" can be obtained when coalition C is able to impose conditions (1), (2), and (3) of proposition 2, which, if adequately used, can ensure that a project will be "cleared," even if it is irrational.

†System capacity had indeed to expand. But all solutions other than the new track were discarded without thorough examination.

$C = (1, 2, 4)$ to have the new track Paris-Lyon project cleared (the T.G.V. global concept is not formally mentioned). But again intersection of $Y^C(2)$ and $y^C(2)$ with $C = (1, 2, 4)$ is empty because the main costs of T.G.V. are represented by the track. The attempt to define a new decision procedure fails.

STAGE 4. $t = 3$

This period follows the 1973 Yom Kippur war in the Middle East. Because of the tremendous increase in oil prices and the fear of being energy dependent on Middle East oil, it is quite easy for coalition $(1, 4)$ to include the energy-saving nuclear electricity people, whom we will denote by $j = 5$. Up to this point the T.G.V. project had nothing to do with the goal of energy saving and nothing to do either with electricity. Indeed, the locomotives had been designed with turbo-diesel motors, i.e., were designed to use diesel oil! Nevertheless, $C = (1, 4, 5)$ can easily form because the shorter, straighter track will save energy; and trains save energy with respect to planes. This time it is possible on the new grounds to convince top authorities. $Y^C(3)$ and $y^C(3)$ intersect, with $C = (1, 2, 4, 5)$. A new decision procedure is set forth. The H.E.G. put by $C = (1, 2, 4, 5)$ in charge of examining the trace project clears it (*limited consensus*).

STAGE 5. $t = 4$

The new track having been built, a *point of no return* has been reached. Engineers and ministry of transportation authorities (say $j = 6$) form a coalition $(1, 3, 6)$ with urban planners ($j = 3$). Locomotives are redesigned as electricity using, and it is then easy to convince S.N.C.F. top authorities ($j = 2$) that the "technically optimal" use of the new track implies having the new locomotives and cars. The latter subproject is then cleared by the H.E.G. put in charge of it by $C = (1, 2, 3, 6)$.

The "global decision" to proceed with the T.G.V. Paris-Lyon never took place as such. Yet the system is operating. The decision has been effective* for all six players, but we hardly could say that it has been a rational decision for players 2 and 6 (S.N.C.F. and ministry of transportation top authorities):

1. These authorities did not actively define their own values and goals, but largely reacted to other players.
2. From *stage 4* on, conditions of proposition 2 were met. Limited consensus on the track implied clearance of the whole T.G.V. system, but not all players involved realized that.

*Following Shakun and Sudit (1983), a project is effective with respect to a goal target for a coalition C if the intersection of the goal target $Y^C(t)$ and the technological feasible performance $y^C(t)$ of the project is nonempty. Thus the subproject cleared at time $t = 4$ (stage 5), in effect implementing the T.G.V. project, is effective for coalition $(1, 2, 3, 6)$.

3. Alternatives taken into consideration were very limited, if any. Airport facilities were developed in Lyon at the same time, although the air traffic could be expected, with the T.G.V. project cleared, to be reduced by 50 percent. An alternative more traditional track project along the existing right-of-way was too quickly dismissed. As for substitutes for travel (new telecommunication facilities), they were simply forgotten.

From a broader viewpoint such a procedural irrationality can result in an ineffective decision for the S.N.C.F. and ministry of transportation in the sense of Shakun and Sudit (1983): The T.G.V. project *as a whole* might not have intersected a goal target resulting from a rational procedure.

15.5 Concluding Remarks

We have thus offered a new explanation as to why one encounters effective and irrational decisions, especially concerning major operations of public agencies.

We feel that our contribution is complementary to that of Sfez (1981), whose main emphasis is on the ambiguity of preferences of subsystems and on their interplay. March's (1978) emphasis is on ambiguity of preferences of the organization as a whole. Our emphasis is on organizational procedures and strategy. We rather see it as a contribution to the domain of "procedural" rationality or irrationality in Simon's (1978) sense.

REFERENCES

Berge, C. 1970. *Graphes et hypergraphes*. Dunod, Paris.

Duru, G. 1980. *Contribution à l'étude des structures des systèmes complexes dans les sciences humaines*. Thèse de Doctorat d'Etat, Départément de Mathématiques, Université de Lyon I.

Egea, M. 1984. "La préuniformité en théorie des coalitions." *Note de Recherche GRASCE*, No. 84.10, Faculté d'Economie Appliquée d'Aix-en-Provence.

Gazier, P. 1983. "D'un septennat à l'autre: Genèse du T.G.V." *Revue Transports*, pp. 507–514.

March, J. G. 1978. "Bounded Rationality, Ambiguity, and the Engineering of Choice. *Bell Journal of Economics*, 9, No. 2, Autumn, pp. 587–608.

Munier, B. 1982. "Systèmes de décision complexes et coordination des activités économiques." *Communication au colloque N.Y.U.-Dauphine*, 16–17 Juin, Paris.

______, and Panariello, A. 1982. "Coordination économique et infrastructures de transport." *Cahiers Transport et Espace*, Mars, Lyon.

______, and Walliser, B. 1982. "Décision et organisations socio-économiques: Contributions françaises récentes." *Note de Recherche GRASCE*, 82.04, Faculté d'Economie Appliquée, Aix-en-Provence.

Offner, J.-M. 1982. "L'extension de la ligne C du métro de Lyon." *Etudes de l'Institut de Recherche des Transports*, Paris.

Simon, H. 1978. "Rationality as Process and as Product of Thought." *American Economic Review*, 68, No. 2, May, pp. 1–16.

Sfez, L. 1981. *Critique de la décision*. Presses de la Fondation Nationale des Sciences Politiques, Paris, 3è édition.

Shakun, M. F. 1981a. "Formalizing Conflict Resolution in Policy Making." *International Journal of General Systems*, 7, No. 3, pp. 207–215. *See also* Chapters 1 and 2 of the present volume.

______. 1981b. "Policy Making and Design of Purposeful Systems." *International Journal of General Systems*, 7, No. 4, pp. 235–251. *See also* Chapters 1 and 2 of the present volume.

______, and Sudit, E. F. 1983. "Effectiveness, Productivity and Design of Purposeful Systems: The Profit-Making Case." *International Journal of General Systems*, 9, No. 3, pp. 205–215. *See also* Chapter 14 of the present volume.

CHAPTER 16

An Evolving Conclusion and Wall As All

A book on evolutionary systems design (ESD) has an evolving conclusion. As design methodology, ESD—the evolutionary design of evolutionary systems—is cybernetics/self-organization that can be modeled by relations (5), (6), (7), (8), (9), as discussed in Section 1.3. Collectively, these relations stand for an evolving coalition (group) problem representation. The ESD methodology with its expression as this representation depicts formalized artificial intelligence (AI)—decision support for problem definition and solution in complex, self-organizing contexts. In other words ESD is an AI framework for GDSS. In turn GDSS implement and provide concrete expression and operational meaning for ESD. This has been actualized in Chapters 8 through 13.

Although not undertaken here, ESD as an AI framework for GDSS should lend itself to a fuzzy set formulation—e.g., the group problem representation and the concepts of group goal target, group technologically feasible performance and their intersection could be modeled using fuzzy sets in the spirit of Negoita (1985).

ESD may also be viewed as evolutionary operations research (OR)/management science (MS). This new evolutionary OR/MS, while including the classical field of well-structured problems, focuses on ill-structured (evolving) problems—in general, multiplayer, multicriteria, ill-structured, dynamic problems that at least initially may have no feasible solution—and self-organizing phenomena. This is the domain of ESD—policy-making under complexity.

Thus the new evolutionary OR/MS is identified with ESD as desig methodology—AI framework—for and implemented by GDSS, i.e., it i identified with GDSS. The new evolutionary OR/MS is also characterized b the ESD paradigm (Section 1.2), of which ESD as design methodology is a integral part. To complete our discussion of the ESD paradigm, it remains t discuss the wall-as-all metaphor, which emphasizes the experiential aspect o ESD.

6.1 Wall As All

The ESD paradigm is scientific, philosophic, and experiential—we experience it, and it is derived from experience. In general all three aspects—scientific, philosophic, and experiential—are involved in designing purposeful systems. The preceding parts of the book have focused on the scientific and philosophic aspects. Here we emphasize the experiential component by looking at some simple systems—the design problems hardly need formalization—where one vividly lives (experiences) the system and ESD metaphors of Section 1.2 as a local space/time expression of all there is that involves its direct experience. By chance—or is it?—each of the systems to be described involves a wall—hence the metaphor "wall as all." When all the metaphors stop, we "see," overcome separateness from all there is—myself, internally, there is all there is. The meaning of this section is enhanced by Sections 1.2 and 1.3 and Chapter 2. Each of these portions mutually enhances each other's meaning.

AVALANCHE AND MOZART

Self-organization:
"Everything Happening All Over,
All the Time, At Once."
Graffiti on post office wall in New York City.

Champagny-en-Vanoise, France
January 20, 1981

The new President of the United States in being inaugurated in Washington—"un spectacle" the French radio reports. At Champagny in the Alps we are tunneling through nearly 2 meters of snow. Most of it has fallen since Norma and I arrived six days ago to go skiing. Too much snow and fog —we've had only skiing hors d'oeuvres, although delicious ones. Today, because of the weather, instead of skiiing the "mur" (wall) on the glacier, we are looking up at the wall of snow above our abode. It is still snowing. We are on avalanche watch. Eighty years ago the wall of snow let go, tearing down the "couloir" (corridor) just to the right of our igloo. Our bedroom on ground level faces up the mountain. "Don't worry," we are told. "Stay home; don't go out. If it comes, the avalanche will pass to the right." Oops, the electricity has gone out.

Suddenly and unexpectedly the electricity comes on again. All at once there is heat and Mozart and warm lunch. The wall of snow doesn't let go. Tao is in these mountains. "Everything is happening all over, all the time, at once." Self-organization.

RANDONNÉE ON SKIIS

Champagny-en-Vanoise, France
10 P.M., April 4, 1980

Dear Everyone (David, Laura, Paul, Julie, Joshua, Kara, Jessica, Mom, Dad, Elaine, Lou, Walt, and Esther),

We went on a fantastic "randonnée" (tour) on skiis today. Chance would have it so, as you will see. The Huttners (Théo, Eric, Bruno, Olivia—Sophie doesn't arrive until tonight), two Huttner friends (Hervé and Carole—friends of Bruno), and two Shakuns (Norma and Mel, of course)—eight in all—set out at 9 A.M. this morning for a vigorous day of skiing. Our backpacks, as usual, were filled with lunch, hats, sweaters, rain gear, etc., as anything, weatherwise, can happen in the mountains.

Actually, it was snowing lightly when we reached the Champagny ski area after about a 20-minute ride up the chairlift from Champagny village where we are staying. A network of poma lifts and some fast skiing brought us to La Plagne-Bellcôte ski area. There are several interconnected ski areas—over 300 kilometers of ski slopes and trails.

We took another chairlift to the Montchavin ski area and did some skiing there. Theo had previously made a descent through a forest to a poma lift below, near Montchavin village. This involved "hors pistes" (off the normal marked slopes) skiing—also called ski "sauvage" (skiing in the wild). We decided to try it. The initial descent before we reached the forest was "un rêve" (a dream) of unbroken powder (just deep enough, not too deep to cause any difficulty). It was a dream but a wide-awake one: an immense, open, dropping-with-oh-such-beautiful-curves natural slope, unbroken snow. I felt such smoothness, such softness beneath my skiis—each turn, carving, carving in the untracked snow.

At the edge of the forest we stopped for lunch. Have you ever backpacked buttered matzos and Beaufort (a French cheese)? Well, they were delicious together. It's Passover you know! A couple of oranges and a swig of ice cold water and we were into the forest of pines, skiing around tress, descending, descending, descending, descending. Eric and Bruno were in front and spontaneously they decided to try a new way down. Down, down... suddenly, without warning, Eric came to the brink of a sudden drop-away—a wall. It was quite extensive—we would have to go around it. The problem was a matter of altitude. We were low and had to go around this canyon. We would have to climb on our skis up very steep terrain. Hard climbing became impossible. There was a narrow passage up—a wall—too steep to climb even sideways on our skis. Fortunately the snow was soft. We took off our skis, being extremely careful that they didn't slide away forever down the canyon. We formed a human chain ladder, passing the skis and poles upward. Then we climbed the narrow snow wall (fortunately it was short) in our boots, remounted our skis above, and continued.

The forest was beautiful, but one thing was clear—we would never arrive at the poma lift at Monthavin. Where exactly we would arrive no one knew, but we did know we had to go down. We finally reached a snow-covered trail used in summer for hiking. We followed it down, skied across a field, came up against a forest that was too patchy with snow to ski, took off our skis, carried them on our shoulders through the forest, put them on again, skied as far as we could on summer trails covered with snow, fields, until, oops, the snow ran out! We had dropped from 2,400 meters (about 7,200 feet) in altitude to 900 meters (2,700 feet).

Now we saw a village below at 700 meters. We had to hike down, finally, on a dirt road, passing through a lovely farm, with the skis on our shoulders. At various degrees of exhaustion we filed into the village, stopping at the village pump for the best ice-cold water I ever drank. (I always say that about mountain water from melting snow.)

The village was Sangot near the larger village of Macot. It was clear that we had to phone for a taxi—two taxis, actually, for eight people. From Sangot to La Plagne—Bellcôte the taxi ride cost 90 Francs (about $22). From there we could take a poma lift back over the mountain and after half an hour of skiing could arrive at the Champagny chair lift for a ride down to Champagny village. A taxi directly to Champagny village would cost over 200 Francs. So in two taxis at 90 Francs each we drove to La Plagne-Bellcôte, had a coffee, and made the trip back home over the mountain, arriving at 5 P.M. After eight hours of mountaineering, we were home again.

What would have been a fine descent through a forest to the Montchavin poma lift—the planned descent—became through chance a fantastic experience in endurance and team cooperation, a challenging randonnée on skiis—cooperative control to all there is. Chance was good to us today (although at one point Norma said that she'd had it—she changed her mind later after a hot bath). By the way, because she was so tired, as chance and human endurance would have it, I cooked the dinner myself. The menu: paté, left-over gefilte fish from our sedar here, veal scallopine in a creme fraiche sauce, salad, wine (St. Emilion, 1976), fruit salad, French pastry (I didn't make it), and coffee. Norma pooped out at the end of the meal, so I finished it off myself while she relaxed on the bed. I did the dishes, took a bath—Norma had aken one before while I did the shopping in the nearby town of Bozel—picked up this pen, wrote it all down, and now I am finished!

Love,
Mel

THE WALL: RUNNING

I thought 787 was a good number. Suddenly 10 and 50 km championship races would be run practically in my backyard in Vermont. I had been jogging but never with the thought of entering a race. The most I had jogged was 5 miles. The 10 km race was 6.2 miles—still I probably could do it. If I ran, I'd be late for a political meeting but could still make it.

Over 800 of us were off on the 10 km run. After the first mile the field spread out. The leaders were fast—way ahead. At 2 miles, where the course funneled from downhill concrete to uphill dirt, I was tiring with two-thirds of the race to go. I had been going faster than my usual to keep up with the slows. At 2, the times were being called out from the side of the road. "24: 37" for 2 miles meant more than 1 hour for the run. I hit the dirt uphill where work was hard. I stopped thinking about the runners. My goal was to finish. My legs ached and my breathing strained uphill. Could I finish under 1 hour? Are you kidding? First make it up the hill! Perhaps I had hit a kind of wall. Not possible—in marathons the "wall" is normally encountered at about 20 miles. I kept going painfully.

The forest along the dirt road was green with lovely dancing sunlight and wetness of the brook. I tuned in and stopped time. My aching body told me I would make it. At 4 miles I clocked a surprising 38: 42—I had a chance of breaking 1 hour. The course changed to concrete downhill. I let myself go. Whew!

Maybe downhill was my thing. On the flat I held things slower but moving. Guess a half mile to go. Give it all you've got—let go totally. Faith. Risk dropping at the end—you know you won't. "57: 25." Placed 559th. I finished! Values and time. The process of all there is.

At the Vermont Democratic Party Platform Convention that afternoon I introduced a motion to support a ban on all smoking in all buses in Vermont. It passed—it was close.

ON THE WALL: L'ESCALADE

It is summer. These Alp mountains. I am climbing. "L'escalade" (rock climbing) it is called. Paul, "guide de haute montagne" (mountain guide), is the other player in the coalition. Assume the values are: v_1, overcoming separateness from all there is, and v_2, Paul earning a living. The goals are: y_1, Mel climbing the wall to the top and rapelling down the cord, and y_2, Paul

Table 16.1. Goals / Values Matrix for "On the Wall." $\lambda^j_{ki} = 1, 0, X$ indicates player j is respectively, "for," "against," "neutral or does not perceive" value v_k being delivered by goal y_i.

Values v_k \ Goals y_i		y_1 *Mel climbing the wall to top and rapelling down on the cord*	y_2 *Paul receiving fee for his guide services*
v_1,	overcoming separateness from all there is	1,1	X, X
v_2,	Paul earning a living	X, X	1,1

receiving a fee for his guide services. The goals/values bimatrix (see Section 1.3) shown in Table 16.1 is simple.

I climb the wall to the top and rest for a few minutes. For the descent I decide to step down to a toehold near the top of the wall before leaning out for the rapel on the cord. Suddenly, I slip while negotiating the toehold. I find myself dangling in midair as Paul holds me on the safety cord. When I say "thanks," he replies with an old French saying: "Chacun à son métier, Et les vâches sont bien guardées," which translates to "Each to his profession, And the cows are well guarded." I remount the wall to the top and this time successfully rapel down.

The players, the sets of values and goals and their relation, the technology, and controls constitute a system—a local space/time expression of all there is that involves its direct experience.

TOUR RONDE, 3,798 METERS

It is summer. The Alp mountains. I am climbing the wall—an incredibly steep snow-covered "couloir" (corridor) leading to the "arête" (ridge) of Tour Ronde. My legs have passed through pain beyond pain. "Bonjour," a passing descending team calls out as if we were strolling down the Champs-Elysées. Waking at 4 A.M., driving two hours over the pass from France to Italy, taking the "téléferic" (cable car) up to the Valée Blanche, putting on our "crampons" (spikes), climbing, leaping over a "crevass" (crevice), and now the three of us—our guide, Reggis, my son, David, and I—on the snow wall, "encordés" (roped together), a system. Cooperative control to all there is—our goal the summit of Tour Ronde, 3,798 meters altitude. Faith, control, faith, control, faith, control... perfect action, nonaction... faith, control... there she is—the bronze-green Madonna with the shiny head at the top of the world. Shine that shiny head, eat lunch, don't trip over your "crampons" and fall off (two have already died that day). Then directly over the top this time—bypassing the "arête"—straight down. A warm bath hours later at Tignes.

The design of a dynamical system of participants, values, goals, technology, and controls constitutes an act of cooperative control (faith)—a local space/time expression of all there is that involves its direct experience. Design constitutes meaning. It is also policy-making. Wall as design problem—wall as all.

COLLEGE FOOTBALL

I slam into the line and go down. Next time I find a hole in the wall and carry the ball through.

The players, the sets of values and goals and their relation, the technology, and the controls constitute a football team (system)—a local space/time expression of all there is that involves its direct experience.

LOVE AS SYSTEM: TO NORMA

A single rose upon a wall,
A single rose doth telleth all,
Of sweet-wild kisses ... (1957).
Love as system—always evolving,
Manifested by increasing consciousness ... (1985).

CABRIES: ALL

The frontispiece is my painting ("Cabries: All") of the hill town of Cabries, France. The wall is right there. The painting was done in 1980 while I was lecturing on evolutionary systems design at the University of Aix-Marseille. The arts are in part goal-oriented and in part relationship-oriented systems where controls and technology deliver values directly (goals may be imputed). In painting, as in other systems design processes, after the present control (paint stroke) is implemented, a redesign of the painting may be undertaken at the next present, one time period later. Painting is all the ESD metaphors of Section 1.2 and expresses all (One).

When all the metaphors stop, we "see," overcome separateness from all there is—myself, internally, there is all there is. The metaphors have stopped–à ce moment.

REFERENCES

Negoita, C. V. 1985. *Expert Systems and Fuzzy Systems*. Benjamin/Cummings, Menlo Park, CA.

Index

Aggregation/disaggregation, 121, 154
Agreement, binding and enforceable, 99
Alternatives, 171, 176
Ambiguity, 266, 270, 275
Analytical judgments, 154
Anticipation, 37, 52
Artificial intelligence, 1, 4, 196, 277
Autopoiesis, 18, 39
Awareness, 40, 51
Axioms
  average utility, 144
  Nash, 141, 159
  negotiation, 134, 136, 140, 156, 197
  Shapley, 141

Bargaining, 227
  game, 84
Behavior concepts, 84
Boundary
  conditions, 87
  Pareto, 141

Coalition, 5, 39, 42, 49, 136, 155, 159, 233, 248, 265
  formation, 272
  grand, 7, 43, 50
  single, 11
Competitive, 227
Compromise, 153, 180, 189, 197, 214
Concession, 51, 83
  on control space, 146
  on goal target, 145
  by a player, 83, 98
Conditional target expansion, 161
Conflict, 153, 155, 197, 232, 248
  absence of, 160
  of expertise, 185, 188
  geometry of, 28, 35
  of legitimacy, 225
  of policy-making, 185
  resolution, 1, 25, 41, 50, 51, 116, 249
Consensus, 153, 158, 197, 227
  limited, 269
Control, 1, 7, 10, 25, 200, 282, 283
  act of, 38
  adaptive, 51
  admissible, 13
  cooperative, 2, 3, 5, 9, 31, 39, 41, 51
  definition of space, 167, 183
  process of, 16, 39
  theory, 2, 29
Cooperation, 5, 11, 39, 197, 227
Criteria, 172
  definition, 168, 173, 176, 184
  objective, 121
  subjective, 116, 121
  unused, 202
  values, 173
Cybernetics, 2, 3, 5, 29, 51, 240, 277
  cycle, 29
  paradigm, 29
  system, 31, 51

Database, 167, 183
  approach, 153, 166
  records, 171
  selection, 175
Decision support systems, 53, 115, 166, 168
  group, 1, 3, 4, 141, 197
  individual, 197
  negotiation, 134, 152
  for new products, 3, 182, 196
  single-user, 136
Delphi technique, 87, 185
Design, 1, 5, 283
  evolutionary systems, 117, 134, 202, 254
  system, 30, 31, 248
Disturbance, 37

Effectiveness, 3, 249, 251, 254, 262, 264
  in public decision making, 264
Electronic mail, 166
Evolutionary system design, 1, 25, 58, 69, 183, 196, 219, 231, 279
  Arab-Israeli conflict, 50, 58
  bank-women's group conflict, 19, 21, 48
  dog run conflict, 19, 21, 44, 46, 50
  effectiveness-productivity example, 247
  freeing hostages example, 219
  mandatory retirement age example, 50, 64
  skiing example, 27, 50
  space industrialization example, 70

Face-saving, 228
Faith, 52, 282

easibility
  controls, 10, 13
  region, 28, 30
  set, 134
  solution, 30, 50, 61
  values, 31
Feedback, 5
  negative, 18, 39
  output, 12, 18, 37, 52
  positive, 18, 39
Feedforward, 5, 32, 38
  disturbance, 12, 18, 38, 52
  negative, 18, 39
  positive, 18, 39
Flexibility, 33

Game
  cooperative/noncooperative, 99
  difference, 35
  dynamic, 5, 38, 248
  prisoner's dilemma, 39
  theory, 13, 93, 97, 98
Gestalt, 41
Goal
  dimension, 14, 43, 45, 62, 232
  nonoperational, 26
  operational, 26, 42, 48, 60, 232, 240
  space, 28, 42, 48, 184, 245
  target agreement, 43, 119, 136, 154, 156, 250, 252, 254
Goals/values
  hierarchy, 13, 45, 49, 254
  matrix, 14, 45, 48, 62, 236, 287
  referral process, 2, 5, 14, 21, 25, 27, 28, 41, 45, 48, 50, 52, 58, 61, 117, 203, 233, 235, 240, 249
  tree, 28

Heuristics, 2, 8, 204, 219, 235–243
Hierarchy, 7, 8, 28, 250
Holograms, 40
Homogeneous examination group, 269

Incidence matrix, 26
Independence
  of irrelevant alternatives, 141
  of linear transformations, 141
Information
  exchange of, 153, 197
  incomplete, 4, 99
  substantive, 219, 245
Input, 10, 248, 252
  admissible, 154
Intuition, 8

Justice, Rawls' theory of, 143

Learning process, 120, 228
Level
  of aspiration, 26, 28, 50, 93, 119
  nonoperational, 30
  operational, 30
  of utility, 119
Linear
  number of pieces, 125
  piecewise, 154

Maintenance, self-referential, 39
Manager
  communications, 169
  data, 169
  dialog, 169
  model, 169
Mapping, 11, 15, 20, 135, 153, 159, 186, 197, 198
  expansion/contraction, 140, 149, 156
Matrix
  alternate, 14
  decision, 171, 173
  graphical and relational data, 140
  incidence, 248
Maximin, 13, 51, 143, 162, 165
Meaning, 3, 7, 41, 51, 53, 265, 283
Measures
  of goals, 29
  of performance, 29
MEDIATOR, 20, 152, 196, 219
Metaphor, 1, 2, 7, 283
  mutually explanatory, 3
  stabilization of, 41
  wall-as-all, 4, 7, 278, 280, 281, 282, 283
Model
  concession making, 145
  descriptive/normative, 81
  Fogelman-Soulie, Munier and Shakun, 51
  Rao-Shakun, 51, 81
Multicriteria, 1, 3, 51, 152, 182, 277
Multivariate, 93
  analysis, 189
  bivariate negotiations, 3, 51, 96
  utility functions, 96

Negotiation
  behavior, 86
  break off, 91
  deadlocked, 21, 48, 99, 227
  under hostility, 227
  international, 225
  phase, 168, 178
  sessions, 113

Negotiation support system, 3, 134, 152, 183, 196, 219

Operations research, 53
  evolutionary, 279
Optimization, 36
  Pareto optimality, 141, 206
Outcomes, 154
  table of, 37
Output, 10, 37
  admissible, 50, 119, 154, 156, 248
  error, 37
  possible, 154

Paradigm
  ESD, 3, 278
  OR, 53
Payoff
  conflict, 83, 84, 99
  probability distribution, 145
PC123, 189
Performance, 29, 43, 52
  technologically feasible, 5, 12, 43, 48, 49, 117, 154
Planning, 25, 31, 33
Policy-making, 1, 3, 4, 5, 25, 52, 283
Predictions, 13, 37
PREFCALC, 131, 135, 154, 159, 169, 171, 182, 185, 189, 198, 205, 206, 213, 214
Preference
  compromise model, 125
  definition, 15, 185
  global, 188
  initial rank order, 121, 124
  wholistic, 120, 121, 124
Probability
  conditional, 91, 101
  distributions, 16, 22, 106
  subjective, 83
Problem
  adaptive control, 51
  dynamic, 1, 3, 53, 152, 182, 277
  evolving representation, 153, 159, 162, 277
  ill-structured, 1, 3, 4, 53, 152, 182, 277
  joint representation, 20, 153, 167, 179, 197
  multiparticipant, 1, 51, 53, 152, 182, 277
  multiperiod, 31
  new product, 182, 196
  sequential decision, 51
Process
  evolutionary, 39
  negotiation, 51
  search, 60
Productivity, 252, 254, 262
Programming, dynamic, 3, 51, 86, 87, 103
Purposeful, 6, 26
  system, 1, 3, 7, 25, 26, 41, 51, 53, 247, 278

Q-analysis, 2, 27, 62, 69

Rationality, 8
  bounded, 8, 97, 264
  individual, 141, 265
  irrationality, 3, 264
  mutually expected, 97
  procedural, 1, 275
Reachability, 35
Regulator, 30, 37
Relation, 26, 58, 62
  controls/goals/values, 14, 58, 62, 232, 233, 235, 283
  false, 27
  inverse, 27
  recursive, 88, 103
  structure, 27, 62
  true, 27
Resilience, 19, 33
Response capacity, 40, 51
Return, maximum expected, 87
Risk
  limit, 84
  minimax, 84
Robustness, 39

Satisficing, 31
Search, 2, 8, 60
  iterative process, 50, 58, 61
  purposeful, 27
  systematic, 50
Self-organization, 2, 6, 19, 277
  dissipative, 3, 6, 7, 9, 41, 51, 53
  response capacity, 51
Semantic disagreements, 167
Separateness, 5, 40, 47, 49, 52, 278, 281
Set, 25
  of admissible outputs, 50, 121
  a priori admissible, 136
  of constraints, 121
  discrete, 143
  expansion/contraction of, 156
  Pareto optimal, 142
  of players, 49
  of strategies, 154
  subset, 62
Simulation, 113, 183
Solution
  bargaining, 51
  feasible, 28, 61

Kalai-Smorodinsky, 141
Nash equilibrium, 93, 141, 189
Nash-Zeuthen, 97, 144
Shapley value, 97, 141
single-point, 5, 140, 155, 250
starting, 61
ocial welfare function, 144
preadsheet, 183
Lotus, 189
Symphony, 189
tability, 50
evolutionary, 39
tate
admissible reachable, 12, 43, 249
memory, 12, 18, 36
of negotiations, 51
of nonequilibrium, 39
of the system, 248, 266
terminal, 87, 101, 112
transition function, 10, 34, 248
Stochastic terminal control, 3, 51, 87, 96
Structure, 7, 62
dissipative, 39
psychological, 41
semi-structured buying decision, 3, 115
Syntactic disagreements, 167
System, 25
adaptive, 1, 26, 31
composite, 38
design of, 5, 26, 53
dynamical, 4, 33, 36, 52
formulation of, 26
general, 30
implementation, 52
Markovian, 12
open, 5
relationship-oriented, 14, 39
of relationships, 93

Target, 29
expansion/contraction of, 137
intermediate, 30, 41
ultimate, 30, 41
Technology, 137, 156, 283
Thinking, 42

Uncertainty, 52, 228
Utility
assessment procedure, 135
expected, 51, 83, 84
function, 13, 119, 120, 125, 135, 154
ideal, 141
interpersonal comparisons, 144
marginal, 137, 158, 205, 212
minimum, 51, 84
multiattribute theory, 96

Values, 4, 5, 26, 45
additional, 60
common, 240
dynamical, 52
future, 14
general, 28, 62
maximizing expected, 92
operational, 4, 42
system, 226
ultimate, 5, 49, 50, 52
underlying, 117, 232
View integration, 168, 175

What-if analysis, 178, 179, 214
Wholistic judgments, 154